METHODS IN CELL BIOLOGY

VOLUME IX

Contributors to This Volume

Harris Busch

F. C. Charalampous

Michael Crerar

H. A. Crissman

Donald J. Cummings

W. C. Dewey

Sam T. Donta

Emmanuel Farber

Edwin V. Gaffney

Enore Gardonio

N. K. Gonatas

Martin A. Gorovsky

Kari Hemminki

Eliseo Manuel
   Hernández-Baumgarten

Roger R. Hewitt

D. P. Highfield

Ronald M. Humphrey

Josephine Bowen Keevert

Toshio Kuroki

Raymond E. Meyn

P. F. Mullaney

Ronald E. Pearlman

J. Hinrich Peters

Gloria Lorick Pleger

M. Robert

H. W. Rüdiger

George Russev

J. A. Steinkamp

Andrew Tait

Charles W. Taylor

Roumen Tsanev

Robert S. Verbin

Harry Walter

Meng-Chao Yao

Lynn C. Yeoman

Methods in
Cell Biology

Edited by

DAVID M. PRESCOTT

DEPARTMENT OF MOLECULAR, CELLULAR AND
DEVELOPMENTAL BIOLOGY
UNIVERSITY OF COLORADO
BOULDER, COLORADO

VOLUME IX

1975

ACADEMIC PRESS • New York San Francisco London
A Subsidiary of Harcourt Brace Jovanovich, Publishers

ACADEMIC PRESS, INC.
111 Fifth Avenue, New York, New York 10003

United Kingdom Edition published by
ACADEMIC PRESS, INC. (LONDON) LTD.
24/28 Oval Road, London NW1

LIBRARY OF CONGRESS CATALOG CARD NUMBER: 64-14220

ISBN 0–12–564109–5

PRINTED IN THE UNITED STATES OF AMERICA

CONTENTS

9. *Growth of Functional Glial Cells in a Serumless Medium*
   Sam T. Donta

10. *Miniature Tissue Culture Technique with a Modified (Bottomless)
    Plastic Microplate*
    Eliseo Manuel Hernández-Baumgarten

11. *Agar Plate Culture and Lederberg-Style Replica Plating of
    Mammalian Cells*
    Toshio Kuroki

12. *Methods and Applications of Flow Systems for*
     Analysis and Sorting of Mammalian Cells
     H. A. Crissman, P. F. Mullaney, and J. A. Steinkamp

13. *Purification of Surface Membranes from Rat Brain Cells*
     Kari Hemminki

14. *The Plasma Membrane of KB Cells; Isolation and Properties*
     F. C. Charalampous and N. K. Gonatas

15. *The Isolation of Nuclei from Paramecium aurelia*
Donald J. Cummings and Andrew Tait

16. *Isolation of Micro- and Macronuclei of Tetrahymena pyriformis*
Martin A. Gorovsky, Meng-Chao Yao, Josephine Bowen Keevert,
and Gloria Lorick Pleger

17. *Manipulations with Tetrahymena pyriformis on Solid Medium*
Enore Gardonio, Michael Crerar, and Ronald E. Pearlman

18. *The Isolation of Nuclei with Citric Acid and the Analysis of Proteins by Two-Dimensional Polyacrylamide Gel Electrophoresis*
Charles W. Taylor, Lynn C. Yeoman, and Harris Busch

LIST OF CONTRIBUTORS

Numbers in parentheses indicate the pages on which the authors' contributions begin.

HARRIS BUSCH, Nuclear Protein Labs, Department of Pharmacology, Baylor College of Medicine, Houston, Texas (349)

F. C. CHARALAMPOUS, Departments of Biochemistry and Pathology, University of Pennsylvania School of Medicine, Philadelphia, Pennsylvania (259)

MICHAEL CRERAR,[1] Department of Biology, York University, Toronto, Ontario, Canada (329)

H. A. CRISSMAN, Biophysics and Instrumentation Group, Los Alamos Scientific Laboratory, University of California, Los Alamos, New Mexico (179)

DONALD J. CUMMINGS, Department of Microbiology, University of Colorado Medical Center, Denver, Colorado (281)

W. C. DEWEY, Department of Radiology and Radiation Biology, Colorado State University, Fort Collins, Colorado (85)

SAM T. DONTA, Department of Medicine, University of Iowa, and Veterans Administration Hospital, Iowa City, Iowa (123)

EMMANUEL FARBER, Fels Research Institute, Temple University School of Medicine, Philadelphia, Pennsylvania (51)

EDWIN V. GAFFNEY, Department of Biology, The Pennsylvania State University, University Park, Pennsylvania (71)

ENORE GARDONIO, Department of Biology, York University, Toronto, Ontario, Canada (329)

N. K. GONATAS, Department of Pathology, Division of Neuropathology, University of Pennsylvania School of Medicine, Philadelphia, Pennsylvania (259)

MARTIN A. GOROVSKY, Department of Biology, The University of Rochester, Rochester, New York (311)

KARI HEMMINKI, Department of Medical Chemistry, University of Helsinki, Helsinki, Finland (247)

ELISEO MANUEL HERNÁNDEZ-BAUMGARTEN, Research Project on Bovine Paralytic Rabies (Derriengue), Instituto Nacional de Investigaciones Pecuarias (National Institute of Livestock Research), México-Toluca, Mexico D. F. (139)

ROGER R. HEWITT, Department of Biology, The University of Texas System Cancer Center, M. D. Anderson Hospital and Tumor Institute, Houston, Texas (103)

D. P. HIGHFIELD,[2] Department of Radiology and Radiation Biology, Colorado State University, Fort Collins, Colorado (85)

RONALD M. HUMPHREY, Department of Physics, The University of Texas System Cancer Center, M. D. Anderson Hospital and Tumor Institute, Houston, Texas (103)

JOSEPHINE BOWEN KEEVERT, Department of Biology, The University of Rochester, Rochester, New York (311)

TOSHIO KUROKI, Department of Cancer Cell

[1] *Present address:* Department of Medical Genetics, University of Toronto, Toronto, Ontario M5S 1A8, Canada.

[2] *Present address*: Department of Human Biological Chemistry and Genetics, Cell Biology Section, University of Texas Medical Branch, Galveston, Texas 77550.

Research, Institute of Medical Science, University of Tokyo, Tokyo, Japan (157)

RAYMOND E. MEYN, Department of Physics, The University of Texas System Cancer Center, M. D. Anderson Hospital and Tumor Institute, Houston, Texas (103)

P. F. MULLANEY, Biophysics and Instrumentation Group, Los Alamos Scientific Laboratory, University of California, Los Alamos, New Mexico (179)

RONALD E. PEARLMAN, Department of Biology, York University, Toronto, Ontario, Canada (329)

J. HINRICH PETERS,[3] Max-Planck-Institut für Biologie, Tübingen, Germany (1)

GLORIA LORICK PLEGER, Department of Biology, The University of Rochester, Rochester, New York (311)

M. ROBERT, Institut für Genetik, Universität des Saarlandes, Saarbrücken, West Germany, and Department of Biology, Johns Hopkins University, Baltimore, Maryland (377)

H. W. RÜDIGER, I. Medizinische Klinik, Universität Hamburg, Hamburg, West Germany (13)

GEORGE RUSSEV, Institute of Biochemistry, Bulgarian Academy of Sciences, Sofia, Bulgaria (115)

J. A. STEINKAMP, Biophysics and Instrumentation Group, Los Alamos Scientific Laboratory, University of California, Los Alamos, New Mexico (179)

ANDREW TAIT,[4] Department of Microbiology, University of Colorado Medical Center, Denver, Colorado (281)

CHARLES W. TAYLOR, Nuclear Protein Labs, Department of Pharmacology, Baylor College of Medicine, Houston, Texas (349)

ROUMEN TSANEV, Institute of Biochemistry, Bulgarian Academy of Sciences, Sofia, Bulgaria (115)

ROBERT S. VERBIN, Departments of Pathology, University of Pittsburgh Schools of Medicine and Dental Medicine, Pittsburgh, Pennsylvania (51)

HARRY WALTER, Laboratory of Chemical Biology, Veterans Administration Hospital, Long Beach, California; and the Department of Biological Chemistry, University of California at Los Angeles, School of Medicine, Los Angeles, California (25)

MENG-CHAO YAO, Department of Biology, The University of Rochester, Rochester, New York (311)

LYNN C. YEOMAN, Nuclear Protein Labs, Department of Pharmacology, Baylor College of Medicine, Houston, Texas (349)

[3] *Present address:* Max-Planck Institut für Virusforschung, Spemannstrasse 35, D-74 Tübingen, West Germany.

[4] *Present address:* Institute of Animal Genetics, University of Edinburgh, Scotland.

PREFACE

In the years since the inception of the multivolume series *Methods in Cell Physiology*, research on the cell has expanded and added major new directions. In contemporary research, analyses of cell structure and function commonly require polytechnic approaches involving methodologies of biochemistry, genetics, cytology, biophysics, as well as physiology. The range of techniques and methods in cell research has expanded steadily, and now the title *Methods in Cell Physiology* no longer seems adequate or accurate. For this reason the series of volumes known as *Methods in Cell Physiology* is now published under the title *Methods in Cell Biology*.

Volume IX of this series continues to present techniques and methods in cell research that have not been published or have been published in sources that are not readily available. Much of the information on experimental techniques in modern cell biology is scattered in a fragmentary fashion throughout the research literature. In addition, the general practice of condensing to the most abbreviated form materials and methods sections of journal articles has led to descriptions that are frequently inadequate guides to techniques. The aim of this volume is to bring together into one compilation complete and detailed treatment of a number of widely useful techniques which have not been published in full detail elsewhere in the literature.

In the absence of firsthand personal instruction, researchers are often reluctant to adopt new techniques. This hesitancy probably stems chiefly from the fact that descriptions in the literature do not contain sufficient detail concerning methodology; in addition, the information given may not be sufficient to estimate the difficulties or practicality of the technique or to judge whether the method can actually provide a suitable solution to the problem under consideration. The presentations in this volume are designed to overcome these drawbacks. They are comprehensive to the extent that they may serve not only as a practical introduction to experimental procedures but also to provide, to some extent, an evaluation of the limitations, potentialities, and current applications of the methods. Only those theoretical considerations needed for proper use of the method are included.

Finally, special emphasis has been placed on inclusion of much reference material in order to guide readers to early and current pertinent literature.

DAVID M. PRESCOTT

CONTENTS OF PREVIOUS VOLUMES

Volume I

Volume II

Volume III

Volume IV

Volume V

Volume VI

Volume VII

Volume VIII

Chapter 1

Preparation of Large Quantities of Pure Bovine Lymphocytes and a Monolayer Technique for Lymphocyte Cultivation

J. HINRICH PETERS[1]

Max-Planck-Institut für Biologie,
Tübingen, Germany

I.   Introduction

In recent years lymphocytes have become one of the main systems for the study of activation and differentiation of mammalian cells. In the case of immunology, mouse and man have been the best studied animals but in both of these cases, the amount of available lymphocytes is minimal.

Nowell (1960) found that high percentages of small lymphocytes are activated *in vitro* by the plant lectin phytohemagglutinin to an increased

[1] *Present address:* Max-Planck-Institut für Virusforschung, Spemannstrasse 35, D-74 Tübingen, West Germany.

RNA, protein, and DNA synthesis and finally to mitosis. Since this finding lymphocytes have become a widely used model for the study of the biochemical mechanisms of cell activation. Lymphocyte activation by mitogenic substances, including a series of agglutinating and nonagglutinating plant lectins as well as inorganic substances, e.g., Zn (Rühl et al., 1971) periodate (Novogrodsky and Katchalski, 1971, 1972), and Hg (Schöpf et al., 1967) appear to be independent of the immune specificity of lymphocytes, thus favoring the use of lymphocytes from donors other than mouse and man. Small lymphocytes from healthy donors, however, cannot be cultivated for long periods of time or as permanent cultures. They therefore have to be freshly prepared and used as primary cultures which respond to mitogenic stimuli for only a few days. There are some advantages in using a primary culture of lymphocytes, and two of these are that laboratory infections such as mycoplasma contamination are reduced or excluded and unwelcome genetic selections which alter the biological behavior of the cells cannot occur. Another advantage is that experiments are repeated with cultures from different animals, thus excluding the possibility of detecting phenomena characteristic of only one selected cell line which are not generally true for the cell class used.

Taking the above factors into consideration, we decided to use bovine lymph node lymphocytes (Hausen et al., 1969) as a suitable source for large quantities of small lymphocytes. Bovine lymph nodes are readily obtained from slaughtered cows, thus making it unnecessary to keep laboratory animals as lymphocyte donors. It is posssible to obtain large quantities of pure lymphocytes from the bovine lymph nodes (10^{10} cells/2 lymph nodes from one animal) which respond to mitogenic stimuli as well as blood or lymph node lymphocytes from other species. The ease of preparing large quantities of these cells together with the fact that the lymphocytes do not attach to surfaces under standard culture conditions make this system suitable for cell biological as well as analytical and preparatory biochemical studies.

There is a close connection between lymph node and thymus lymphocytes. Calf thymus is an approved source of a homogeneous population of mammalian cells with a comparatively low cytoplasmic content. Since thymus cells do not respond to mitogens, the introduction of bovine lymph node lymphocytes as a cell type susceptible to mitogenic stimulation is an attractive alternative to the calf thymus system.

Advantages of the Lymphocyte System

The advantages of this system can be summarized as follows:

1. With a simple technique, large amounts of lymphocytes can be prepared from bovine lymph nodes.

2. Under serum-free conditions lymphocytes are susceptible to stimulation, thus offering a unique system for keeping cells at defined medium conditions (see Section III).

3. Lymphocytes remain as single cells unattached to surfaces under standard conditions, whereas under serum-free conditions bovine lymphocytes can be cultivated as a monolayer.

4. On addition of the mitogen, lymphocytes switch over from a resting state to a highly activated state irrespective of prior treatment or conditions.

5. A high percentage of lymphocytes respond to the mitogenic stimulus, leading to striking biochemical changes.

6. Metabolic changes are synchronized within the first 2 days after stimulation.

7. The observed biochemical changes caused by the stimulation are reproducible; lymphocytes do not have any special requirements with respect to the composition of media and are comparatively insensitive to gaseous and pH changes.

8. Mitogenic stimulation leads to the expression of differentiated functions: Depending on the type of mitogen and the method of application (Greaves and Bauminger, 1972), immunoglobulins (Parenti et al., 1966; Parkhouse et al., 1972), lymphokines (Dumonde et al., 1969), interferon (Wheelock, 1965), and cytotoxic factors (Holm and Perlman, 1965), are synthesized.

The procedure for obtaining lymphocyte suspensions from bovine lymph nodes was developed earlier and has been published in a brief form (Hausen et al., 1969). Since then we have successfully worked with this procedure and recommend it as a simple and reproducible method, which we will describe in more detail here. In Section IV a technique used to cultivate bovine lymphocytes as a monolayer and to stimulate them in the attached form is described.

II. Preparation of Bovine Lymphocytes

A. Lymph Node Preparation

Within 30 minutes after slaughtering, the lymph nodes are cut out of the animal by the veterinary surgeon. Normally as a control against infection they are opened, but for this preparation they remain unopened in order to keep them sterile. The *lymphonodi retropharyngeales mediales* are the most convenient lymph nodes for use since they exhibit a lower degree of inflammation foci and are easier to prepare than tonsils or udder lymph nodes. Lymph nodes from 4 to 5 animals are collected so that the less reactive ones can be selected. If necessary, *lymphonodi tracheales* and *bifurcales* are included. The viability of the lymphocytes remains remarkably high provided the lymph nodes remain intact and are kept at room temperature. Even when the lymph nodes are stored for 3 hours after slaughtering, the vitality of the prepared lymphocytes as measured by their response to phytohemagglutinin appears to be normal.

The lymphocyte preparation is begun by removing the tissue surrounding the lymph node with a pair of surgical scissors. The lymph nodes are then inspected macroscopically for the presence of inflammation foci which appear as livid areas of a hard consistency as compared to the brown-yellow color and soft consistency of less reactive lymph nodes. The lymph node is then fixed with two needles at both of the small ends to a stanniol covered cork plate and swabbed once with 70% ethanol. Under sterile conditions the connective tissue capsule is opened with one lengthwise cut on the convex side of the lymph node. The capsule is held open with forceps and it is undermined with the scalpel until the *hilus* of the lymph node is reached on both sides. The *parenchyma* is separated from the connective tissue of the hilus using a scalpel while holding and elevating the *parenchyma* with forceps. The *parenchyma* is minced into fine pieces which are then with a plastic pestle squeezed through a stainless steel sieve (pore size approximately 0.5 mm) into a petri dish containing Eagle's medium plus 10% inactivated calf serum. The resulting suspension from the two lymph nodes is made up to 200 ml with Eagle's medium plus 10% inactivated calf serum. Prior to further purification this single cell suspension contains lymphocytes, reticulum cells, and negligible amounts of red cells and granulocytes.

B.   Glass Fiber Purification of Lymphocytes

For futher lymphocyte purification the technique developed by Rabinowitz *et al.* (1968) is adapted to our system. It is based on the principle that lymphocytes do not attach to glass surfaces when incubated in a serum-containing medium, whereas reticulum cells and granulocytes stick to the glass surface, thus being retained by the glass fiber-containing column.

Careful cleaning of the glass wool (chemical grade, Merck, Germany) is a necessary prerequisite in order to remove toxic material. This is achieved by heating the glass wool in a glass flask containing concentrated sulfuric acid (which may be used several times, even if it becomes brown) to 100°C. After overnight cooling, most of the acid is poured off and replaced with distilled water. The glass wool can be stored for long periods of time in this dilute sulfuric acid. The second step for purification is carried out when the glass wool has been filled into a glass column. The glass column is 40 cm in height, 5 cm in diameter, and tapers off on the downward side which ends in a glass tap or an adapted silicone rubber tube. The upper end of the column is widened to form a sphere 15 cm in diameter with a 5 cm opening at the top.

After the column is half filled with distilled water, it is filled with glass wool to a height of 12 cm, corresponding to a volume of approximately 200 ml. In the following washing steps the accumulation of air bubbles in the wool is prevented by allowing the solutions to outflow only until the upper limit of the glass wool is reached. The acid is removed from the glass wool by washing with distilled water until the outflow becomes neutral. Alkaline-soluble contaminations are removed by a single wash with 0.1 N KOH, followed by a distilled water wash to return to neutral pH. The column is filled with 0.9% NaCl solution, the upper and lower ends are covered with aluminum foil, and then the column is autoclaved in an upright position.

Lymphocyte purification is carried out at 37°C. The saline solution in the column is replaced by 200 ml of Eagle's medium plus 10% inactivated calf serum and then 200 ml of cell suspension is added. The cells are allowed to interact with the glass fibers for 30 minutes, after which the cell suspension is slowly removed (3 min) from the column. After the cells have been spun at 400 g for 10 minutes, the supernatant containing subcellular particles is removed and the cells are resuspended in Eagle's medium plus 10% inactivated calf serum to give a final concentration of 5×10^6 cells/ml.

The resulting cell suspension consists of more than 95% small lymphocytes, the remaining nonlymphocytes being mainly reticulum cells. Another advantage of this purification is that dead lymphocytes are completely removed by the glass fibers. Therefore when the cells are tested with trypan blue immediately after preparation, a 100% viability is found. Between 1000 and 3000 ml of purified cell culture (5×10^6 cells/ml) are obtained from two retropharyngeal lymph nodes, depending on the size of the lymph nodes and the method of preparation. This high yield enables us to use lymphocytes from only one animal so that the immunological cross reactions which occur when lymphocytes from different donors are cocultivated ("mixed lymphocyte reaction") can be avoided. Excluding the preparation of the column, the described lymphocyte preparation and purification require no more than 2 hours.

C. Culture Conditions

From 200 to 250 ml of lymphocyte culture are contained in a tightly stoppered 750-ml Erlenmeyer flask. The culture is incubated at 37°C under permanent gassing with a CO_2 and air mixture adjusted to keep the pH of the medium as close to pH 7.2 as possible. The lymphocytes settle down but do not attach to the glass surface as long as the medium contains 10% inactivated calf serum.

During the following days there is a considerable decrease in cell number, viability, and susceptibility to mitogenic stimulation since these are primary cultures of nondividing cells. Therefore stimulation experiments are started within the next 2 days after the cultures are set up. That this cell aging is partly due to insufficient culture conditions is shown by the fact that a medium exchange always enhances the stimulation yield. A single addition of 10^{-4} M 2-mercaptoethanol, originally found by immunologists to enhance in $vitro$ immune reactions (Chen and Hirsch, 1972; Click et $al.$, 1972), leads to a significantly prolonged life time of the cultured cells and sometimes to a doubling of the stimulation as measured by the rate of incorporation of labeled uridine into acid-insoluble material after a 24-hour and 48-hour stimulation. In addition, the reproducibility of the quantitative stimulative effect is enhanced by the use of 2-mercaptoethanol. It is not known at which level 2-mercaptoethanol acts to favor the lymphocyte viability.[2]

We have found that lymphocyte stimulation strongly depends on cell contact (Peters, 1972). Usually this is without practical importance, since the most-used mitogens, phytohemagglutinin and concanavalin A, cause strong agglutination which enables the cells to form optimal cell contact. However, when mitogens are used that have low agglutinating capacity, e.g., pokeweed mitogen or staphylococcal filtrate, culture conditions have to be chosen to favor cell contact between lymphocytes during the stimulation. This is done by stimulating the cells in round-bottom test tubes of 1.2 cm diameter, which leads to the formation of a small cell button. The tubes are tightly stoppered and incubated in a 45° (angle) tilted position.

III.   Cultivation of Bovine Small Lymphocytes in a Monolayer

A.   Introduction

Although a large amount of work has been done on mitogenic and immunogenic stimulation, the cellular mechanism for the activation process remains unsolved. Among the different metabolic and cell physiological alterations that occur at the beginning of stimulation, membrane functions

---

[2] After this manuscript was submitted, other more detailed reports appeared showing that several sufur-containing substances share similar properties (Broome and Jeng, 1973; Heber-Katz and Click, 1972; and Peck et $al.$, 1973).

may have a control function in the transmission of the stimulus. This is indicated by the fact that a mitogen is able to stimulate the cell by direct cell membrane contact without entering the cell (Greaves and Bauminger, 1972). Furthermore, the first reaction of the cell to the mitogenic stimulus, is activation of a large number of membrane functions (Peters and Hausen, 1971a,b). Among these functions, the altered cell contact behavior which is registered at the beginning of the stimulation has become our central interest. Lymphocyte–lymphocyte cell contact appears to be a necessary prerequisite for the induction of the stimulation (Peters, 1972). Intercellular communication is formed between lymphocytes immediately after the addition of phytohemagglutinin, and this has led to the hypothesis that the exchange of cytoplasmic factors between lymphocytes may be the inductive stimulus for the lymphocyte nucleus (Hülser and Peters, 1972).

In order to investigate the topological aspects of lymphocyte stimulation, methods were developed to cultivate lymphocytes in a fixed localization. In the literature only the method of Brody et al. (1968), which involved the use of a gelatin film, was found for lymphocyte immobilization. For unknown reasons, however, the lymphocytes in our experimental system did not adhere sufficiently to the gelatinized surface. Therefore we initially developed a method to immobilize lymphocytes in an agar film (Peters, 1972). This method was rather complicated, and it was restricted to shorter incubation periods, since the viability of the lymphocytes was reduced in agar. Although it was possible to extract nucleic acids from the agar and to count the incorporated radioactivity in a scintillation counter, morphological observations of the cells were restricted and autoradiographic studies of the agar-covered cells were impossible.

During our experiments, we found that lymphocytes strongly adhered to the inside surface of glass tubes when they were incubated in buffer or medium without serum during certain washing steps. At the same time it was found that mitogenic stimulation of lymphocytes does not require serum (Gazit and Harris, 1972; Steinmann et al., 1972), and it was shown that the production of immunoglobulins after a mitogenic stimulus was more easily detectable under serum-free conditions (Mann and Falk, 1973).

When cells are incubated in vitro, rather complicated interactions occur between cell membrane, glass or plastic surface, serum components, and extracellular macromolecular factors, and these differ from one cell type to another (Nordling, 1967). For practical use, however, it is important that lymphocytes are stimulated independently on serum in the medium, this being in contrast with the majority of uninfected tissue culture cells, which require serum for the initiation of DNA synthesis. This does not exclude the possibility that small amounts of serum proteins remain adherent to the lymphocyte surface even after repeated washings of the cells.

B.   Conditions of Serum-Free Lymphocyte Cultivation

Our experiments showed that:

1. Unstimulated bovine lymphocytes adhered to glass and plastic surfaces when serum was omitted from the medium. The adherence was enhanced by the addition of a mitogen, and this was in agreement with older observations that mitogens were capable of changing the adhesive properties of lymphocytes (Killander and Rigler, 1965; Forsdyke, 1968).

2. Lymphocytes were well stimulated by mitogens under serum-free conditions.

3. Unfortunately this technique could not be adapted to human peripheral blood lymphocytes. We defibrinated human blood and separated the lymphocytes in a Ficoll–Isopaque solution (Lymphoprep, Nyegaard and Co., Oslo) according to the method of Böyum (1968). The obtained lymphocytes were washed once and incubated with serum-free Eagle's medium in glass and plastic petri dishes from different manufacturers. Lymphocytes were found to attach to approximately 60% to the dishes, and this was insufficient for our purposes. We did not test whether this different behavior was due to species differences or to differences between lymph node and blood lymphocytes.

4. When bovine lymphocytes stimulated under serum-free conditions were pulse-labeled with uridine-^{14}C, a higher incorporation rate was found than in cultures containing 10% inactivated calf serum. This effect could be attributed to the phenomenon of agglutination: The immediate adherence of bovine lymphocytes to surfaces under serum-free conditions prevented the formation of very large agglutinates which occurred in normal serum containing media. We therefore assumed that the diffusion of label and nutrients in large agglutinates was reduced. To confirm this, we centrifuged (400 g, 5 minutes) different amounts of nonagglutinated, pokeweed mitogen-stimulated cells to form a dense button prior to pulse labeling with uridine-^{14}C for 1 hour. It appeared that the reduction in incorporation of label was dependent on the thickness of the cell sediment. Another possible explanation for the reduced labeling in serum containing cultures was that serum factors might compete for the uridine molecules at the membrane carrier site. This possibility was excluded in an experiment where lymphocyte cultures stimulated without serum and receiving 10% inactivated calf serum immediately before labeling with uridine-^{14}C showed no reduction in the rate of uridine incorporation.

5. If serum was added to the cultures after attachment and stimulation, the cells detached slowly during the next day. When the cultures were not disturbed during a 24-hour incubation period, however, the majority of cells remained moderately attached.

According to these preconditions the technique for cultivating lymphocytes in a monolayer was developed for studying the phenomenon of lymphocyte "contact cooperation."

C. The Method for Monolayer Cultivation

Suspended lymphocytes from the stock cultures containing 10% inactivated calf serum are sedimented at a low speed (400 g, 5 minutes), and the supernatant is discarded. The pelleted cells are resuspended in Eagle's medium without added serum and transferred to 5-cm plastic petri dishes (tissue culture quality, Greiner, Nürtingen, Germany). The cells are allowed to settle down and to attach during 1 hour under normal incubation conditions, and then they are stimulated by the addition of a mitogen. If it is necessary to test for serum effects, serum is added 1 hour after the mitogen. After 24 or 48 hours of mitogen treatment the cultures are labeled with the desired radioactive precursor.

1. Preparation for scintillation counting. The culture medium is sucked off, and phosphate-buffered saline (PBS) (pH 7.4, 5 ml) is added to the cells. They are then scraped off the surface with the aid of a plastic spatula or a "rubber policeman" and transferred to a centrifuge tube, this procedure being repeated twice. The cells are collected on Sartorius membrane filters (Göttingen, Germany), washed three times with 3 ml of 4% trichloroacetic acid, dried, and submitted to scintillation counting.

2. Preparation for morphological observations: The culture medium is removed from the petri dishes and the cells are gently rinsed once with PBS. Without drying the cells then are fixed with either concentrated methanol for 5 minutes or with a mixture of 96% ethanol and acetic acid (3 + 1 v/v) for 10 minutes. They are stained with the May-Grünwald and Giemsa stains according to the normal blood smear staining technique.

3. Autoradiography. The cultures are stopped and fixed (see previous paragraph) after which they are rinsed twice with 4% trichloroacetic acid followed by a 30-minute wash in distilled water and then dried. When the Ilford K5 dipping film technique is used, the bottom of the petri dish is completely covered by the gelatin solution, which is dried, exposed, and developed under autoradiographic standard conditions. For use in the Kodak AR 10 stripping film technique, round disks (25 mm diameter) are cut out from the bottom of the petri dish by using a heated brass tube. These disks are mounted with the cell side up on gelatinized slides with a polymerizing plastic glue. The glue is dried, then the plastic disks are covered by an autoradiographic film under standard conditions.

After development of the autoradiographies the cells are stained in the following way: dry autoradiographic preparates are preincubated for 5

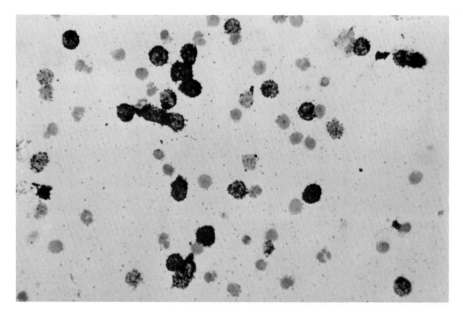

FIG. 1. Result of a 24-hour lymphocyte stimulation with phytohemagglutinin under suboptimal cell density conditions (Peters, 1972). With a few exceptions only the lymphocytes that are in cell contact with other lymphocytes are stimulated, as seen by uridine-³H incorporation, thus demonstrating the phenomenon of lymphocyte "contact cooperation."

minutes in 20% phosphate-buffered saline (pH 7.2) and then stained for 25 minutes in an identical buffer solution containing 20% May-Grünwald solution (Merck, Germany). They are briefly washed in water and then dried (Fig. 1).

REFERENCES

Boyum, A. (1968). *Scand. J. Clin. Lab. Invest.* **21**, Suppl. 97, 77–89.

Brody, J. A., Harlem, M. M., and Plank, C. R. (1968). *Proc. Soc. Exp. Biol. Med.* **129**, 968–972.

Broome, J. D., and Jeng, M. W. (1973). *J. Exp. Med.* **138**, 574–592.

Chen, C., and Hirsch, J. G. (1972). *Science* **176**, 60–61.

Click, R. E., Benck, L., and Alter, B. J. (1972). *Cell. Immunol.* **3**, 156–160.

Dumonde, D. C., Wolstencroft, R. A., Panayi, G. S., Matthes, M., Morley, J., and Howson, W. T. (1969). *Nature (London)* **224**, 38–42.

Forsdyke, D. R. (1968). *Biochem. J.* **108**, 297–302.

Gazit, E., and Harris, T. N. (1972). *Proc. Soc. Exp. Biol. Med.* **140**, 750–754.

Greaves, M. F., and Bauminger, S. (1972). *Nature (London), New Biol.* **235**, 67–70.

Hausen, P., Stein, H., and Peters, H. (1969). *Eur. J. Biochem.* **9**, 542–549.

Heber-Katz, E., and Click, R. E. (1972). *Cell Immunol.* **5**, 410–418.

Holm, G., and Perlman, P. (1965). *Nature (London)* **207**, 818.

Hülser, D. F., and Peters, J. H. (1972). *Exp. Cell Res.* **74**, 319–326.

Killander, D., and Rigler, R. (1965). *Exp. Cell Res.* **39**, 701–704.

Mann, P., and Falk, R. E. (1973). Abstr. in *Seventh Leucocyte Culture Conference*, p. 18.

Nordling, S. (1967). *Acta Pathol. Microbiol. Scand., Suppl.* **192**.

Novogrodsky, A., and Katchalski, E. (1971). *FEBS (Fed. Eur. Biochem. Soc.) Lett.* **12**, 297–300.

Novogrodsky, A., and Katchalski, E. (1972). *Proc. Nat. Acad. Sci. US.* **69**, 3207–3210.

Nowell, P. C. (1960). *Cancer Res.* **20**, 462–466.

Parenti, F., Francescini, P., Forti, G., and Cepellini, R. (1966). *Biochém. Biophys. Acta* **123**, 181–187.

Parkhouse, R. M. E., Janossy, G., and Greaves, M. F. (1972). *Nature (London), New Biol.* **235**, 21–23.

Peck, A. B., Heber-Katz, E., and Click, R. E. (1973). *Eur. J. Immunol.* **3**, 516–519.

Peters, J. H. (1972). *Exp. Cell Res.* **74**, 179–186.

Peters, J. H., and Hausen, P. (1971a). *Eur. J. Biochem.* **19**, 502–508.

Peters, J. H., and Hausen, P. (1971b). *Eur. J. Biochem.* **19**, 509–513.

Rabinowitz, Y., Schimo, I., and Wilhite, B. A. (1968). *Brit. J. Haematol.* **15**, 455–464.

Rühl, H., Kirchner, H., and Borchert, G. (1971). *Proc. Soc. Exp. Biol. Med.* **137**, 1089–1092.

Schöpf, E., Schulz, K. H., and Gromm, M. (1967). *Naturwissenschaften* **54**, 568–569.

Steinmann, H. G., Fowler, A. K., and Hellman, A. (1972). *Proc. Soc. Exp. Biol. Med.* **140**, 48–53.

Wheelock, E. F. (1965). *Science* **149**, 310.

Chapter 2

Methods to Culture Diploid Fibroblasts on a Large Scale

H. W. RÜDIGER[1]

I. Medizinische Klinik, Universität Hamburg,
Hamburg, West Germany

I. Introduction

Human diploid fibroblast cell cultures are now a widely used tool in biomedical research because they are genetically stable and in many important aspects (chromosome number, enzyme pattern, immunologic properties, and others) they resemble the body cells of the donor *in vivo* (Mellman, 1971). In particular their properties of growth largely correspond to those of normal cells *in vivo* as, for example, limited life-span and contact inhibition. Normal diploid cells are needed in mass scale for several purposes, such as the production of vaccines and biochemicals in human use, where heteroploid cancerous cells cannot be used although they are easier to culture.

[1] The author's work on the subject is supported by the Deutsche Forschungsgemeinschaft.

Growth Properties of Diploid Fibroblasts

Normal diploid fibroblasts can grow on surfaces only, to which they stick firmly while forming a confluent monolayer. On one hand, this provides an excellent experimental condition for cell procedures which include the need for rapid exchange of the culture medium inasmuch as the cells can be washed exhaustively *in situ* without centrifugation. On the other hand, growth in monolayer yields only very small amounts of cells even when large bottles are used. Proteolytic enzymes, such as trypsin or Pronase, in concentrations between 0.02 and 0.25% are necessary to suspend the attached cells, an effect that can be enhanced by combination with 0.02% ethylene-diamine tetraacetate (EDTA). The action of trypsin and EDTA is abolished immediately when culture medium and fetal calf serum is added, because this complete medium contains excess of α_1-antitrypsin and calcium ions.

After the suspended cells have been seeded into a culture vessel, it takes about 3 hours until they are firmly attached to the bottom and 8 hours until they completely regain their spindle-shaped form. If the material of the vessel does not allow attachment of cells (e.g., siliconized glass and some kinds of plastics will not), they cannot grow and divide and will die within a short time.

II.   Establishing a Human Fibroblast Culture

A small piece of skin of about 2 × 4 mm is taken with forceps and scissors under sterile conditions, and submerged immediately into culture medium. There it can be maintained at least 20 days at room temperature before

FIG. 1.   T-15 flask from Bellco with plastic net as used for the primary culture of fibroblasts from skin explants.

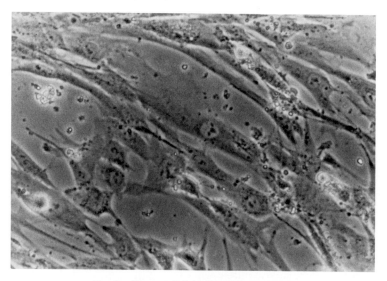

FIG. 2.   Human diploid fibroblasts in culture.

culturing. To establish a primary culture, the skin is cut into very small pieces with a disposable surgical blade. These pieces are sucked off with a pipette and placed together with a small volume of culture medium into a small flask (we use T-15-flasks from Bellco, Fig. 1) and covered with a plastic film which has many small perforations. No more culture medium should be used than is needed to fill the narrow space between the bottom and the plastic film lest the inserted pieces of skin begin to float. The cultures are fed carefully every 48 hours, and within 2 to 5 days the first cells begin to grow out from the pieces of skin (Fig. 2); 10–20 days after the skin explants have been set into culture, cells can be subcultured.

When a culture has been established, the fibroblasts divide every 15 to 20 hours as long as they do not become confluent. The final cell density which can be reached depends on several factors, such as the type of culture medium, pH, age of the culture, and the concentration of fetal calf serum in the culture medium; it varies considerably between cells from different donors. As a rule, a confluent monolayer has about 10^5 cells per square centimeter, which yields about 0.05 mg of cell protein. Subcultures are usually done by dividing the cells from one culture flask into two, a procedure that can be repeated about 40–70 times per cell strain until cells stop growing and begin to die (depending on the age of the individual from whom the cells have been taken). A wide variety of culture media have been described, which do not differ substantially from each other with respect to their ability to support growth of fibroblasts (Eagle, 1965; Morton, 1970). We use Dulbecco's modified Eagle's Medium, which is one of the richest media, supplemented with 16% fetal calf serum, penicillin, streptomycin, and neomycin.

III. An Automated Method for Growing Cells on Glass Beads (Perlacell)

A. Design and Function of the Apparatus

The surface available for cell attachment can be greatly enhanced when a culture vessel is filled with glass beads (Gey, 1933; Robineaux et al., 1970). For example a 1-liter-flask containing 800 ml of glass beads with an average diameter of 3.5 mm provides a total inner surface of 1 m². This area is theoretically sufficient for about 10^9 fibroblasts. Such a large number of cells in a relatively small volume will exhaust the culture medium between the glass beads within a few hours, thus necessitating to refeed the culture. An automated medium exchange can easily be afforded when the culture medium is recirculated by a pump under constant monitoring of the pH. The pH becomes acid when the culture medium is depleted and a 3-way valve under the control of an electronic level sensor is switched to discard the medium which previously had been recirculated. The threshold can be adjusted. The culture vessel is connected with a bottle of fresh medium from which the discarded medium is replaced in exactly the same amount. Figure 3 gives a schematic representation of a construction we have used (Wöhler et al., 1972). In addition, it shows the waterjacket around the culture vessel used to maintain 37°C without need of an incubator. The culture vessel itself should be of conical shape in at least its lower part to guarantee an even flow of medium through the vessel. Two glass rods cover the lower outlet and support the beads. The complete apparatus includes the culture vessel, glass beads 4–5 mm in diameter, a 3-way valve, a peristaltic pump, a level sensor, a pH meter, an autoclavable pH electrode, and a thermostat (Figs. 3 and 4). Instead of a peristaltic pump a membrane pump can be used which can be autoclaved in part (New Brunswick Scientific Co.); this facilitates filling and evacuating the vessel and enforces the effect of trypsinization. In this case the 3-way valve with its small connections should be bypassed with a larger tubing. As a 3-way valve, we used successfully Model 11300 from LKB Stockholm, which is magnetically switched. The part in contact with liquid can be removed entirely and autoclaved separately. This valve is

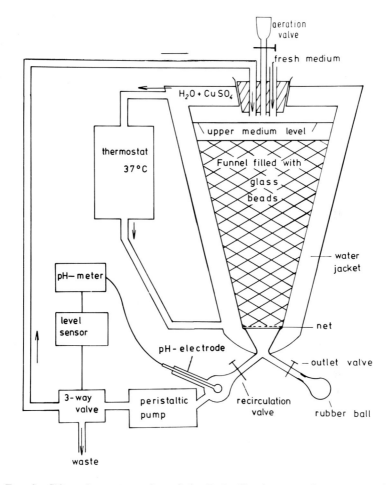

FIG. 3. Schematic representation of the Perlacell culture vessel with accessories as described in the text. From W. Wöhler *et al.*, *Exp. Cell Res.* **74**, 571 (1972). Reproduced with permission of Academic Press, Inc., New York.

plugged into a level sensor Model 11300 from LKB. The whole set was originally built to control column eluates. Any commercially available pH meter can be used which has a separate output plug of 50 mA. All tube connections have to be made according to Fig. 3 with silicon rubber tubings prior to autoclaving. The culture vessel is autoclaved for at least 2 hours at 121°C. We found dry heat to be insufficient even for 12 hours at 150°C; higher temperatures destroy the silicon rubber tubings and seals.

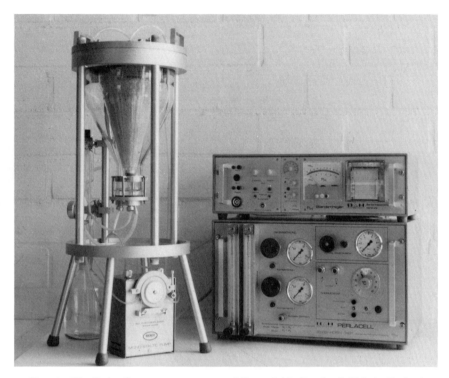

FIG. 4. The Perlacell apparatus as commercially available (Bender & Hobein, Karlsruhe, W. Germany). Culture vessels of different sizes can be used from 1 liter to 5 liters. The vessel is equipped with a pneumatic valve and, in addition to the apparatus described in the text, with a pH recorder and a device for continuous aeration with a sterile gas mixture.

B.   The Operation Procedure

Fibroblasts are suspended in culture medium corresponding in volume to the capacity of the culture vessel (a 3-liter vessel contains 1300 ml of medium). For each 100 ml of medium, at least 10^6 fibroblasts are needed to ensure sufficient initial cell density. We find DME buffered with 20 mM HEPES (Eagle, 1971) instead of sodium bicarbonate to be optimal for growth because it dispenses with CO_2 equilibrium. The medium is supplemented with 16% fetal calf serum and antibiotics and adjusted to pH 7.7 at room temperature prior to sterile filtration. The culture vessel is filled with cell suspension through the outlet for discarding the medium by running the pump in opposite direction. About 24 hours after inoculation, the peristaltic pump is started, and the flow is set at 1 liter per hour (3-liter culture vessel). The threshold of the level sensor is set at pH 7.1; the volume of the discarded medium should be recorded daily. Usually, within 8–10 days, it no longer increases when the cells have become confluent. At this time the vessel is ready for harvest. The medium is removed through the outlet, and the cells are washed twice with normal saline. A prewarmed solution containing trypsin 0.125% and EDTA 0.01% in normal saline is then pumped through the vessel. It is circulated with the pump at maximal speed in order to suspend all cells. After 10 minutes the cell suspension is drawn out through the outlet and residual cells are washed out with fresh culture medium. This also stops undesirable continued action of trypsin on the suspended cells. All these procedures must be carried out under sterile conditions if the cells will be cultivated further. From a 3-liter chamber about 2×10^9 cells can be harvested from an initial inoculum of 5×10^7 cells. Control of CO_2 and O_2 tension in the medium has not been necessary in our experience. The reason could be either that the cells grow even better at low oxygen tension (Cooper et al., 1958) or that the exchange of gas through the walls of the silicon tubings compensates a disequilibrium to some extent.

Often it is advantageous to trypsinize the cells incompletely, and refill the vessel with fresh culture medium thereafter. By this method about 10^9 cells can be drained out of a 3-liter vessel every 3 days for several weeks without any additional work. If the same cells are repeatedly needed in larger amounts for continued studies, this is nearly the ideal device for culturing.

IV. Other Methods for Mass-Scale Culturing of Fibroblasts

Many methods for mass culturing have been described for cells which grow in suspension, but there is only a very limited number of large-scale methods for monolayer cultures (Paul, 1970). The following synopsis is not complete but will give a brief description of either well established or interesting fundamental techniques that have come to my attention.

A. Roller Bottles

This technique is based on the idea that not only the bottom, but the entire wall, of a cylindrical culture bottle can be used for cell attachment. The method was originally described in detail by Gey in 1933 and has since become the most widely used method for culture of large numbers of diploid cells. A cylindrical bottle is filled to about one-tenth of its volume with cells suspended in culture medium. It is then rolled at a rate of 2–4 rpm in a specially designed automated apparatus (Fig. 5). The cells will spread evenly

FIG. 5. A roller bottle apparatus (Bellco, New Brunswick Scientific Co.). The speed of rotation can be adjusted. The whole machine must be placed in a walk-in incubator.

over the inner surface of the bottle. They can be resuspended by replacing the culture medium with trypsin. This technique has been extended further by combining 18 two-liter bottles in a cylindrical wire rack; 108 of these racks containing altogether 2052 bottles are rotated together in a roller mill yielding a total amount of about 1.6×10^{12} BHK cells within 6 days (Bachrach and Polatnik, 1968). The surface of an individual roller bottle can be enhanced further by a plastic foil arranged helically in the direction of rota-

FIG. 6. Disposable 2-liter plastic roller bottle. The surface available for cell attachment has been augmented by a coiled plastic film. This enhances the surface available for cell growth about 10-fold to 8500 cm². The vessel can be gassed through a pipe mounted in the center and an excentric fixed, stoppered gas outlet pipe. The vessel is filled and emptied by a separate outlet.

tion (Fig. 6). This bottle is commercially available in plastic as a disposable vessel (Greiner, Nürtingen, West Germany). Although widely used, roller bottles are in my opinion far from being the ideal tool for large-scale culturing. The roller machine and special bottles are expensive, and usually the whole apparatus has to be placed in a walk-in incubator. In addition, large individual bottles are laborious to handle and to clean, especially as compared with the Perlacell apparatus, which can be operated continuously simply by draining out about half the cell mass every 3 days.

B. Plastic Films

It has been demonstrated that cells in monolayer attach to different kinds of plastic films just as well as to glass. At least the following materials have been used: thermoplastic fluoroethylene–propylene copolymer (FEP-Teflon, Du Pont), cellophane, polycarbonate, polyethylene polyester and polypropylene (Hösli, 1972; Munder et al., 1971). The material can either be autoclaved or UV-sterilized. For mass-scale culturing on membranes, pieces of the desired size are cut from a roll, soaked overnight in 7 × washing compound (Serva, Heidelberg), and rinsed the next day with distilled water six times. Then the membranes are folded once, and the open edges are sealed by diathermy. Cell suspension is then inoculated through a hypodermic needle at one corner of the bag, which is diathermically sealed thereafter to prevent leakage (not necessary if the material is elastic). The

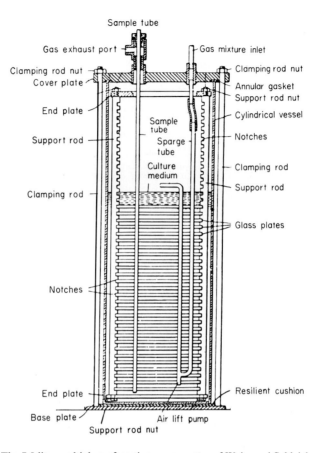

FIG. 7.   The 7.5-liter multiple surface tissue propagator of Weiss and Schleicher. A sterile gas mixture is bubbled through a U-shaped pipe. Its one side reaches from the cover plate to the botton of the jar. The upward side is open at its lower end and leads to the surface of the medium. The rising gas bubbles create a decrease of hydrostatic pressure and cause a constant rise of fluid in the pipe, which is discharged at the top.

bag contains fluid layers 2–10 cm thick placed on a tray in an incubator. After 1–2 hours under culture, depending on the diameter of the fluid layer, the bag is inverted to obtain cell growth on the opposite membrane surface. Some membranes will permit gas exchange, which must be taken into consideration when a particular CO_2 equilibrium of the medium has to be maintained. In this case a CO_2 incubator is used. Cells can be harvested either by folding and stretching the membrane or by trypsinization (Munder *et al.*, 1971).

C. Multilayers of Glass Plates

Weiss and Schleicher (1968) have described an apparatus consisting of multiple layers of window glass which are stacked on top of each other 0.6 cm apart. They used successfully vessels of 1 to 200 liters which resemble one another in basic details, differing only in size. The medium is both recirculated and aerated by a simple and effective arrangement (see Fig. 7). It also serves to distribute the cells evenly after inoculation. Cells are harvested after trypsinization.

D. DEAE-Sephadex

The basic idea is to grow cells on particles which can be kept suspended in the culture medium. The method offers the advantage of homogeneous culture to diploid cells. DEAE-Sephadex has been found to be appropriate because the negatively charged cells adhere to the positively charged matrix and form a confluent monolayer (van Wezel, 1967). DEAE-Sephadex is equilibrated with phosphate-buffered saline and autoclaved. Cells and carrier are suspended together in culture medium and slowly stirred to establish a culture. When the stirring is stopped, the Sephadex settles, and the supernatant fluid can be replaced by fresh medium. However, as yet harvest of the cells seems to be difficult.

REFERENCES

Bachrach, H. L., and Polatnik, J. (1968). *Biotechnol. Bioeng.* **10**, 589.
Cooper, P. D., Burt, A. M., and Wilson, J. N. (1958). *Nature (London)* **182**, 1508.
Eagle, H. (1965). *Science* **148**, 42.
Eagle, H. (1971). *Science* **174**, 500.
Fedoroff, S. (1967). *J. Nat. Cancer Inst.* **38**, 607.
Gey, G. O. (1933). *Amer. J. Cancer* **17**, 752.
Hösli, P. (1972). "Tissue Cultivation on Plastic Films." Technomara AG, Zürich.
Mellman, W. J. (1971). *Advan. Hum. Genet.* **2**, 259.
Morton, H. J. (1970). *In Vitro* **6**, 89.
Munder, P. G., Modolell, M., and Wallach, D. F. H. (1971). *FEBS (Fed. Eur. Biochem. Soc.) Lett.* **15**, 191.
Paul, J. (1970). "Cell and Tissue Culture," 4th ed. Livingstone, Edinburgh.
Robineaux, R., Lorans, G., and Beaure d'Angères, C. (1970). *Eur. J. Clin. Biol. Res.* **15**, 1066.
van Wezel, A. L. (1967). *Nature (London)* **216**, 64.
Weiss, R. E., and Schleicher, G. B. (1968). *Biotechnol. Bioeng.* **10**, 601.
Wöhler, W., Rüdiger, H. W., and Passarge, E. (1972). *Exp. Cell Res.* **74**, 571.

Chapter 3

Partition of Cells in Two-Polymer Aqueous Phases: A Method for Separating Cells and for Obtaining Information on Their Surface Properties

HARRY WALTER

*Laboratory of Chemical Biology, Veterans Administration Hospital, Long Beach, California;
and the Department of Biological Chemistry, University of California at Los Angeles
School of Medicine, Los Angeles, California*

I. Introduction

A. Usefulness

This chapter describes an extremely sensitive method for the separation
and subfractionation of cell populations based on differences in their mem-

brane surface properties (primarily surface charge). In addition, subtle changes in membrane surface properties can be traced that occur as a function of cell differentiation, maturation, age, growth and proliferation, metabolism, *in vivo* and *in vitro* treatments.

B.   Background

Separation of soluble materials by differential partition between two immiscible phases is one of the classical purification procedures. If one introduces a mixture of substances (e.g., A and B) into such a phase system and A is more soluble in the top phase while B is more soluble in the bottom phase, a separation of A and B can be effected. If A is almost exclusively in the top and B in the bottom, the separation is virtually complete in one extraction. If the differences in the relative solubilities of A and B in the top and bottom phases are not so extreme, then separation of the two substances can be achieved only by a multiple extraction procedure (e.g., countercurrent distribution). Separation of cells by partition has, until relatively recently, not been feasible because at least one of the phases has had to be an organic solvent.

When aqueous solutions of two different polymers (e.g., dextran, polyethylene glycol) are mixed above certain concentrations, they form immis-

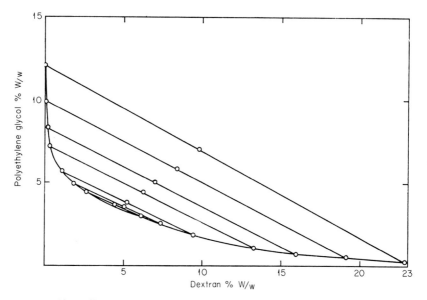

Fig. 1.   Phase diagram and phase compositions of the dextran T500-polyethylene glycol 6000 system at 20°C. From Albertsson (1971).

cible liquid two-phase systems (see phase diagram, Fig. 1) even though each of the phases may contain more than 90% water (Albertsson, 1971). Such systems, being aqueous, can be buffered and rendered isotonic and have proved useful for the partition and separation by countercurrent distribution of cells, particles, and membranes (Albertsson and Baird, 1962; Walter and Selby, 1966, 1967; Walter et al., 1969a,b, 1970, 1971, 1973a,b,c; Miller and Walter, 1971; Brunette and Till, 1971).

The partition of cells depends on the relative affinity of the membrane surface to the polymers, salts, and water constituting the top or bottom phases and to the cells' adsorption at the interface.

Brønsted (1931) has deduced that the partition coefficient, K (ratio between the quantities of particles in the top and bottom phases), may be represented by

$$K = \exp \left[\frac{A\lambda}{RT} \right]$$

where A is the surface area of the particle, R is the gas constant, T the absolute temperature, and λ is a constant characteristic of the particle–phase system interaction. Cell properties that influence the cell partition would be included in the exponential λ term. Thus partition is shown to provide an extremely sensitive cell fractionation method. As an example, surface charge (as will be outlined below) is a major determinant of cell partition, and partition is, hence, a far more sensitive indicator of cell charge than is electrophoresis in which mobility is related linearly to particle surface charge.

C.   Properties of the Phases

Dextran-polyethylene glycol aqueous two-phase systems consist of an upper polyethylene glycol-rich phase and a bottom dextran-rich phase. These systems are extremely mild and generally display a protective effect on the materials partitioned in them. No deleterious effect has been reported, and fully viable cells are recovered after partition and countercurrent distribution (e.g., Brunette et al., 1968).

Albertsson (1971) was the first to use two-polymer aqueous phases for the partition of suspended biological materials and to study some of the parameters affecting partition in a systematic manner. Of major importance in determining the partition coefficient (see below) are the polymer composition and concentration and the ionic composition, concentration, and ratio between the different ions. The manipulation of these variables to obtain a partition suitable for the separation of cell populations of one's interest or for their subfractionation will be outlined.

1. DESIRABLE PARTITION FOR CELLS

The partition of soluble materials take place according to their relative solubilities between two phases and is expressed in terms of a partition coefficient, K, defined as the concentration of the material in the top phase/concentration of the material in the bottom phase. K is independent of the relative top and bottom phase volumes.

The partition of cells usually takes place between one of the phases and the interface. In the case of all cells to be discussed here, the partition is between the top phase and the interface (Fig. 2). Surprisingly, the concentration of cells in the top phase changes inversely with changing top phase volume. Thus, for example, when the top phase volume is doubled, the concentration of cells is halved. This means that it is the quantity of cells in the top phase that is constant. For this reason, the partition of cells is presented as the *quantity of cells in the top phase* (percent of total cells added).

A useful partition for cells for subfractionation of a cell population by countercurrent distribution is in the range of 20 to 80% cells in the top phase. For the separation of two (or more) cell populations one can use phase compositions that give few or no cells in the top phase for one population

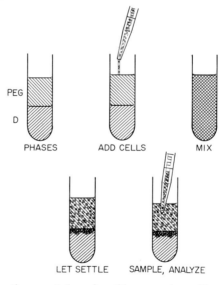

FIG. 2. Diagrammatic presentation of partition procedure with cells. PEG denotes the polyethylene glycol-rich phase; D, the dextran-rich phase. A known quantity of cells is added to the phases, which are then mixed and permitted to settle by the clock. At the end of this time an aliquot is withdrawn from the top phase, and the cell quantity is determined. Partition is expressed as the quantity of cells in the top phase (percent of total cells added). From Walter (1969).

and a higher partition for the second population or partitions that are sufficiently different for the different cell populations (preferably one population having a partition below 50% and the other above 50% cells in the top phase).

2. POLYMER COMPOSITION AND CONCENTRATION

In all the work to be described, the polymers used are dextran T500 (D) molecular weight about 500,000, from Pharmacia Fine Chemicals, Piscataway, New Jersey, and polyethylene glycol (PEG: Carbowax 6000 or 4000) molecular weight about 6000 or 4000, respectively, from Union Carbide, New York.

Aqueous two-phase systems are obtained only above certain "critical" polymer concentrations (Fig. 1). As one goes farther from the "critical point" (i.e., as one increases the polymer concentrations) the partition of all cells decreases. Conversely, as one gets closer to the "critical point" the partition of cells increases, but, as one gets quite close, the reproducibility of the partitions diminishes, and the considerable ionic effects on the partition of cells (see below) no longer pertain. A system containing 5% (w/w) D and 4% (w/w) PEG 6000 has been found suitable for most of the experiments to be described. These concentrations of polymers combine the advantage of being close enough to the critical point to give optimal partition while far enough away to allow appropriate manipulation of the partition of cells by changing ionic parameters.

3. IONIC COMPOSITION AND CONCENTRATION

Salts and buffer are incorporated in two-polymer aqueous phases for several reasons. First, one can render the system isotonic in this manner (a 5% D–4% PEG system in the absence of salt is only about 40 mOsM). Second, one can manipulate the partition of cells by appropriate choice of ionic composition and concentration. Finally, one can often choose salts that are most suitable for preserving the viability of the cells of interest.

It was established quite early that ions fall into series (analogous to the Hofmeister series) in which the substitution of one ion for another in the phase system would increase (or decrease) the partition of *all* tested cells (Albertsson, 1971). Thus, as one example, when a given sodium salt is substituted for the analogous potassium salt the partition of cells increases. The substitution of the corresponding lithium salt increases the partition still further.

The ratio of ions is similarly of great importance. If one partitions cells in a series of 5% D–4% PEG two-polymer phases containing diminishing concentrations of phosphate buffer and increasing quantities of sodium chloride [such that the combined phosphate buffer–sodium chloride con-

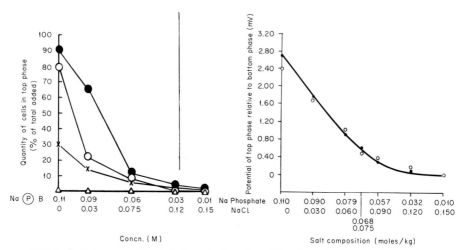

Concn. (M)   Salt composition (moles/kg)

FIG. 3   *Left*: The quantity of erythrocytes from four different species found in the top phase
of a system containing 5% dextran, T500–4% polyethylene glycol (carbowax 6000) at different sodium chloride and phosphate buffer concentrations. ●, dog; ○, human; ×, sheep;
△, rabbit. From Albertsson and Baird (1962).

Right: Electrical potential between top and bottom phases as a function of salt composition. Polymer composition is same as in left part of the Figure. The electrodes contained either
1.5 *M* KCl (○——○) or 3 *M* KCl (○——○). From Reitherman *et al.*, *Biochim. Biophys. Acta*
297, 193 (1973).

centrations are always isotonic for the (red) cells], the partition of cells
will be greatest in the phosphate buffer alone and diminish with decreasing
phosphate/chloride ratios (Fig. 3, left). If one repeats such a series of partitions in a phase system composed of 5% D–3.5% PEG, which is closer to the
critical point (see discussion above) changes in ionic composition have
no effect (Albertsson and Baird, 1962). Changes in the ratios of mono-
to dibasic phosphates also have an appreciable effect on the partition of
cells.

The concentration of ions is another factor in the partition of cells. If the
phosphate concentration is decreased (and the isotonicity is maintained
by incorporating sucrose into the system), the partition of cells will increase.
An increase in pH will also generally tend to increase the partition.

4.   POTENTIAL BETWEEN THE PHASES

The generalities enumerated in the section above hold for all negatively
charged materials (e.g., cells). However, Walter *et al.* (1968a) found that when
positively charged materials were partitioned in these phases the sequence
of the ionic series is reversed and ion substitution, which increases the

partition of cells, decreases the partition of the positive substance. As an example, the substitution of a sodium salt for the analogous potassium salt decreases the partition of the positive material.

The basis for this phenomenon appears to be the unequal distribution of some salts between the phases (Johansson, 1970a). That is, even though the polymers themselves are nonionic, the concentration of some salts (notably phosphate) is not the same in the top and bottom phases. This in turn leads to an electrical and to a zeta potential between the phases. Seaman and Walter (1971) demonstrated the charge between the phases by measuring the electrophoretic mobility of top phase droplets in continuous bottom phase and of bottom phase droplets in continuous top phase. The sign and magnitude of the charge was found to depend markedly on the salt composition of the phases. Reitherman et al. (1973) subsequently measured the electrical potential between the phases in a series of systems containing diminishing phosphate buffer and increasing sodium chloride. The potential was found to decrease with decreasing phosphate/chloride ratio (Fig. 3, right). The decrease in cell partition with decreasing potential between the phases can clearly be seen by comparing Fig. 3, left and right.

5. MANIPULATION OF PHASE VARIABLES TO OBTAIN A DESIRED PARTITION FOR CELLS

Procedurally the partition of a cell population of interest is determined in an arbitrarily selected phase system (e.g., 5% D–4% PEG containing 0.11 M sodium phosphate buffer, pH 6.8). A useful partition is then obtained by changing the ionic composition, ionic concentration, as well as polymer concentration, in a manner as to predictably give a higher or lower partition (as desired).

If the partition in the selected phase system is too low, one can increase the pH, change from sodium phosphate to lithium phosphate, reduce the phosphate concentration and maintain isotonicity by incorporating sucrose into the system, move the system closer to the critical point (e.g., change the polymer concentration from 5% D–4% PEG to 5% D–3.5% PEG). One can also incorporate small quantities of a positively charged polymer into the phase system (Walter and Selby, 1967). A particularly useful charged polymer is trimethylamino-PEG, which tends to collect with the PEG in the top phase and "pull" negatively charged cells "up" (Johansson, 1970b). All these procedures tend to increase the partition of cells.

If the partition of the cells of interest is too high in the phase system tried, one can lower the pH, change from sodium phosphate to potassium phosphate, reduce the phosphate concentration and maintain isotonicity by incorporating K, Na, or Li chloride into the system, move the system farther from the critical point (e.g., change the polymer concentration from 5% D–4% PEG to 7% D–4.4% PEG). All these procedures tend to decrease the partition of cells.

D.   Properties of the Cells Measured by Partition

1.   Membrane Surface Charge

Even before the unequal distribution of some salts in the phase systems had been discovered and before anything was known about the charge between the phases a correlation (with some exceptions) had been reported (Fig. 4) between the partition of red blood cells from a number of different species and their relative electrophoretic mobilities (Walter *et al.*, 1967). This was the first indication that surface charge (or surface charge-associated properties) influenced cell partition. At about the same time it was found, by a combination of radioisotope techniques and countercurrent distribution of erythrocytes, that red blood cells of different ages could be separated by this method (Walter and Selby, 1966; also see Walter *et al.*, 1967). The youngest erythrocytes have the highest partition, and cells of

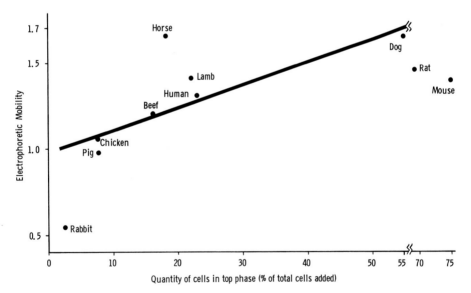

Fig. 4.   Relationship between the electrophoretic mobilities of red blood cells from 10 different species and the quantities of these cells found in the top phase of a system containing 5% (w/w) D–4% (w/w) PEG (carbowax 6000), 0.09 M sodium phosphate buffer, and 0.03 M NaCl. From Walter *et al.*, *Biochim. Biophys. Acta* **136**, 148 (1967).

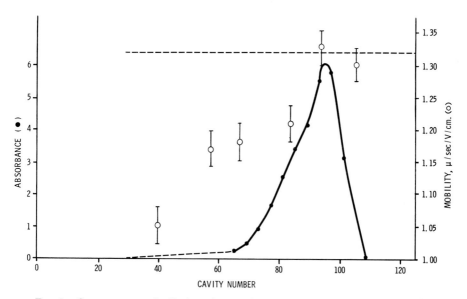

FIG. 5.   Countercurrent distribution of rat erythrocytes, in a two-polymer aqueous phase system, and the electrophoretic mobilities of these cells taken from different cavities along the extraction train. The broken line indicates mobility of normal, unfractioned rat red blood cells. From Brooks *et al.* (1971).

increasing age have ever-decreasing partitions. This finding appeared to be in line with the reported diminution of electrophoretic mobilities of erythrocytes as a function of increasing age (Danon and Marikovsky, 1961). Subsequently, the electrophoretic mobilities of cells from different cavities of the extraction train after countercurrent distribution were measured, both in the case of red cells (see Fig. 5, Brooks *et al.*, 1971) and liver cells (Fig. 8, top, Walter *et al.*, 1973c), and it was found that cells with lower partition have a lower mobility. Increasing mobility and increasing partition were concomitant.

2.   NATURE OF MEMBRANE CHARGE MEASURED BY PARTITION

As mentioned above, there are exceptions to the correlation between electrophoretic mobility and partition of erythrocytes from different species (Fig. 4), indicating that factors in addition to or instead of electrokinetic charge determine the partition of cells in some cases.

An interesting series of experiments served to shed some light on the nature of the membrane charge measured by partition. Partition of beef erythrocytes from a large number of animals reveals that these fall into three partition classes: high, intermediate, and low partition. Neuraminidase

TABLE I

PARTITION AND MEMBRANE SIALIC ACID OF BEEF ERYTHROCYTES[a,b]

Class of cell	Partition in phase system[c]		Sialic acid released[d] by	
	I	II	Neuraminidase	Trypsin
1	37 ± 13(10)	13 ± 2(8)	88 ± 8(8)	47 ± 9(8)
2	62 ± 9(11)	17 ± 3(8)	91 ± 8(5)	49 ± 9(5)
3	90 ± 11(14)	27 + 3(11)	137 ± 11(9)	105 ± 11(5)

[a]From Walter *et al.* (1972).

[b]Results are presented as the mean ±SD with the number of experiments indicated in parentheses.

[c]Partition is expressed as quantity of cells found in top phase (percent of total cells added). Phase system I contained 5% (w/w) dextran, 4% (w/w) polyethylene glycol, and 0.11 M sodium phosphate buffer, pH 6.8. Phase system II had the same polymer composition but 0.09 M sodium phosphate buffer, pH 6.8, and 0.03 M sodium chloride.

[d]Membrane sialic acid released by treatment with neuraminidase or trypsin (γ-sialic acid/10^{10} cells).

or trypsin treatment of the beef red cells releases far more sialic acid (the main charge-bearing component of the red cell membrane) from those cells having a high partition than from the other two partition classes (Table I). However, all beef red blood cells have identical electrophoretic mobilities. It thus appears that partition measures membrane charge deeper into the cell membrane than does electrophoresis, which reflects only the charge at the plane of shear (Walter *et al.*, 1972).

A number of results to be discussed below will make use of these findings as a probable interpretation.

3. OTHER PROPERTIES MEASURED BY PARTITION

Surface charge is undoubtedly the major determinant in cell partition but other factors may play a role as well in the partition of cells. This area of investigation is being actively pursued at present.

It must be borne in mind that the greater sensitivity of partition to charge-associated properties when compared to electrophoresis (see Brønsted formula discussion earlier) together with the fact that partition measures an additional membrane charge (i.e., deeper into the membrane) may make it appear at times that factors other than charge are involved in determining partition even when this is not the case.

II.   Methods

A.   Preparation of Cells for Partition

Since partition reflects most sensitively the differences and changes in the membrane surface properties of cells, the methods chosen for the preparation of single cells must be carefully considered. If the phase systems are to be used solely to isolate a given cell population, one can prepare the cells by any suitable means (e.g., enzymic or mechanical) so long as the method is not detrimental to the biological parameters to be studied on the separated cells. In such a case one may be willing to waive potential information on the original cell surface properties (since these may have been altered during the preparatory steps) and on the relative surface properties of cells of primary interest and those from which they were separated. Since modification of membrane surface properties as a consequence of some cell preparatory procedures may be different for different cell populations (and is not predictable) one might find unchanged, enhanced, or diminished separability between different cell populations following such treatments.

If one wishes to obtain information on the cell membrane properties of a given cell population as a function of differentiation, maturation, aging, etc., one must exert effort to obtain single cells in a manner that will cause the least possible alteration in the membrane surface. With some cell populations this presents little problem (e.g., blood cells, bone marrow cells, intestinal epithelial cells, tissue culture cells grown in suspension, bacterial cells, *Chlorella*) whereas with others (e.g., liver cells, tissue culture cells grown in monolayers) no choice is wholly satisfactory. For example, mechanical procedures with liver cells (Pertoft, 1969) cause membrane damage visible on electron microscopy while enzyme procedures cause no such damage but most likely change the surface properties (e.g., charge, permeability) of the cells (Seaman and Uhlenbruck, 1962; Seaman and Heard, 1960; Schreiber and Schreiber, 1971). Such alterations can markedly distort the original membrane properties as measured by partition (Walter and Coyle, 1968).

B.   Preparation of Stocks and Phase Systems

1.   PREPARATION OF STOCKS

a. Dextran (D).   Dextran T 500 is obtained from Pharmacia Fine Chemicals, Piscataway, New Jersey. While most batches give the same partition results with cells, we have, on occasion, found a batch which gives

significantly different results with some (but not necessarily all) cell populations. It is therefore advantageous to use the same batch of dextran in a
given series of experiments or, if another batch is to be used, to check
whether the new batch gives the same results as previously obtained before
ordering a larger quantity of it.

Dextran is hygroscopic. In order to make a 20% (w/w) stock solution we
generally weigh out about 220 gm of dextran and make it up to 1000 gm with
water. The dextran slowly dissolves at room temperature with occasional
shaking. After it has dissolved it is advisable to heat the solution for
15 minutes at 90–95°C to avoid subsequent microbiological contamination.
Five grams of the dextran solution is weighed out and diluted 5-fold with
water. Concentration of the dextran is then determined by polarimetry,
$[\alpha]_D^{25} = +199°$, and calculated as percent (w/w). Dextran stock solutions
can be stored in the cold for several weeks.

b. Polyethylene Glycol (PEG).   Polyethylene glycol, under the trade name
Carbowax 6000 (or 4000) is obtained from Union Carbide, New York. It is
not hygroscopic, and a 40% (w/w) stock solution is simply made by weighing
out 400 gm and making it up to 1000 gm with water. Such solutions keep
almost indefinitely in the cold.

c. Salts, Sugars, and Buffers.   Salts and sugars, analytical grade, are
obtained from any standard supply house. Stock solutions of salts, sugars,
and buffers that are a convenient number of times more concentrated than
desired in the final phase system are then prepared. We generally use 0.44 M
sodium phosphate buffer (equimolar mixture of mono- and dibasic phosphates), 0.60 M NaCl, 1.0 M sucrose, etc.

2.   PREPARATION OF PHASE SYSTEMS

One must first decide how much of a phase system is required. For single-
tube partition experiments we usually use 5 or 10 gm of a phase system.
For a countercurrent distribution experiment (on our apparatus, see description below) we usually require 225 gm. To make up 225 gm of a phase system

Stocks available	Quantities needed (gm) (to make above phase system)
20.72% (w/w) dextran	54.30
40% (w/w) PEG 6000	22.50
0.44 M Na phosphate, pH 6.8	46.02
0.60 M NaCl	11.25
Water	90.93
	225.0

composed of 5% (w/w) dextran, 4% (w/w) PEG 6000, 0.09 M sodium phosphate buffer, pH 6.8, and 0.03 M NaCl calculate the dilutions of the available stock solutions needed in 225 gm of total phase system to give the desired composition (see tabulation).

C.   Partition Procedure

The desired phase system(s) is made up and permitted to come to the temperature at which the partition of cells is to be studied (e.g., room or cold-room temperature). The phase system is mixed and poured into a calibrated 10-ml tube. The tube may then be centrifuged (at the same temperature) so as to speed phase settling, and the top phase volume is recorded. If the top and bottom phase volumes are greatly different (which depends on the particular phase composition chosen) enough top or bottom phase should be carefully removed so as to have the top and bottom phase volumes nearly equal. This results in a faster subsequent settling time and yields more reproducible results. Of the washed, packed cells to be partitioned 0.1 ml is pipetted into the tube (representing 1% of total phase volume) and an equal aliquot into another, empty, tube ("whole"). The tube with the phase and cells is inverted several times so as to completely mix the contents. Phase settling is permitted to proceed for a given time by the clock. With 5%D–4%PEG we have found that 20 minutes is usually an adequate time (both at room temperature and at 5°) although if sucrose is present as one of the phase components settling proceeds somewhat more slowly, and 25 or 30 minutes should be used. At the end of this time a known aliquot of the top phase is removed, diluted with saline, and analyzed for cell concentration (turbidity, Coulter count, microscopic cell count, or, in the case of red cells, by lysis and measurement of hemoglobin absorbance) and, if desired, for specific activity of a biochemical marker of interest. The partition is calculated (taking into account cell dilution and total top phase volume) and is expressed as the quantity of cells in the top phase (percent of total cells added). The *total* cell quantity added is obtained by analysis of cells in the tube which received an aliquot equal to that in the partition tube (see above).

For cells that cannot be pipetted when they are packed by centrifugation an alternate way of partitioning is as follows: The phase system to be used is allowed to settle, and top and bottom phases are separated. A suitable aliquot of cells is washed once with top phase of the system in which the cells are to be partitioned. The cells are centrifuged at low speed, the supernatant is discarded, and the cells are suspended in 3 ml of top phase. One milliliter of the suspension (a) is removed (for calculation of total cell quantity) and 2 ml of bottom phase are added to the other 2 ml of top phase.

The phase system is mixed and permitted to settle for a given time by the clock as above. One milliliter of top phase is then carefully pipetted from the phase system (b); a and b are centrifuged as above, the supernatant solution is discarded, and the cells in each case are suspended in 2 ml of aqueous isotonic sodium chloride or other suitable solution. The absorbance (or Coulter count, etc.) is now measured on a and on b. From this, the partition of cells is calculated.

For cell populations that cannot readily be centrifuged and resuspended (e.g., they adhere too strongly after centrifugation, there are too few cells to handle) one can incorporate the cell suspension, in appropriate medium, directly into the phase system. For example, if 10 g of a phase system for partition has the composition of the one described under preparation of phase system, above, and the cells of interest are in 0.09 M sodium phosphate buffer +0.03 M NaCl (e.g., 0.5 ml) one can weigh out the appropriate amounts of D and PEG (needed for 10-gm phase system), calculate the quantity of 0.44 M phosphate buffer and 0.60 M NaCl needed on the basis of 9.5 gm (instead of 10 gm), make up with water to 9.5 gm, and add the 0.5 ml of cell suspension. Partition analysis proceeds as above.

It is clear from these examples that a variety of procedures for cell partition can be followed and one should adopt or adapt one most suitable for the cells of one's interest.

D. Phase Selection

After determining the quantity of cells in the top phase of an arbitrarily selected phase system (e.g., 5%D–4% PEG containing 0.11 M sodium phosphate buffer, pH 6.8), one proceeds to modify (if necessary) the polymer concentration and the ionic composition and concentration of the phase system in order to obtain a desirable partition for separation purposes for the cell population(s) in question. An outline of these procedures is given in Sections I,C,1 and 5.

E. Countercurrent Distribution (CCD)

1. APPARATUS

In the current work we used an automatic, thin-layer CCD apparatus (Albertsson, 1970) manufactured by Incentive Research and Development AB, Bromma, Sweden, and sold by Buchler Instruments Division, Fort Lee, New Jersey. The advantage of using a "thin-layer" unit is that phase settling proceeds more rapidly when the height of the phase column is reduced and cell separations can thereby be appreciably speeded. The extraction train is constructed of two circular plexiglass plates with 120 concentric cavities. The bottom (or stator) plate cavities have a capacity of 0.7 ml. There is a hole above each cavity in the top (rotor) plate for loading of phase and cells.

2. LOADING THE APPARATUS FOR CCD OF CELLS

Of the phase system selected for CCD, 225 gm is prepared, allowed to come to the temperature at which the CCD is to be run, shaken up in a separatory funnel and permitted to settle (usually overnight). Top and bottom phases are separated (a small volume of interface is discarded since small quantities of insoluble material from the dextran accumulate there).

Each of 5 adjacent cavities receives 0.5 ml of bottom phase, 0.1 ml of packed cells of interest, and 0.8 ml of top phase (or 0.5 ml of bottom phase and 0.9 ml of top phase containing 0.1 ml of suspended cells). Loading in this manner assures that the interface will be treated as part of the bottom phase. All other cavities receive 0.6 ml of bottom phase and 0.8 ml of top phase. (Actual loading details vary somewhat in the different experiments.) CCD on our automatic apparatus then proceeds. The plates are shaken in a rotary motion, usually 20 seconds, during which time the phases are mixed in each cavity. A separation in load cavities (i.e., those containing cells) is analogous to that described for single-tube partition above. Some cells prefer the top phase and some the interface (Fig. 2). After the phases settle (usually 5–6 minutes), the top plate rotates in a clockwise manner through 3° (while the bottom plate remains stationary). Thus, the top phase in each cavity is carried forward to the next cavity in the bottom plate. Those cells that were suspended in the top phase are carried to the next cavity, where they are reextracted with fresh bottom phase; those cells that were at the interface and remained behind in the bottom plate are reextracted with fresh top phase. In this manner the cycle is repeated as many (usually 60 or 120) times as desired: mixing, settling, transfer, etc. Cell populations which differ only slightly in surface properties require more transfers for a separation than do those which differ significantly. As a rule of thumb one can load as many as 10% of the total number of cavities as there are transfers made without significant sacrifice of resolution (e.g., 12 cavities for 120 transfers, 6 for 60, etc.). One can load as few cells as the analytical method used after cell separation will permit. However, the quantity of cells loaded can alter the effective partition coefficient, G, of the distribution obtained (see caption to Fig. 10 for definition of G).

3. COLLECTION OF CELLS AFTER CCD AND THEIR ANALYSIS

At the end of the CCD run, a circular fraction collector containing 120 tubes is placed over the countercurrent plates. The latter are "flipped over," and the 120 cavities are emptied simultaneously into the centrifuge tubes.

After addition of saline (or other solution), which converts the two-phase system into a single phase (by bringing the polymer concentrations below the critical point), the cell concentration can be determined directly or the cells can be centrifuged and the supernatant discarded. They can be suspended in a suitable medium (e.g., buffer, saline) and analyzed both for cell quantity and for those biochemical or biophysical properties of interest (e.g., electrophoretic mobility of cells in different cavities, microscopic examination of separated cells, protein, enzyme activities, isotope incorporation).

III. Results

Of the numerous studies on cells undertaken with the aid of two-polymer aqueous phase systems, only a few representative samples can be cited here. They are selected to illustrate the usefulness of partition for (A) separation of cells, (B) subfractionation of cell populations and tracing of membrane changes as a function of *in vivo* processes, (C) examining altered membrane properties as a function of *in vitro* treatments, and (D) gaining information on "inner" vs "outer" surface membrane charge.

It might be useful to note that CCD curves obtained with cells are usually broader than theoretical indicating the heterogeneity of cell populations which, by other applied criteria, may appear to be homogeneous.

A. Separation of Cells

1. BLOOD CELLS

The partition data obtained with red and white blood cells from a number of different species in two selected phase systems are presented in Table II (Walter *et al.*, 1969b). These partitions are species specific and the separability of red from white blood cells depends both on the species and the phase system selected. Figure 6 depicts an experiment in which a mixture of rat red and white blood cells were subjected to CCD. The heterogeneity of the white cell population is indicated by the "bumpy" nature of the distribution curve.

An attempt was made to separate polymorphonuclear leukocytes from lymphocytes (Walter *et al.*, 1969a). A buffy coat preparation from horse was examined after CCD, the percentage of PMN and lymphocytes in the different cavities of the extraction train being determined by microscopic count. It is clear (Fig. 7) that a separation of these cells was, at least partially, effected with the lymphocytes having the higher partition.

TABLE II

PARTITION OF ERYTHROCYTES AND LEUKOCYTES FROM DIFFERENT SPECIES IN
DEXTRAN-POLYETHYLENE GLYCOL PHASE SYSTEMS[a,b]

Species	Partition[c] of erythrocytes in phase system indicated		Partition[c] of leukocytes in phase system indicated	
	I	II[d]	I	II
Rabbit	<5	<5	34 ± 14(7)	—
Pig	6 ± 1(6)	8	14 ± 5(6)	—
Beef	39 ± 9(11)[e]	16[e]	19 ± 4(6)	—
Lamb	41 ± 6(10)	22	39 ± 6(5)	—
Human	53 ± 6(8)	24	81 ± 11(5)	54 ± 10(4)
Horse	54 ± 8(8)	18	32 ± 7(8)	11 ± 6(10)
Rat	73 ± 7(7)	67	44 ± 8(5)	22 ± 10(5)
Dog	83 ± 5(9)	54	55 ± 9(5)	19 ± 4(5)

[a]From Walter et al. (1969b); the details for the preparation of leukocytes used in these partitions are also given.

[b]Phase system I contained 5% (w/w) dextran, 4% (w/w) polyethylene glycol, and 0.11 M sodium phosphate buffer (pH 6.8). Phase system II contained the same polymer composition, but 90 mM phosphate buffer and 30 mM NaCl. Partition was carried out at 22–24°. Phase-settling proceeded for 20 minutes with the tubes in vertical position. Results are presented as the mean ±SD with the number of experiments indicated in parentheses.

[c]Partition is expressed as quantity of cells found in top phase (% of total cells added).

[d]Partition of erythrocytes in the phase system II is taken from Walter et al. (1967).

[e]Beef erythrocytes belonging to partition class I (see Table I).

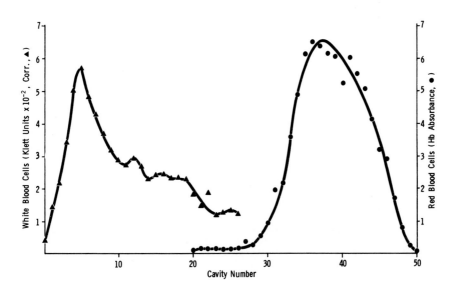

FIG. 6. Countercurrent distribution of a mixture containing approximately 0.4 ml white blood cells and 0.1 ml red blood cells from the rat. Phase system II (see Table II) was used, and 50 transfers were completed at room temperature. From Walter et al. (1969b).

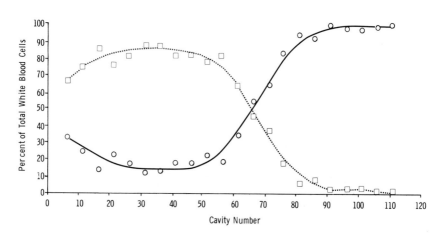

FIG. 7.   Granulocytes (□----□) and lymphocytes (O——O) expressed as percentage of total leukocytes found in the different cavities after countercurrent distribution of a horse buffy coat cell preparation. One hundred twenty transfers were completed in a phase system containing 5% (w/w) D–4% (w/w) PEG (carbowax 6000) and an isotonic sodium phosphate buffer, pH 7.5, at 5°C. From Walter *et al.* (1969a).

2.   LIVER CELLS

Liver cells prepared by the mechanical dispersion method of Pertoft (1969) were examined by CCD. These cells have a rather low partition (reflecting their relatively low surface charge). CCD was carried out in two phase systems, one containing D–PEG, sucrose, and a low concentration of phosphate buffer and the other containing in addition to D–PEG also the positively charged polymer trimethylamino-PEG. (The exact phase compositions are given in the captions to the Figs. 8 and 9.)

The partition and electrophoretic mobility of regenerating liver cells is higher than that of normal liver cells. CCD of liver cells in the uncharged phase system gives rise to two peaks: the left one being larger than the right one for normal cells, and the reverse holding for regenerating cells. Liver cells show a correlation between increasing electrophoretic mobilities and increasing partition coefficients through the bulk of the distribution curve (Fig. 8). In the charged phase system, CCD of liver cells gives single curves. Liver cells from rats injected with India ink yield a separation of hepatocytes from cells containing India ink particles (histiocytes). The latter are at the extreme right end of the distribution curve (Fig. 9). Mono- and binucleated liver cells also appear to have different surface properties since their partial separation is effected by CCD. Incubation of liver cells with valine-^{14}C followed by CCD indicates that the more highly radioactively labeled cells are under the right and left ends of the distribution.

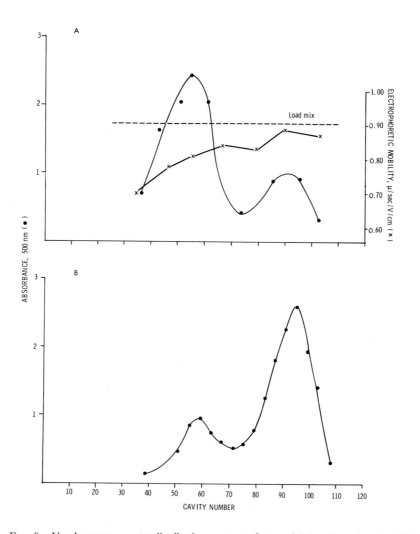

FIG. 8. Usual countercurrent distribution pattern of normal (A) and regenerating (B) rat liver cells. Regenerating liver cells were from a rat partially hepatectomized 7 days earlier. Distribution of liver cells is in terms of cell absorbance at 500 nm. The electrophoretic mobility (μm/sec/V/cm) of normal liver cells was also measured in different cavities after countercurrent distribution. Phase system contained 5% (w/w) D–3.8% (w/w) PEG (carbowax 6000), 0.01 M sodium phosphate buffer (pH 6.8), and 0.25 M sucrose. One hundred twenty transfers were completed at 4°C. From Walter *et al.* (1973c). *Exp. Cell Res.* **82**, 15.

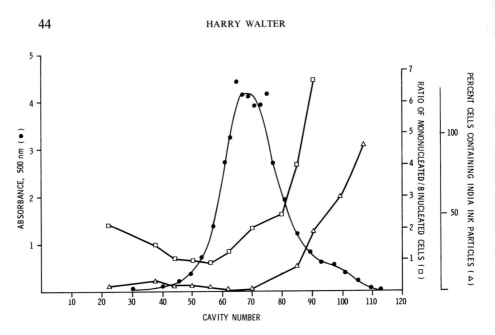

FIG. 9.   A rat was injected intravenously with India ink on 3 successive days. On the fourth day its liver cells were prepared and subjected to countercurrent distribution (120 transfers) in a phase system containing 7% (w/w) D–4.2% (w/w) PEG 4000–2.8% (w/w) trimethyl-amino–PEG (carbowax 4000), 0.055 M sodium phosphate buffer (pH 6.8), and 0.125 M sucrose. The distribution curve is depicted in terms of cell absorbance (as in Fig. 8). In addition, slides were prepared of cells in different cavities along the extraction train and the precentage of cells containing India ink particles was determined. The ratio of mono-/binucleated cells was also obtained by microscopic count. From Walter *et al.* (1973c).

B.   Subfractionation of Cell Populations and Tracing Membrane Changes That Occur as a Function of Normal (or Abnormal) *in Vivo* Processes

One of the greatest strengths of this method (and one that cannot be duplicated by any other presently available procedure) is its ability to trace certain subtle alterations in membrane surface properties as a function of *in vivo* processes.

1.   RED BLOOD CELLS

CCD of red blood cells gives rise to a single, essentially symmetrical curve (Fig. 5). Combination of isotopic labeling techniques and CCD reveals that red blood cells of different ages are under different parts of such a distribution curve (Walter and Selby, 1966; also see Walter *et al.*, 1967). If one injects animals (e.g., rats) with radioactive iron and bleeds them at different times after injection, labeled red blood cell populations of different ages will be obtained. If these are then examined by CCD it is found that, while the distribution of the entire red cell population is always the same, the distribution of the radioactivity shifts depending on the age of the red cell subpopulation that is labeled. If a rat is bled 18 hours after isotope injection the labeled cells are displaced markedly to the left of the whole red cell distribution curve indicating that these young cells (i.e., the youngest reticulocytes) have a lower partition than any other red cells in the peripheral blood. Rapidly over the next few hours this partition increases so that, at the end of about 40 hours, the cells (now the oldest reticulocytes in the blood) have the highest partition of any red cells present (Walter *et al.*, 1971). These cells then become mature erythrocytes and their partition diminishes over the entire life-span of the cell in the blood (i.e., 55 days in the case of rats). Near the end of the life-span, the oldest erythrocytes have a partition very close to that of the youngest reticulocytes. These partition changes are summarized in terms of the effective red blood cell partition coefficients at different cell ages in Fig. 10.

We have recently also examined the partition of hemoglobin-containing cells in the bone marrow and found that the nucleated precursor cells and the reticulocytes in the bone marrow have still lower partitions than the red cells first released into the peripheral blood (Walter *et al.*, 1973a).

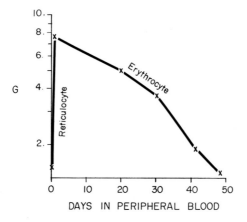

FIG. 10. The effective partition coefficient, G^*, of red blood cells of different ages in the peripheral blood of rats. The data were obtained by a combination of isotopic and countercurrent distribution techniques as outlined in the text. Phase system II (see Table II) was used.

$^*G = r_{max}/(n - r_{max})$, where r_{max} is the number of the peak cavity of the distribution curve and n is the total number of transfers (Craig and Craig, 1956).

2. Intestinal Epithelial Cells

Rat intestinal epithelial cells prepared by the method of Weiser (1973) were examined by CCD (Walter and Krob, 1974). Weiser's method yields, sequentially, cell fractions from villus tips to crypt base. CCD on the first fractions obtained have a peak to the left and a smaller peak to the right; last fractions have a peak only to the right. When all fractions are pooled prior to CCD two well-defined peaks are obtained with the right peak sometimes showing additional heterogeneity. Isotope experiments with methylthy-midine-^{14}C or glucosamine-^{14}C (which, at short times after rat injection, label, respectively, the crypt and villus cells) indicate that the highest thymidine specific activity is between the two peaks after 4–5 hours, moves to the right over the next 11–24 hours, and then moves to the left, whereas the highest glucosamine specific activity is over the left half of the distribution at 4–5 hours after injection and moves to the left over the next 24 hours. These data, which are schematically presented in Fig. 11, indicate that the young (most highly labeled) crypt cells have a partition between the two distribution peaks. Their partition increases during further differentiation, and (possibly as a consequence of the appearance of brush borders) the cells', partition then diminishes as they become villus cells. Their partition is the lowest just before being sloughed off.

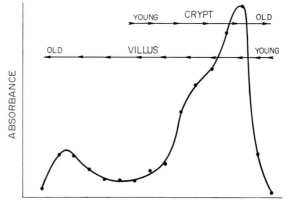

Fig. 11. Schematic diagram depicting the countercurrent distribution pattern obtained with rat intestinal epithelial cells prepared by the method of Weiser (1973). CCD was in a phase system containing 5% (w/w) D–3.8% (w/w) PEG (carbowax 6000), 100 mOsM sodium phosphate buffer, pH 7.4, and 0.17 M sucrose. Use of radioisotopic techniques (see text) permit one to trace alterations in partition of cells during differentiation from the crypt base to the villus tips. From Walter and Krob (1975).

C. Changed Partitions Due to *in Vitro* Treatment

1. ENZYME TREATMENT OF RED BLOOD CELLS

Treatment of red cells with neuraminidase or with trypsin reduces their partition (Walter and Coyle, 1968) because the membrane surface charge is diminished by the removal of sialic acid. Enzyme treatment of cells over different time periods followed by CCD indicates that the essentially symmetrical normal cell distribution becomes broader and much more heterogeneous as the distribution moves to the left (i.e., as the partition decreases). A possible explanation is that enzyme treatment does not remove equal amounts of charged surface groups from all cells over the same time span (H. Walter, unpublished observation).

2. STORED RED BLOOD CELLS

Storage of human (but not rat) red blood cell leads, progressively, to increased partitions (Walter *et al.*, 1968b). No change in cell electrophoretic mobility accompanies cell storage, however (Brooks *et al.*, 1971). Recently it has been found that if stored red cells are exposed to phase (or polymer solution) and then washed prior to cell electrophoresis, an increased electrophoretic mobility is observed for stored cells when compared to normal similarly treated cells. Apparently some of the polymer is irreversibly bound to the stored cells and causes the increase in mobility (Brooks, 1971). It is not at all clear what changes in the cell membrane are measured by partition in the case of stored cells, nor what property might be responsible for the irreversible binding of polymer to the stored cell surface.

D. Information on "Inner" vs "Outer" Surface Membrane Charge

1. BEEF RED BLOOD CELLS

The results with beef red blood cells have been described in Section I,D,2 (Table I). Suffice it to repeat that the results (Walter *et al.*, 1972) indicate that partition measures membrane charge deeper into the cell membrane than does electrophoresis.

2. RED BLOOD CELLS FROM DIFFERENT SPECIES

There is, as mentioned earlier, a correlation between the partition and relative electrophoretic mobilities of erythrocytes from a number of different species (Fig. 4; Walter *et al.*, 1967). However, a number of exceptions to the correlation exist including the red cells from horse, rat, and mouse. It is interesting to speculate on the basis for these differences. It is known, for example, that the sialic acid on the horse red cells is different from other species in that it contains only *N*-glycolylneuraminic acid. In the case of rat erythrocytes a difference in the binding of their membrane sialic acid is indicated by the fact that, while practically all sialic acid can be removed by neuraminidase treatment of other species' red cells, this cannot be done with rat. It may also be that, since partition measures charge more deeply into the membrane than does electrophoresis, the three species' red cell partitions reflect some such difference.

3. ALDEHYDE-FIXED ERYTHROCYTES

Acetaldehyde-fixed red blood cells have the same partition and electrophoretic mobilities as the normal erythrocyte from which they are derived. Lipid extracted, acetaldehyde-fixed red cells have an unchanged electrophoretic mobility but greatly increased partition (Table III). Again, one may conjecture that lipid removal from the cell surface exposes charge groups which affect partition but, being too deep in the membrane, do not affect electrophoretic mobility (Walter *et al.*, 1973b).

TABLE III

HUMAN GLUTARALDEHYDE AND ACETALDEHYDE-FIXED RED BLOOD CELLS: PARTITION IN DEXTRAN-POLYETHYLENE GLYCOL PHASE SYSTEMS AND ELECTROPHORETIC MOBILITY[a]

Material	Partition[b] in phase system		Electrophoretic mobility[c] (μm/sec/V/cm)
	I	II	
Red blood cells	$67 \pm 8(10)$	$22 \pm 4(10)$	-1.08 ± 0.03
Glutaraldehyde-fixed red cells	$90 \pm 7(5)$	$62 \pm 5(9)$	-1.16 ± 0.04
Glutaraldehyde-fixed red cells, lipid extracted	$94 \pm 5(3)$	$90 \pm 6(5)$	-1.15 ± 0.06
Acetaldehyde-fixed red cells	$71 \pm 9(27)$	$28 \pm 9(8)$	-1.09 ± 0.04
Acetaldehyde-fixed red cells, lipid extracted	$91 \pm 8(3)$	$63 \pm 7(3)$	-1.08 ± 0.05

[a] From Walter *et al.* (1973b).
[b] Partition is expressed as quantity of cells found in top phase (% of total cells added). Phase system I contained 5% (w/w) dextran, 4% (w/w) polyethylene glycol, and 0.11 M Na-phosphate buffer, pH 6.8. Phase system II contained the same polymer composition but 0.09 M sodium phosphate buffer, pH 6.8, and 0.03 M NaCl. Results are presented as the mean $\pm$ SD with the number of experiments indicated in parentheses.
[c] Electrophoretic mobility as measured in 0.15 M NaCl, pH 7.4, at 25°C.

IV.   Concluding Remarks

It is clear from the foregoing examples that partition and CCD of cells in two-polymer aqueous phases is an extremely useful and sensitive method for separating cell populations, tracing membrane alterations that occur as a function of *in vivo* processes or *in vitro* treatment and uncovering suspected or unexpected heterogeneities in cell populations.

It is important, when attempting to devise a phase system useful for the partition of cells of interest, to change the available variables in a systematic manner (as outlined in Section I). Single tube partitions are an important first step in finding a phase system that can be subsequently used in CCD. At the same time it should be noted that single tube partitions can be somewhat misleading in (a) determining the separability of cell populations when these are highly heterogeneous (since the single tube partitions give only *average* values for the partitioned cells) and (b) that the CCD does not always agree quantitatively with the partition obtained in a single tube. Hence, some systematic adjustment of the phase system composition (again as outlined in Section I) may therefore be required after results of an initial CCD are obtained.

REFERENCES

Albertsson, P. Å. (1970). *Sci. Tools* **17**, 53.
Albertsson, P. Å. (1971). "Partition of Cell Particles and Macromolecules," 2nd ed. Wiley (Interscience), New York.
Albertsson, P. Å., and Baird, G. D. (1962). *Exp. Cell Res.* **28**, 296.
Brønsted, J. N. (1931). *Hoppe-Seyler's Z. Physiol. Chem., Ser. A*, 257.
Brooks, D. E. (1971). Ph.D. Thesis, University of Oregon Medical School, Portland.
Brooks, D. E., Seaman, G. V. F., and Walter, H. (1971). *Nature (London), New Biol.* **234**, 61.
Brunette, D. M., and Till, J. E. (1971). *J. Membrane Biol.* **5**, 215.
Brunette, D. M., McCulloch, E. A., and Till, J. E. (1968). *Cell Tissue Kinet.* **1**, 319.
Craig, L. C., and Craig, D. (1956). *In* "Techniques of Organic Chemistry" (A. Weissberger, ed.), 2nd ed., Vol. 3, Part I, p. 149. Wiley (Interscience), New York.
Danon, D., and Marikovsky, Y. (1961). *C. R. Acad. Sci.* **253**, 1271.
Johansson, G. (1970a). *Biochim. Biophys. Acta* **221**, 387.
Johansson, G. (1970b). *Biochim. Biophys. Acta* **222**, 381.
Miller, A., and Walter, H. (1971). *Fed. Proc., Fed. Amer. Soc. Exp. Biol.* **30**, 1312a.
Pertoft, H. (1969). *Exp. Cell Res.* **57**, 338.
Reitherman, R., Flanagan, S. D., and Barondes, S. H. (1973). *Biochim. Biophys. Acta* **297**, 193.
Schreiber, G., and Schreiber, M. (1971). *J. Biol. Chem.* **247**, 6340.
Seaman, G. V. F., and Heard, D. H. (1960). *J. Gen. Physiol.* **44**, 251.

Seaman, G. V. F., and Uhlenbruck, G. (1962). *Biochim. Biophys. Acta* **64**, 570.

Seaman, G. V. F., and Walter, H. (1971). *Fed. Proc., Fed. Amer. Soc. Exp. Biol.* **30**, 1182a.

Walter, H. (1969). *In* "Red Cell Membrane, Structure and Function" (G. A. Jamieson and T. J. Greenwalt, eds.), p. 368. Lippincott, Philadelphia, Pennsylvania.

Walter, H., and Coyle, R. P. (1968). *Biochim. Biophys. Acta* **165**, 540.

Walter, H., and Krob, E. J. (1974). *Fed. Proc., Fed. Amer. Soc. Exp. Biol.* **33**, 1320.

Walter, H., and Krob, E. J. (1975). *Exp. Cell Res.* (in press).

Walter, H., and Selby, F. W. (1966). *Biochim. Biophys. Acta* **112**, 146.

Walter, H., and Selby, F. W. (1967). *Biochim. Biophys. Acta* **148**, 517.

Walter, H., Selby, F. W., and Garza, R. (1967). *Biochim. Biophys. Acta* **136**, 148.

Walter, H., Garza, R., and Coyle, R. P. (1968a). *Biochim. Biophys. Acta* **156**, 409.

Walter, H., Garza, R., and Selby, F. W. (1968b). *Exp. Cell Res.* **49**, 679.

Walter, H., Krob, E. J., and Ascher, G. S. (1969a). *Exp. Cell Res.* **55**, 279.

Walter, H., Krob, E. J., Garza, R., and Ascher, G. S. (1969b). *Exp. Cell Res.* **55**, 57.

Walter, H., Edgell, M. H., and Hutchison, C. A., III. (1970). *Biochim. Biophys. Acta* **204**, 248.

Walter, H., Miller, A., Krob, E. J., and Ascher, G. S. (1971). *Exp. Cell Res.* **73**, 145.

Walter, H., Tung, R., Jackson, L. J., and Seaman, G. V. F. (1972). *Biochem. Biophys. Res. Commun.* **48**, 565.

Walter, H., Krob, E. J., and Ascher, G. S. (1973a). *Exp. Cell Res.* **79**, 63.

Walter, H., Krob, E. J., Brooks, D. E., and Seaman, G. V. F. (1973b). *Exp. Cell Res.* **80**, 415.

Walter, H., Krob, E. J., Ascher, G. S., and Seaman, G. V. F. (1973c). *Exp. Cell Res.* **82**, 15.

Weiser, M. M. (1973). *J. Biol. Chem.* **248**, 2536 and 2542.

Chapter 4

Synchronization of Cell Division in Vivo through the Combined Use of Cytosine Arabinoside and Colcemid[1]

ROBERT S. VERBIN AND EMMANUEL FARBER[2]

Departments of Pathology, University of Pittsburgh Schools of Medicine and Dental Medicine, Pittsburgh, Pennsylvania and Fels Research Institute, Temple University School of Medicine, Philadelphia, Pennsylvania

I. Introduction

A. Historical Perspective

Many of the early studies on proliferation of both prokaryotic and eukaryotic cells were mainly concerned with attempts to illucidate the physico-

[1] This research was supported in part by Grants CA-11390 and CA-12218 from the National Cancer Institute, Grant AM-14882 from the National Institute of Arthritis and Metabolic Diseases, and Grant BC-7N from the American Cancer Society.
[2] Dr. Farber is an American Cancer Society Research Professor of Pathology and Biochemistry.

chemical factors associated with division. With the introduction of radioactive precursors and the development of autoradiographic and liquid scintillation techniques, the emphasis was shifted toward the delineation of the molecular events during interphase, i.e., the period between two successive nuclear divisions. The first major contribution in this area was made by Howard and Pelc (1953). Through the use of thymidine-^{3}H these investigators were able to provide convincing evidence that the replication of the genome and mitotic division occurred at two different and distinct times during cell proliferation. From these observations, the concept of the "cell cycle" was subsequently formulated. As a result, it is now generally recognized that continuously dividing cells pass through four relatively well-defined stages: G_1, the interval between the completion of mitosis and the beginning of DNA synthesis; S, the period of DNA replication; G_2, the interval between the completion of DNA synthesis and the onset of mitosis; and M, the stage of actual nuclear and cellular division.

Originally, the G_1 and G_2 compartments were referred to as "gaps" and, as such, accurately reflected the paucity of information relative to the biosynthetic activities and regulatory mechanisms that were presumably operating during these particular intervals. Furthermore, it was apparent that, in most instances, the passage of cells through the various stages took place in an asynchronous manner. Therefore it became obvious that the eventual identification and analysis of the various metabolic interractions characteristic of interphase was intimately dependent upon the availability of more sophisticated experimental models which would permit detailed investigations of the biochemical and molecular features unique to each specific compartment of the cell cycle. Consequently, a concerted effort was made to devise systems that would simulate the synchronous cellular division that occurred in certain natural situations such as the simultaneous cleavage of marine eggs. As cited by James (1966), among the first cells to be artificially synchronized were the prokaryotic *Chlorella*, *Amoeba*, bacteria, and *Tetrahymena*.

B.   Synchronization *in Vitro*

More recently, partial synchronization of division in eukaryotic cells *in vitro* has been obtained with considerable success through the implementation of a variety of techniques (Newton and Wildy, 1959; Terasima and Tolmach, 1963; Robbins and Marcus, 1964; Kim and Perez, 1965; Mitchison and Vincent, 1965; Sinclair and Morton, 1965; Stubblefield and Klevecz, 1965; Whitfield and Youdale, 1965; Lindahl and Sorenby, 1966; Morris *et al.*, 1967; Ramsdahl and Deanen, 1968; Rao, 1968). One method generally used is the exposure of randomly proliferating cells to agents that interfere with specific biosynthetic activities, e.g., DNA replication. The cells thus blocked in the S phase are subsequently released by simply being washed in control media, followed in some instances by the addition of exogenous nucleotides, thereby permitting their simultaneous wavelike movement into succeeding segments of the cell cycle (Eidenoff and Rich, 1959; Rueckert and Mueller, 1960; Littlefield, 1962; Xeros, 1962; Erickson and Szybalski, 1963; Mueller, 1963; Schindler, 1963; Till *et al.*, 1963; Bootsma *et al.*, 1964; Petersen and Andersen, 1964; Kasten *et al.*, 1965; Firket and Mahieu, 1966; Galavazi *et al.*, 1966; Tobey *et al.*, 1966; Adams and Lindsay, 1967; Priest *et al.*, 1967; Puck, 1967; Steffen and Stolzmann, 1969; Kuroki and Sato, 1970). For a more complete discussion of the techniques and other aspects of synchronization in *in vitro* systems, the reader is referred to the excellent reviews by Engelberg and Hirsh (1966), Frindel and Tubiana (1971), and Nias and Fox (1971).

C. Synchronization *in Vivo*

Despite the encouraging results obtained from *in vitro* systems, attempts to attain comparable degrees of synchronization in the intact animal have been, for the most part, substantially less rewarding. The basis for the difficulties encountered *in vivo* can be attributed to a large extent to the inherent diversity in the proliferative capacity of mammalian cells. On the basis of the differences in the growth kinetics of their cell populations, all tissues of the body can be placed into either of three major categories: (a) permanent nonproliferating cells, i.e., those that have lost the ability to proliferate under any circumstances, e.g., neurons; (b) "resting" cells; and (c) cells programmed for continuous replication.

1. STUDIES IN "RESTING" CELLS

Certain mammalian cells are programmed for occasional but self-limiting proliferative activity that can be triggered by the application of an appropriate stimulus. A common prototype for this category of cells is represented by the population of hepatocytes which are stimulated to enter the cell cycle following partial hepatectomy. Although the parenchymal cells of the undamaged mammalian liver are mitotically inert, removal of two-thirds of its mass results in the active synthesis of nucleic acids followed by a burst of cell division. Between 22 and 24 hours after surgery 55–60% of the liver cells are stimulated to engage in DNA synthesis (Bucher and Swaffield, 1964) while the peak of mitotic activity (approximately 4.0%) is reached 28 hours postoperatively (Bucher, 1963; Verbin *et al.*, 1969). Baserga (1968), in his review of the maximal levels of DNA synthesis obtained in other *in vivo* systems, reported the following figures: 80% in salivary gland cells stimulated by isoproterenol, 20% in folic acid stimulated kidney, regenerating pancreas and lactating mammary gland, and 50% in the basal layer of the epidermis after wounding. The response of the thyroid and bone marrow to thiouracil (Doniach and Logothetopoulos, 1955) and erythropoietin (Guzman and Lajtha, 1970), respectively, represent additional instances of stimulated cellular proliferation. However, as pointed out by Nias and Fox (1971), "In each case, only a cohort of the population is triggered to proliferate and the degree of synchrony is relatively low in proportion to the total population."

2. STUDIES IN CONTINUOUSLY REPLICATING CELLS

A further characteristic of the aforementioned cells is their ability to traverse the cell cycle for only a relatively short period of time before reverting to their usual "resting" state. In contrast, there is a variety of tissues in the body that, even under unstimulated conditions, demonstrate a high and sustained rate of replication. These particular tissues are best exemplified by hematopoietic and lymphoid elements as well as by the intestinal crypts of Lieberkuhn, all of which consist in part of populations of continuously dividing cells. This feature, together with their accessibility, quite naturally have made them attractive model systems in which to test a variety of agents which might induce a state of synchronous division *in vivo*.

The synchronization action of X-irradiation, as demonstrated by dose fractionation, has been well established (Kallman, 1963; Till and McCulloch, 1963; Frindel *et al.*, 1966; Withers, 1967; Elkind *et al.*, 1968). However, the utilization of chemical antimetabolites to duplicate *in vivo* the synchronous movement of cells noted *in vitro* presents a much more complex situation. In the intact animal, intricate enzymic reactions determine not only the rapidity by which the inhibitory agent exerts its influence, but also the rapidity by which it is catabolized. This latter factor is obviously a prerequisite both for the resumption of movement of the previously arrested cells and for establishing the degree to which synchronization is attained. Consequently, this approach necessitates the careful selection of suitable inhibitors with structural configurations and chemical characteristics that render them susceptible to the intrinsic intracellular regulatory mechanisms. In an attempt to identify such a compound, we have been examining the effects of various antimetabolites on the crypts of the small intestine. Thus far the agent that seemed to possess the greatest potential for meeting the metabolic criteria outlined above was cytosine arabinoside (Ara-C), a fraudulent nucleoside and potent inhibitor of DNA synthesis (Doering *et al.*, 1960; Pizer and Cohen, 1960; Chu and Fischer, 1962, 1968; Buthala, 1964; Bader, 1965; Kim and Eidinoff, 1965; Silagi, 1965; Estensen and Baserga, 1966; Kimball *et al.*, 1966; Kit *et al.*, 1966; Peterkofsky and Tomkins, 1967; Atkinson and Stacey, 1968; Creasey *et al.*, 1968; Smith, 1968; Hirschman *et al.*, 1969; Inagaki *et al.*, 1969; Lenaz *et al.*, 1969; Graham and Whitmore, 1970; Lenaz and Philips, 1970; Lieberman *et al.*, 1970; Bertalanffy and Gibson, 1971; Verbin *et al.*, 1972a). This tentative conclusion was formulated on the basis of the following data: (a) An almost complete inhibition of DNA synthesis occurs in the crypts of Lieberkuhn within 10–15 minutes (Lieberman *et al.*, 1970). In apparent contrast to the findings of other investigators (Estensen and Baserga, 1966; Lenaz *et al.*, 1969; Lenaz and Philips, 1970; Lieberman *et al.*, 1970), this interruption in DNA synthesis is not associated with the concomitant development of epithelial cell death (Verbin *et al.*, 1972a). (b) The subsequent restitution of DNA replication in this tissue occurs within a reasonable period of time (Lenaz *et al.*, 1969; Lenaz and Philips, 1970; Lieberman *et al.*, 1970).

A number of studies have also demonstrated that Ara-C inhibits mitotic activity in mammalian cells both *in vitro* (Kim and Eidinoff, 1965) and *in vivo* (Lenaz *et al.*, 1969; Bertalanffy and Gibson, 1971). Recent experiments conducted in this laboratory (Verbin *et al.*, 1972a) revealed a similar effect in the proliferative compartment of rat intestinal mucosa and, in addition, disclosed that recovery from the initial interference with cell division was characterized by a pronounced rebound 12 hours after the administration of the antimetabolite. At this particular time interval, the population of dividing cells in the intestinal crypts increased from the 3 to 5% levels normally occurring in control animals to approximately 15% in rats treated with Ara-C. Additional data obtained from autoradiograms indicated that the cells contributing to this rebound were those originally blocked in the S phase of the cell cycle (Verbin *et al.*, 1972a).

These observations raised the possibility that an even greater accumulation of proliferating cells and, consequently, a considerable degree of synchrony could be achieved by temporarily preventing the subsequent passage of these particular cells through mitosis. Such an approach has been made feasible by the availability of agents, such as colchicine and Colcemid (CCM), both of which arrest mitotic cells in metaphase (cf. Brinkley *et al.*, 1967). The utilization of the former compound was deemed impractical because it induces epithelial cell degeneration in the crypts of Lieberkuhn (Hooper, 1961). Since CCM has been reported to be less cytotoxic (Schär *et al.*, 1954), this agent was used in the current investigations. The method of synchronization of a large proportion of the small intestinal crypt cells with the use of Ara-C plus CCM is here described.

II.   Materials and Methods

Male white Wistar rats (Carworth Farms, Inc., New City, New York), weighing 165–200 gm after a 16- to 18-hour period of fasting, were utilized in all experiments. Freshly prepared solutions of the following compounds were injected intraperitoneally: (a) Ara-C (250 mg/kg) (Cancer Chemo-therapy Service Center of the National Cancer Institute, Bethesda, Mary-land), as the hydrochloride, was made up in distilled water (50 mg/ml) and adjusted to pH 5.5–6.0 with a few drops of 10% NaOH; and (b) CCM (0.05 mg/100 gm) (Ciba Pharmaceutical Company, Summit, New Jersey) in a con-centration of 1 mg/ml (CCM, 0.01 gm; 95% ethanol, 0.65 ml; propylene glycol, 1.0 ml; $Na_2HPO_4 \cdot 7 H_2O$, 0.015 gm; $NaH_2PO_4 \cdot H_2O$, 0.0133 gm; and distilled water, a sufficient quantity up to 10 ml). Control animals received either vehicle (i.e., the solvent mixture for CCM) or 0.9% NaCl solution by a similar route.

The rats were sacrificed by decapitation, and the first 15-cm segment of the small intestine was removed and fixed in Stieve's solution. For histological examination, sections of paraffin-embedded tissue were stained with hema-toxylin and eosin. Mitotic activity, expressed as percentage of mitoses, was established in each animal by counting the number of dividing and non-dividing nuclei in at least 30 longitudinally sectioned intestinal crypts.

III.   Results

A.   Effects of Cytosine Arabinoside on Mitotic Activity

One control and 2 experimental animals were killed at each of the fol-lowing intervals: 0.5, 1.5, 2.5, 3.5, 8, 12, 14, 16, and 24 hours after the administration of Ara-C. As shown in Fig. 1, mitotic activity, markedly reduced at 1.5 hours, was completely abolished within 2.5 hours. Sub-sequently, mitoses began to reappear and were appreciable, but were still below the normal value at 8 hours. After this time they progressively in-creased in number, reaching a peak at 12 hours when they showed a 3-fold increase over control values. Cell division subsequently decreased and returned to the range of the controls by 16–24 hours.

Examination of autoradiograms prepared from tissues obtained from animals pretreated with thymidine-^{3}H and killed 12 hours after the injection of Ara-C revealed the presence of silver grains over virtually 100% of the mitotic figures (Verbin et al., 1972a). This particular observation precludes the possibility of a selective destruction of S-phase epithelial cells by the analog (Young and Fischer, 1968; Karon and Shirakawa, 1969) or a mobili-zation of reserve G_2 cells (Gelfant, 1962; Post and Hoffman, 1969; Pederson and Gelfant, 1970) for entry into mitosis.

B.   Effects of a Single Injection of Colcemid on Mitotic Activity

This series of experiments was designed to establish: (a) the sequential pattern of mitotic activity and (b) the presence or the absence of necrosis

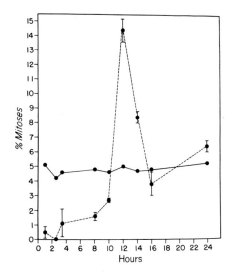

FIG. 1. Effects of Ara-C on mitotic activity in the intestinal crypts. Points in the experimental groups represent the mean of two animals: vertical bars indicate ranges. ●——●, Control; ● --- ●, experimental. From Verbin *et al.* (1972a).

TABLE I

EFFECTS OF A SINGLE INJECTION OF COLCEMID (CCM) ON MITOTIC ACTIVITY IN THE INTESTINAL CRYPTS OF THE RAT[a]

Treatment	Hour after Injection	Mitoses (%)
0.9% NaCl solution (1)[b]	1	4.9
Vehicle (1)		3.6
CCM (2)		9.2 ± 0.5[c]
0.9% NaCl solution (1)	2	4.0
Vehicle (1)		4.2
CCM (2)		15.6 ± 0.3
0.9% NaCl solution (1)	3	5.0
Vehicle (1)		3.0
CCM (2)		8.0 ± 0.5
0.9% NaCl solution (1)	4	4.0
Vehicle (1)		4.0
CCM (2)		4.6 ± 0.8
0.9% NaCl solution (1)	6	4.3
Vehicle (1)		4.9
CCM (2)		3.8 ± 0.4

[a]From Verbin *et al.* (1972b).
[b]Number in parentheses is number of animals treated.
[c]Mean ± average deviation

in the intestinal crypts of animals given a single injection of CCM. Table I demonstrates that there were no significant differences in mitotic activity in control animals treated with either 0.9% NaCl solution or vehicle. The number of dividing cells in these two groups of rats ranged from 3.0 to 4.9%. In contrast, mitoses increased substantially in those animals treated with CCM. This augmentation was apparent within 1 hour, when 9.2% of the crypt cells were in mitosis. After reaching a peak value of 15.6% at the 2-hour interval, mitotic activity progressively declined, returning to control levels within 4–6 hours after the administration of CCM. Also, there was no evidence of cell damage at any of the intervals studied. These observations confirm those of Kleinfeld and Sisken (1966), which indicated that, under certain conditions, cells exposed to CCM are capable of reforming a functional spindle and completing mitosis.

C. Mitotic Response to the Combined Administration of Cytosine Arabinoside and Colcemid

On the basis of the data derived from the preceding experiments, it seemed reasonable to consider that a synergistic effect on mitotic activity might be obtained through the utilization of Ara-C in conjunction with CCM. For a test of this hypothesis, Ara-C was administered at zero time and CCM was administered 12 hours thereafter. Additional rats that served as controls received the following combinations of injections at comparable times: NaCl–NaCl, NaCl–vehicle, NaCl–CCM, Ara-C–NaCl, or Ara-C–vehicle. All animals were killed 2 hours after the second injection. Figure 2 summarizes data obtained from three separate experiments; findings in each of the three independent studies were quite similar and reproducible. The mean mitotic activity in the NaCl–NaCl- or NaCl–vehicle - treated animals ranged between 4.7 and 5.1%, while mitoses increased to approximately 15% in rats receiving 0.9% NaCl solution followed by CCM. In those animals treated with Ara-C–NaCl or Ara-C–vehicle, approximately 8.0–8.5% of the crypt cells were in mitosis. However, in rats given injections of both Ara-C and CCM, 38% of the crypt cells were undergoing division. Of additional significance was the finding that this latter regimen was not necrogenic to the intestinal epithelium.

D. Maintenance of Synchronization

To determine the degree to which this synchronization of cell division was maintained, animals were given either 0.9% NaCl solution or Ara-C at zero time and CCM 12 hours later. Groups of rats were then killed at 2-hour intervals beginning at 14 hours and terminating at 24 hours after the initial

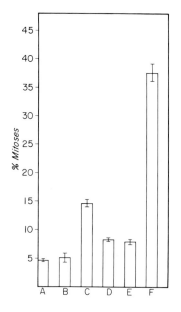

FIG. 2. Effects of cytosine arabinoside–Colcemid (Ara-C–CCM) treatment on mitotic activity in the intestinal crypts. At zero time, 0.9% NaCl solution or Ara-C was injected, followed by the administration of 0.9% NaCl solution, vehicle, or CCM 12 hours thereafter. All rats were killed 2 hours after the second injections. Numbers in parentheses indicate number of animals utilized; vertical lines, SD. A: NaCl + NaCl (3); B: NaCl + vehicle (5); C: NaCl + CCM (6); D: Ara-C + NaCl (7); E: Ara-C + vehicle (6); F: Ara-C + CCM (11). From Verbin *et al.* (1972b).

injections. Table II shows that mitotic activity in animals treated with NaCl–CCM followed a pattern analogous to that found in our initial studies with this agent (Table I). Additionally, Table II reveals a pronounced increase in mitotic cells in Ara-C-treated animals that were killed 2 hours after the subsequent administration of CCM. In these latter rats, 33% of the crypt cells were in mitosis at this particular interval. This value is consistent with the earlier findings reported above. However, after this initial augmentation, the percentage of mitotic cells in Ara-C–CCM-treated animals killed 4 hours after the administration of CCM dropped abruptly to 2.5%, indicating a release from the blockade. Mitosis then returned to the 3–5% levels observed previously in control rats treated with either 0.9% NaCl solution or vehicle.

For a further examination of the kinetics of cellular proliferation induced by the concomitant administration of Ara-C and CCM, additional studies that extended over a longer period of time and utilized a second injection of CCM were conducted. The design and results obtained from

TABLE II

MITOTIC ACTIVITY IN THE INTESTINAL CRYPTS AT VARIOUS INTERVALS AFTER A SINGLE
INJECTION OF CYTOSINE ARABINOSIDE (ARA-C) OR NaCl SOLUTION AND COLCEMID (CCM)[a]

Treatment given at		Time of sacrifice (hours after first injection)	Mitoses (%)
0 Hours	12 Hours		
0.9% NaCl solution (1)[b]	CCM	14	14.9
Ara-C (2)	CCM		33.0 ± 2.5[c]
0.9% NaCl solution (1)	CCM	16	4.6
Ara-C (2)	CCM		2.5 ±0.6
0.9% NaCl solution (1)	CCM	18	43.
Ara-C (2)	CCM		3.5 ±0.5
0.9% NaCl solution (1)	CCM	20	5.2
Ara-C (2)	CCM		4.5 ±0.6
0.9% NaCl solution (1)	CCM	22	5.3
Ara-C (2)	CCM		4.5 ±0.2
0.9% NaCl solution (1)	CCM	24	3.8
Ara-C (2)	CCM		5.0 ±0.4

[a]From Verbin et al. (1972b).
[b]Number in parentheses is number of animals treated.
[c]Mean ± average deviation.

these experiments are outlined in Table III. In control animals receiving NaCl–vehicle–vehicle, from 4.0 to 4.7% of the crypt cells were in mitosis. No significant differences in the percentage of mitoses were noted at any of the intervals in animals receiving either NaCl–CCM–CCM or Ara-C–CCM–CCM. In both instances, the values were essentially the same as those observed in animals killed 2 hours after the injection of CCM alone. Measurements of the incorporation of labeled thymidine into DNA in the total intestinal wall or Ara-C–CCM–treated animals showed a pronounced decrease (80%) at 12, 14, and 18 hours after the initial injection of the analog (E. Farber and H. Liang, unpublished data). DNA synthesis continued to be markedly depressed at 20 hours even though the percentage of mitoses at that particular time had fallen to within the control range (Table II). No evidence of synchronization of the cells during the subsequent S phase was obtained. Thus it appears that the initial synchronization of cell division induced by the combined utilization of Ara-C and CCM is lost once traversal through the cell cycle is reinstituted. A similar dispersion or decay of synchrony has also been noted in many different cell types in vitro (e.g., Till et al., 1963; Nias and Fox, 1971).

TABLE III

Mitotic Activity in the Intestinal Crypts at Various Intervals Following a Single Injection of Cytosine Arabinoside (Ara-C) or NaCl Solution and 2 Injections of Colcemid (CCM)[a]

0 Hours	Treatment given at						Mitoses (%)
	12 Hours	24 Hours	26 Hours	28 Hours	30 Hours	32 Hours	
0.9% NaCl solution (1)[b]	CCM	CCM	Sacrificed	—	—	—	17.2
0.9% NaCl solution (1)	Vehicle	Vehicle	Sacrificed	—	—	—	4.0
Ara-C (2)	CCM	CCM	Sacrificed	—	—	—	15.4 ± 0.6[c]
0.9% NaCl solution (1)	CCM	—	CCM	Sacrificed	—	—	17.7
0.9% NaCl solution (1)	Vehicle	—	Vehicle	Sacrificed	—	—	4.5
Ara-C (2)	CCM	—	CCM	Sacrificed	—	—	18.6 ± 0.8
0.9% NaCl solution (1)	CCM	—	—	CCM	Sacrificed	—	18.7
0.9% NaCl solution (1)	Vehicle	—	—	Vehicle	Sacrificed	—	4.7
Ara-C (2)	CCM	—	—	CCM	Sacrificed	—	15.9 ± 0.7
0.9% NaCl solution (1)	CCM	—	—	—	CCM	Sacrificed	16.7
0.9% NaCl solution (1)	Vehicle	—	—	—	Vehicle	Sacrificed	4.2
Ara-C (2)	CCM	—	—	—	CCM	Sacrificed	14.9 ± 0.1

[a]From Verbin et al. (1972b)

[b]Number in parentheses, is number of animals treated.

[c]Mean ± average deviation.

IV.   Discussion

The application to the entire organism of methods of cell manipulation developed *in vitro* would seem to be of considerable importance in the study of many biological problems. Similarly, a greater understanding of the mechanisms by which certain antimetabolites interfere with the functional activities of replicating cells would seem relevant to a more rational approach to the treatment of neoplastic disorders. Requisite to the accumulation of such information is the availability of an experimental model system that will permit a detailed analysis of the various metabolic interactions occuring at specific loci within the cell cycle. From a practical standpoint these objectives can be achieved most effectively by utilizing synchronous cell populations *in vivo*. The availability of a variety of different methods to achieve this is essential in this context.

Efforts to induce cell synchronization in the intact animal through the introduction of chemical mediators have, for the most part, been hindered by at least two important factors: (a) the inability to collect a sufficiently large number of cells in any given compartment of the cell cycle and (b) the induction of irreversible damage in those cells exposed to the inhibitor. That these difficulties are not insurmountable, however, has been demonstrated by a number of studies which have clearly shown that a considerable degree of synchronization can be achieved *in vivo* in a variety of malignant and normal tissues without a concomitant loss of cell viability.

Nitze *et al.* (1971) studied the effects of 5-fluorouracil (FU) on human tumors. Patients were given 1.0 gm of the antimetabolite in 1000 ml of 5.4% glucose by infusion over an 18-hour period. Small pieces of tumor were removed at various times and subjected to autoradiography after exposure to thymidine-^{3}H. Upon cessation of the infusion (release of the block), the number of labeled cells increased and reached a plateau within 5 hours. A distinct rise in the mitotic index was also noted, the peak occurring at the 9-hour interval. In a somewhat similar investigation (Barranco *et al.*, 1973), patients with malignant melanoma were infused with either 25 mg/day or 15 mg/day of bleomycin for 4 days. Excisional biopsies of subcutaneous nodules were subsequently obtained at 12- to 24-hour intervals and pulse-labeled with thymidine-^{3}H. Determination of the labeling indices by liquid scintillation and autoradiographic techniques revealed a 1.5- to 4-fold increase in the number of cells synthesizing DNA in the bleomycin-treated tumors.

Induction of apparent increased synchrony has also been accomplished in murine melanomas as well as Ehrlich ascites tumors by the utilization of Ara-C. In their initial studies, Bertalanffy and Gibson (1971) administered four injections of 12.5 mg/kg of Ara-C daily on each of 3 days with a day interposed between each series of injections to allow the animals to recuperate from the toxic effects of the drug. Colchicine (0.2 mg/100 gm) was given 4 hours prior to sacrifice. Their results demonstrate that, after the first two recovery periods, the mitotic rates of the melanomas eventually exceeded control levels in both instances. As a corollary study, Ara-C was given in a single 50 mg/kg dose. Thymidine-^{3}H was administered with or at various times after Ara-C over a 24-hour period. Autoradiograms revealed that, in the Ara-C treated animals the thymidine-labeling indices exhibited an overshoot considerably higher than the control levels noted in the untreated melanomas and Ehrlich ascites tumors.

More recently Gibson and Bertalanffy (1972) examined in greater detail the effects of various doses and regimens of Ara-C administration on the uptake of thymidine-^{3}H and the mitotic index in the transplantable B_{16} mouse melanoma. The best results, with respect to synchronization, were achieved in animals receiving 8 injections of Ara-C (12.5 mg/kg) spaced at 2-hour intervals and in those receiving 16 injections of Ara-C (18.8 mg/kg) given at 1-hour intervals. In the former instance the labeling index (LI) and mitotic index (MI) were 45.06% and 3.5%, respectively, and in the latter instance 38.90% and 3.5%, respectively. Corresponding values in untreated control animals were 18.60% (LI) and 2.18% (MI). The synchronization index, as determined by the method of Sinclair and Morton (1965) was 40.7 in those mice receiving the 8 injections and 33.7 in those given 16 injections.

Augmentation of DNA synthesis and mitotic activity has also been demonstrated in rat mammary carcinoma (Rajewsky, 1970). The maximum mean fraction of cells engaged in the replication of DNA was noted at about 8 hours after the injection of a single 0.5 mg/gm dose of hydroxyurea (HU). These values exceeded the normal mean by about 70–100%. Similarly, a 2- to 3-fold increase in the fraction of cells in mitosis occurred about 12 hours after HU. The effects of this particular compound on fetal liver obtained from BDIX rats on day 18 of gestation have been examined by Rajewsky *et al.* (1971). The rate of thymidine-^{3}H incorporation and the fraction of mitotic cells were determined at various intervals for up to 18 hours after the transplacental administration of HU (0.25 mg/rat). After an initial block of about 2.5 hours DNA synthesis progressively increased and attained a maximum level of 120% of control values at about 7 hours. This enhanced synthesis of DNA was followed by a 2-fold increase in the MI at 7–9 hours. Additional investigations (Vassort *et al.*, 1971) have demonstrated that the exposure of bone marrow cells to HU leads to an accumulation of 52% of colony forming units in the S phase 12 hours after the administration of a 0.5 mg/gm dose of the inhibitor. A somewhat higher concentration (57%) of DNA-synthesizing colony-forming units was noted at 16 hours in animals treated with 2.5 mg/gm of HU. In control mice, 20% of colony-forming units were in the S phase.

The studies with human and experimental cancer are interesting but difficult to interpret. It is now widely appreciated that many solid neoplasms have only a minority of cells that are proliferating and that it is possible to change the growth fraction by experimental manipulation. Therefore, an effect of any proposed synchronization regimen on the mobilization of nonproliferating cells into the cell cycle must be ruled out before a bona fide increase in the degree of synchronization can be considered as possible.

The response of intestinal epithelial cells to HU has been reported by Gillette et al. (1970). This agent was administered to BDF_1 mice in an initial dose of 0.5 mg/gm followed by 2 additional doses of 0.2 mg/gm at 45-minute intervals. Thymidine-^{3}H was injected at hourly intervals after the last HU injection and the animals killed 30 minutes later. Autoradiograms revealed that the thymidine-labeling index reached a peak (approximately 48%) at 6 hours after the final injection of HU, fell to about 10% at 11 hours and then reached a second maximum level of about 55% at 13 hours. In a subsequent study Dethlefsen and Riley (1973) measured the radioactivity in mouse duodenal crypts following the administration of IUDR-^{125}I at various times after a single 3 mg/gm dose of HU. Their results showed that there was no incorporation of the isotope until the sixth hour. However, by 8 hours the radioactivity increased sharply to a peak value of 55% of control values. Following a decline to 16%, a second increase in radioactivity with an overshoot of 150% was noted at 20 hours. In parallel autoradiographic studies IUDR-^{3}H or thymidine-^{3}H was injected at selected times after HU and the animals killed 30 minutes later. A mean labeling index of 0.50 was obtained 8 hours after HU. The index then fell to about 0.15 at 11 hours and subsequently showed another rise to 0.47 by the sixteenth hour. These values, when compared to a mean labelling index of 0.26 in control animals, reflected a 2-fold increase in the number of cells engaged in DNA synthesis.

It is apparent from the data obtained from the present day study that partial synchronization can also be achieved in the intestinal crypt epithelial cells via the administration of Ara-C followed, at an appropriate time, by the injection of CCM. The former compound, by virtue of its ability to interrupt DNA synthesis, institutes an S-phase block. The latter agent prevents progression through actual mitosis by impeding migration of the centrioles, thereby interfering with the organization of the mitotic spindle (Brinkley and Stubblefield, 1966; Brinkley et al., 1967). The efficiency of this regimen is readily apparent; treatment with either Ara-C or CCM results in peak mitotic values of approximately 15%. When compared with controls, this represents a 3- to 5-fold increase in the number of mitoses. When these compounds are used sequentially, however, an even more significant elevation of the mitotic index occurs. Under these latter conditions, 33–38% of the crypt epithelial components can be identified as dividing cells, thus reflecting a considerable degree of synchrony.

The magnitude of the synchronization can best be appreciated by examining the various parameters ascribed to the cell cycle of the intestinal crypts. In the rat, the total generation time of the proliferating cells has been reported to be about 12 hours, with the mitotic phase occupying 1 hour (Cairnie *et al.*, 1965). Our observations as well as those of other investigators (Leblond and Stevens, 1948; Lesher and Bauman, 1969), indicate that, in control animals, 3–5% of the epithelial cells are in mitosis at any given time. Consequently, these latter values would represent one-twelfth of the population actually traversing the cell cycle. Therefore, if total synchronization were achieved, one would expect a 12-fold increase in the number of dividing cells, in which case from 36 to 60% of the crypt cells would be in mitosis. The higher of these two estimated values is in good agreement with that determined by Lesher and Bauman (1969). Thus, the 33–38% values obtained by treatment with Ara-C and CCM indicate synchronization of at least two-thirds of the proliferating epithelial cells in the intestinal mucosa. However, our data also indicate that this synchronization is dissipated once the cells originally arrested in mitosis pass into subsequent stages of the cell cycle. Under our experimental conditions, this appears to occur in G_1, presumably as a result of the variability and complexity of the regulatory mechanisms operative during this stage of the cycle (Prescott, 1968).

The mechanism whereby Ara-C leads to the synchronization of a large proportion of the crypt cells is not clear. Since inhibition of DNA synthesis by Ara-C is rapid and almost complete (Lieberman *et al.*, 1970), presumably the effect on DNA replication is pretty general throughout the S phase. The partial synchronization of cells with Ara-C alone would imply that the recovery from the inhibition is not uniform throughout the S phase but must be nonuniform, thereby allowing a significant proportion of cells to accumulate at certain sites in the cell cycle. The nature of this nonuniform affect and its site or sites remains to be established. It would appear that it is in part a function of the nature of the inhibition. Cycloheximide, which inhibits not only DNA but also protein synthesis, exerts a block in both G_2 and in the S phase. This antibiotic induces only a small degree of synchronization of intestinal crypt cells (Verbin and Farber, 1967). Thus, the unpredictable effect of Ara-C would seem to be worthy of an in-depth study, since it might reveal new features about DNA replication and its control.

During the past several years the sensitivity of dividing cells to a number of cytotoxic agents during different phases of the generation cycle has been studied *in vitro* (Terasima and Tolmach, 1961; Walker and Helleiner, 1963; Sinclair, 1965; 1967; Sinclair and Morton, 1966; Kim *et al.*, 1967; Mauro and Elkind, 1968; Sakamoto and Elkind, 1969; Mauro and Madoc-Jones, 1970; Wheeler *et al.*, 1970; Goldenberg *et al.*, 1971; Tobey, 1972; Wheeler *et al.*, 1972; Bhuyan *et al.*, 1973). In the intact animal, investigations have been conducted to test the variability in the age response of cells synchronized with HU or FU to irradiation (Mauro and Madoc-Jones, 1969; Gillette *et al.*, 1970; Madoc-Jones and Mauro, 1970; Chaffey and Hellman, 1971; Nitze *et al.*, 1971), nitrogen mustard (Ash *et al.*, 1972), and vincristine, vinblastine, and a second injection of HU (Madoc-Jones and Mauro, 1970). More recently we have utilized our Ara-C–CCM method of synchronizing crypt epithelial cells in an attempt to delineate the effects of nitrogen mustard on G_2, M, or G_1 cells. This was accomplished by injecting nitrogen mustard at 10, 14, or 16 hours, respectively, after the initial administration of Ara-C. Our results indicate that nitrogen mustard induces distinctive and severe degenerative morphological alternations which are not only virtually identical regardless of the position of the crypt epithelial cells within the generation cycle at the time of drug administration but also indistinguishable from those induced in asynchronous populations (Verbin *et al.*, 1974). This lack of phase specificity has also been reported in cancer cells (Bruce *et al.*, 1966). Thus synchronization of the crypt epithelial cells with Ara-C and CCM may provide an interesting model for analyzing the response of proliferating cells in the intact animal to disturbances in their biosynthetic activities during different, predetermined stages of the cell cycle.

REFERENCES

Adams, R. L. P., and Lindsay, J. G. (1967). *J. Biol. Chem.* **242**, 1314–1317.
Ash, R., Chaffey, J. T., and Hellman, S. (1972). *Cancer Res.* **32**, 1695–1702.
Atkinson, C., and Stacey, K. A. (1968). *Biochem. Biophys. Acta* **163**, 705–707.
Bader, J. P. (1965). *Science* **149**, 757–758.
Barranco, S. C., Luce, J. K., Ramsdahl, M. M., and Humphrey, R. M. (1973). *Cancer Res.* **33**, 882–887.
Baserga, R. (1968). *Cell Tissue Kinet.* **1**, 167–191.
Bertalanffy, F. D., and Gibson, M. H. L. (1971). *Cancer Res.* **31**, 66–71.
Bhuyan, B. K., Fraser, T. J., Gray, L. G., Kuentzel, S. L., and Neil, G. L. (1973). *Cancer Res.* **33**, 888.
Bootsma, D., Budke, L., and Vos, O. (1964). *Exp. Cell Res.* **33**, 301–309.
Brinkley, B. R., and Stubblefield, E. (1966). *Chromosoma* **19**, 28–43.
Brinkley, B. R., Stubblefield, E., and Hsu, T. C. (1967). *J. Ultrastruct. Res.* **19**, 1–18.
Bruce, W. R., Meeker, B. E., and Valeriote, F. A. (1966). *J. Nat. Cancer Inst.* **37**, 233–245.
Bucher, N. L. R. (1963). *Int. Rev. Cytol.* **15**, 245–300.
Bucher, N. L. R., and Swaffield, M. N. (1964). *Cancer Res.* **24**, 1611–1625.

Buthala, D. (1964). *Proc. Soc. Exp. Biol. Med.* **115**, 69–77.

Cairnie, A. B., Lamerton, L. F., and Steel, G. G. (1965). *Exp. Cell Res.* **39**, 528–538.

Chaffey, J. T., and Hellman, S. (1971). *Cancer Res.* **31**, 1613–1615.

Chu, M. Y., and Fischer, G. A. (1962). *Biochem. Pharmacol.* **11**, 423–430.

Chu, M. Y., and Fischer, G. A. (1968). *Biochem. Pharmacol.* **17**, 741–751.

Creasey, W. A., DeConti, R. C., and Kaplan, S. R. (1968). *Cancer Res.* **28**, 1074–1081.

Dethlefsen, L. A., and Riley, R. M. (1973). *Cell Tissue Kinet.* **6**, 3–16.

Doering, A., Keller, J., and Cohen, S. S. (1960). *Cancer Res.* **26**, 2444–2450.

Doniach, I., and Logothetopoulos, J. H. (1955). *Brit. J. Cancer* **9**, 117–127.

Eidenoff, M. L., and Rich, M. A. (1959). *Cancer Res.* **19**, 521–524.

Elkind, M. M., Withers, H. R., and Belli, J. A. (1968). *Front. Radiat. Ther. Oncol.* **3**, 55–87.

Engelberg, J., and Hirsch, H. R. (1966). *In* "Cell Synchrony" (I. L. Cameron and G. M. Padilla, eds.), pp. 14–37. Academic Press, New York.

Erickson, R. L., and Szybalski, W. (1963). *Radiat. Res.* **18**, 200–212.

Estensen, R. D., and Baserga, R. (1966). *J. Cell Biol.* **30**, 13–22.

Firket, H., and Mahieu, P. (1966). *Exp. Cell Res.* **45**, 11–22.

Frindel, E., and Tubian, M. (1971). *In* "The Biochemistry of Disease" (R. Baserga, ed.), Vol. 1, pp. 391–447. Dekker, New York.

Frindel, E., Chairuyer, F., Tubiana, M., Kaplan, H. S., and Alpen, E. L. (1966). *Int. J. Radiat. Biol.* **11**, 435–443.

Galavazi, G., Schenk, H., and Bootsma, D. (1966). *Exp. Cell Res.* **41**, 428–437.

Gelfant, S. (1962). *Exp. Cell Res.* **26**, 395–403.

Gibson, M. H. L., and Bertallanffy, F. D. (1972). *J. Nat. Cancer Inst.* **49**, 1007.

Gillette, E. L., Withers, H. R., and Tannock, I. F. (1970). *Radiology* **96**, 639–643.

Goldenberg, G. J., Lyons, R. M., Lepp, J. A., and Vanstone, C. L. (1971). *Cancer Res.* **31**, 1616–1619.

Graham, F. L., and Whitmore, G. F. (1970). *Cancer Res.* **30**, 2627–2635.

Guzman, E., and Lajtha, L. G. (1970). *Cell Tissue Kinet.* **3**, 91–98.

Hirschman, S. Z., Fischinger, P. J., Zaccari, J. J., and O'Connor, T. (1969). *J. Nat. Cancer Inst.* **42**, 399–411.

Hooper, C. E. S. (1961). *Amer. J. Anat.* **108**, 231–244.

Howard, A., and Pelc, S. R. (1953). *Heredity* **6**, Suppl., 261–273.

Inagaki, A., Nakamura, T., and Wakisaki, C. (1969). *Cancer Res.* **29**, 2169–2176.

James, T. W. (1966). *In* "Cell Synchrony" (I. L. Cameron and G. M. Padilla, eds.), pp. 1–13. Academic Press, New York.

Kallman, R. F. (1963). *Nature (London)* **197**, 557–560.

Karon, M., and Shirakawa, S. (1969). *Cancer Res.* **29**, 687–696.

Kasten, F. H., Strasser, F. F., and Turner, M. (1965). *Nature (London)* **207**, 161–164.

Kim, J. H., and Eidinoff, M. L. (1965). *Cancer Res.* **25**, 698–702.

Kim, J. H., and Perez, A. G. (1965). *Nature (London)* **207**, 974–975.

Kim, J. H., Gelbard, A. S., and Perez, A. G. (1967). *Cancer Res.* **27**, 1301–1305.

Kimball, A. P., Bowman, B., Bush, P. S., Herriot, J., and LePage, G. (1966). *Cancer Res.* **26**, 1337–1343.

Kit, S., deTorres, R. A., and Dubbs, D. R. (1966). *Cancer Res.* **26**, 1859–1866.

Kleinfeld, R. G., and Sisken, J. E. (1966). *J. Cell Biol.* **31**, 369–379.

Kuroki, T., and Sato, H. (1970). *Exp. Cell Res.* **61**, 210–213.

Leblond, C. P., and Stevens, C. E. (1948). *Anat. Rec.* **100**, 357–371.

Lenaz, L., and Philips, F. S. (1970). *Cancer Res.* **30**, 1961–1962.

Lenaz, L., Sternberg, S. S., and Philips, F. S. (1969). *Cancer Res.* **29**, 1790–1798.

Lesher, S., and Bauman, J. (1969). *Nat. Cancer Inst., Monogr.* **30**, 185–198.

Lieberman, M. W., Verbin, R. S., Landay, M., Liang, H., Farber, E., Lee, T. S., and Starr, R. (1970). *Cancer Res.* **30**, 942–951.

Lindahl, P. E., and Sorenby, L. (1966). *Exp. Cell Res.* **43**, 424–434.

Littlefield, J. W. (1962). *Exp. Cell Res.* **26**, 318–326.

Madoc-Jones, H., and Mauro, F. (1970). *J. Nat. Cancer Inst.* **45**, 1131–1143.

Mauro, F., and Elkind, M. M. (1968). *Cancer Res.* **28**, 1150–1155.

Mauro, F., and Madoc-Jones, H. (1969). *Proc. Nat. Acad. Sci. U.S.* **63**, 686–691.

Mauro, F., and Madoc-Jones, H. (1970). *Cancer Res.* **30**, 1397–1408.

Mitchison, J. M., and Vincent, W. S. (1965). *Nature (London)* **205**, 987–989.

Morris, N. R., Cramer, J. W., and Reno, D. (1967). *Exp. Cell Res.* **48**, 216–218.

Mueller, G. C. (1963). *Exp. Cell Res., Suppl.* **9**, 144–149.

Newton, A. A., and Wildy, P. (1959). *Exp. Cell Res.* **16**, 624–6359.

Nias, A. H. W., and Fox, M. (1971). *Cell Tissue Kinet.* **4**, 375–398.

Nitze, H. R., Vosteen, K.-H., and Ganzer, U. (1971). *Acta Oto-Laryngol.* **71**, 227–231.

Pederson, T., and Gelfant, S. (1970). *Exp. Cell Res.* **59**, 32–36.

Peterkofsky, B., and Tomkins, G. M. (1967). *J. Mol. Biol.* **30**, 49–61.

Petersen, D. F., and Andersen, E. C. (1964). *Nature (London)* **203**, 642–643.

Pizer, L. I., and Cohen, S. S. (1960). *J. Biol. Chem.* **235**, 2387–2392.

Post, J., and Hoffman, J. (1969). *Exp. Cell Res.* **57**, 111–113.

Prescott, D. M. (1968). *Cancer Res.* **28**, 1815–1820.

Priest, J. H., Heady, J. E., and Priest, R. E. (1967). *J. Nat. Cancer Inst.* **38**, 61–72.

Puck, T. T. (1967). *Cold Spring Harbor Symp. Quant. Biol.* **29**, 167–176.

Rajewsky, M. F. (1970). *Exp. Cell Res.* **60**, 269–276.

Rajewsky, M. F., Fabricius, E., and Hulser, D. F. (1971). *Exp. Cell Res.* **66**, 489–492.

Ramsdahl, M. M., and Deanen, L. L. (1968). *Exp. Cell Res.* **50**, 463–467.

Rao, P. N. (1968). *Nature (London)* **160**, 774–776.

Robbins, E., and Marcus, P. J. (1964). *Science* **144**, 1152–1153.

Rueckert, R. R., and Mueller, G. C. (1960). *Cancer Res.* **20**, 1584–1591.

Sakamoto, K., and Elkind, M. M. (1969). *Biophys. J.* **9**, 1115–1130.

Schär, B., Loustalot, P., and Gross, F. (1954). *Klin. Wochenschr.* **32**, 49–57.

Schindler, R. (1963). *Biochem. Pharmacol.* **12**, 533–538.

Silagi, S. (1965). *Cancer Res.* **25**, 1446–1453.

Sinclair, W. K. (1965). *Science* **150**, 1729–1731.

Sinclair, W. K. (1967). *Cancer Res.* **27**, 297–308.

Sinclair, W. K., and Morton, R. (1965). *Biophys. J.* **5**, 1–25.

Sinclair, W. K., and Morton, R. A. (1966). *Radiat. Res.* **29**, 450–474.

Smith, C. G. (1968). *Proc. Int. Pharmacol. Meet., 3rd, 1966* Vol. 5, pp. 33–53.

Steffen, J. A., and Stolzmann, W. M. (1969). *Exp. Cell Res.* **56**, 453–460.

Stubblefield, E., and Klevecz, R. (1965). *Exp. Cell Res.* **40**, 660–664.

Terasima, T., and Tolmach, L. J. (1961). *Nature (London)* **190**, 1210–1211.

Terasima, T., and Tolmach, L. J. (1963). *Exp. Cell Res.* **30**, 344–362.

Till, J. E., and McCulloch, E. A. (1963). *Radiat. Res.* **18**, 96–105.

Till, J. E., Whitmore, G. F., and Gulyas, S. (1963). *Biochim. Biophys. Acta* **72**, 277–289.

Tobey, R. A. (1972). *Cancer Res.* **32**, 309–316.

Tobey, R. A., Petersen, D. F., Andersen, E. C., and Puck, T. (1966). *Biophys. J.* **6**, 567–581.

Vassor, F., Frindel, E., and Tubiana, M. (1971). *Cell Tissue Kinet.* **4**, 423–431.

Verbin, R. S., and Farber, E. (1967). *J. Cell Biol.* **35**, 649–658.

Verbin, R. S., Sullivan, R. J., and Farber, E. (1969). *Lab. Invest.* **21**, 179–182.

Verbin, R. S., Diluiso, G., Liang, H., and Farber, E. (1972a). *Cancer Res.* **32**, 1476–1488.

Verbin, R. S., Diluiso, G., Liang, H., and Farber, E. (1972b). *Cancer Res.* 32, 1489–1495.
Verbin, R. S., Diluiso, G., and Farber, E. (1974). *Cancer Res.* 34, 1429–1434.
Walker, I. G., and Helleiner, C. W. (1963). *Cancer Res.* 23, 734–739.
Wheeler, G. P., Bowdon, B. J., Adamson, D. J., and Vail, M. H. (1970). *Cancer Res.* 30, 100–111.
Wheeler, G. P., Bowdon, B. J., Adamson, D. J., and Vail, M. H. (1972). *Cancer Res.* 32, 2661–2669.
Whitfield, J. F., and Youdale, T. (1965). *Exp. Cell Res.* 38, 208–210.
Withers, H. R. (1967). *Radiat. Res.* 32, 227–239.
Xeros, N. (1962). *Nature (London)* 194, 682–683.
Young, R. S. K., and Fischer, G. A. (1968). *Biochem. Biophys. Res. Commun.* 32, 23–29.

Chapter 5

The Accumulation and Selective Detachment of Mitotic Cells

EDWIN V. GAFFNEY

Department of Biology,
The Pennsylvania State University,
University Park, Pennsylvania

I. Introduction

The sequence of events required for mitosis in a synchronous population must be comparable to the sequence observed in randomly dividing cells. Thus, the need for normalcy in the resulting population limits the synchrony technique employed.

Synchronized populations of mammalian cells are frequently established by the selective harvest of mitotic figures from monolayer cultures. This approach avoids the problem of unbalanced growth reported for other procedures involving inhibitors of DNA synthesis (Studzinski and Lambert, 1969; Rueckert and Mueller, 1960; Rosenberg and Gregg, 1969). However, its application is restricted in two ways. First, the yield of mitotic cells is low since a small percentage of most asynchronous populations is dividing at any one time.

This limitation was overcome when Stubblefield and Klevecz (1965) showed that Colcemid could reversibly increase the percentage of mitotic cells in a population by preventing the formation of the spindle apparatus. Second, cells which demonstrate great cell-to-cell adhesion or enhanced attachment to culture vessels may have little advantage for selection at mitosis. Increased agitation of the culture fluids to displace mitotic cells may also remove loosely attached cells in interphase. Thus, only a few of the established lines of cells are routinely utilized in cell cycle studies and rarely are reports found using diploid cell strains.

Preliminary attempts to obtain synchronous populations of a simian virus 40 transformed human cell line (Gaffney et al., 1970) by selective harvest following incubation in Colcemid were not consistent owing to the detachment of interphase cells. It was observed that cells accumulated in metaphase with Colcemid or vinblastine sulfate were preferentially detached from monolayer cultures during exposure to hypotonic solution. Thus, this investigation describes the application of hypotonic salt concentrations in selectively harvesting mitotic populations of many cell types. In addition the data emphasize the need always to examine the effects of Colcemid concentration and incubation time on the reversibility of metaphase arrest prior to conducting cell cycle studies.

II.   Preparation of Mitotic Cells

A244 is an established line resulting from the simian virus 40 transformation of a primary culture of human amnion cells. The line has been maintained in monolayer culture for 5 years, and contains a modal chromosome number in the triploid range. Other cell lines include the L-929 mouse fibroblast line (Sanford et al., 1948), the CV-1 line (Jensen et al., 1964) derived from monkey kidney cultures, and HEL derived from human embryonic lung. Cells are subcultured at weekly intervals in McCoy's medium 5a supplemented with 10% newborn calf serum, penicillin (100 units/ml), and streptomycin (100 mg/ml). HEL cells were also maintained in MEM with 15% fetal calf serum.

Large synchronized populations of A244 cells could be obtained by the following procedure. Thirty-two-ounce prescription bottles seeded with 5×10^6 cells were maintained for 2 days in 30 ml of medium. Culture fluids were renewed and 2 hours later Colcemid (Grand Island Biological Co.) was added to a final concentration of 0.01 μg/ml. After 4 hours of incubation, this was replaced with 20 ml of 1 part Hank's Balanced Salt Solution (HBSS) and 4 parts sterile distilled water delivered to the back of the bottle.

The bottles were rapidly turned over so that all cell sheets were incubated simultaneously. Cultures were gently shaken for 90 seconds and then turned again to remove the hypotonic medium from contact with the attached cells. Hypotonic solutions were quickly emptied into conical centrifuge tubes containing 10 ml of the Colcemid medium. Ten milliliters of HBSS were added at 1.8 times the normal concentration. These tubes were immediately centrifuged and the cell pellets resuspended in prewarmed, conditioned medium.

III. Colcemid and Mitotic Arrest

Since concentrations of Colcemid above that which just inhibit mitosis have been found to cause aberrant cytokinesis and unequal chromosome distribution, it was necessary to establish the amount of Colcemid which would reversibly accumulate A244 cells in metaphase without inducing morphological alterations during cell separation.

Leighton tube cultures were incubated with Colcemid in concentrations of 0.002 to 0.04 μg/ml and cells were accumulated in metaphase for 6 hours. Triplicate cultures incubated at each drug concentration were stained, and the percentage of the population in mitosis was determined from counting approximately 1000 cells per sample. The results showed that a maximum of 12–14% metaphase figures was attained with incubation in Colcemid at a concentration between 0.01 and 0.02 μg/ml. The accumulation of cells in metaphase at any one Colcemid concentration was also dependent on the time of incubation.

The ability of metaphase cells to progress rapidly into interphase following removal of Colcemid was investigated by incubating cultures at concentrations of either 0.01 or 0.02 μg/ml for periods of 1–6 hours. At 1-hour intervals some tubes were stained and others washed and fluid changed with fresh medium. The mitotic index was then determined during 3 hours after removal of the drug. The data shown in Fig. 1 (Gaffney and McElwain, 1973) illustrate that: (1) The number of cells accumulated in metaphase increased during the period of incubation in both concentrations of Colcemid. However, the rate of accumulation appeared less at the lower drug concentration. (2) Cells arrested in metaphase by 0.01 μg of Colcemid per milliliter progressed more rapidly into interphase following removal of the inhibitor than cells arrested in the higher concentration. (3) Cultures treated for 4 hours or less with 0.01 μg Colcemid per milliliter contained a similar number of metaphase cells (4%) as untreated controls 2 hours after fluid change.

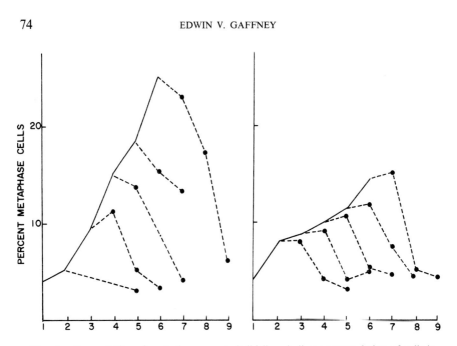

FIG. 1. Reversibility of metaphase arrest. Solid lines indicate accumulation of cells in metaphase with increasing period of exposure to Colcemid (1–6 hours). Dashed lines illustrate the percentage of each population remaining in metaphase during the 3-hour interval following removal of the inhibitor. Cell number, 3.5 × 10⁵ per tube. Control averages 4% mitoses. Concentration: left, 0.02 μg/ml; right, 0.01 μg/ml. From Gaffney and McElwain (1973). Reprinted by permission of the Tissue Culture Association, Inc.

The induction of aberrant mitoses by Colcemid in human cells was studied by incubating cultures in concentrations of 0.01 μg/ml or 0.02 μg/ml for 4, 5, or 6 hours. Colcemid was removed by washing twice with fresh medium and triplicate samples were stained at 1-hour intervals for 3 hours. The percent of the mitotic population which appeared abnormal was determined by counting 500 dividing cells from each coverslip. A mitosis was defined as aberrant when a multipolar division or an unequal distribution of chromosomes between daughter nuclei was observed. Figure 2 (Gaffney and McLwain, 1973) describes the abnormal divisions present in Colcemid-treated populations. Cultures treated with 0.02 μg of Colcemid per milliliter contained a high percentage of aberrant divisions 2 hours after release from the inhibitor. However, the number of mitoses that appeared abnormal with the lower concentration was not much greater than controls regardless of the length of incubation. The high percentage of dividing cells in cultures growing exponentially (3.2–3.8%) and the ease with which it first appeared that dividing cells could be detached from the culture flask suggested that large synchronous populations might be easily obtained. However, pre-

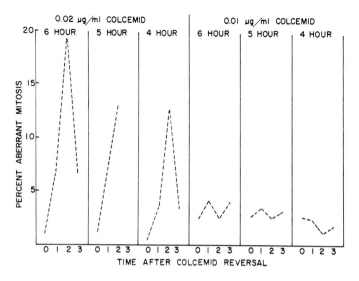

FIG. 2. The percentage of mitotic cells that appear aberrant after removal of Colcemid. From Gaffney and McElwain (1973). Reprinted by permission of the Tissue Culture Association, Inc.

liminary attempts to remove mitotic cells by mechanically shaking mono-layers incubated in 0.01 µg of Colcemid per milliliter were not successful for two reasons. First, the population which was detached was less than 90% synchronous. Second, too few cells were recovered to establish populations for studies of macromolecular systhesis. Increased agitation of the cultures only increased the number of interphase cells removed.

Cell attachment to glass or plastic surfaces is mediated by both cationic bonding and protein or glycoprotein linkage. In addition, cell attachment and growth depend on serum supplements in the medium. Therefore, a series of short-term experiments were designed to manipulate cell bonding so more mitotic cells might be displaced on shaking. Monolayer cultures were grown in either fetal calf serum or newborn calf serum at concentrations of 1, 5, or 10% prior to and during harvest. Also, cultures were agitated in calcium- and magnesium-free Hank's balanced salt solution or in trypsin and/or EDTA at a concentration of 0.02%. These procedures increased the number of interphase cells collected.

Karyotypic analyses of cells requires the separation of chromosomes into distinct visible structures. This is usually accomplished by incubating cultures in a hypotonic salt solution. An attempt was made to diminish cell-to-cell and cell-to-glass bonding and thus enhance the detachment of cells in metaphase by increasing cell volume with a short incubation in

TABLE I

EFFECT OF HYPOTONIC SOLUTIONS ON MITOTIC
RECOVERY[a]

HBSS: H_2O	% Metaphase cells
1:4	96
2:3	62
3:2	67

[a]From Gaffney and McElwain (1973). Reprinted
by permission of the Tissue Culture Association, Inc.

diluted HBSS. The procedure finally adopted for this cell line is outlined
in Section II and consistently yields populations which are 95–98% synchronous with respect to mitosis.

The effect of different hypotonic solutions on the recovery of synchronous
populations was examined. Cells grown in 25-cm² plastic falcon flasks to
a density of 1.3×10^6 per container were incubated with 0.01 µg of Colcemid
per milliliter for 4 hours. The medium was replaced with solutions of diluted
HBSS, and mitotic cells were detached as previously described. Table I
illustrates that HBSS diluted 1:4 with water promoted the detachment of
cells with a high mitotic index (Gaffney and McElwain, 1973). In addition,
it appeared that lower dilutions of salts not only detached fewer mitotic
cells but also resulted in less synchronous populations.

IV. Cell Cycle Analysis

Since populations of cells obtained by this modified selective harvest
procedure had been exposed to both a mitotic inhibitor and a hypotonic
solution, it was necessary to study the effect of these treatments on the
ability of metaphase cells to attach to culture vessels and to progress
normally through a subsequent interphase.

The ability of populations of metaphase-arrested cells to attach to culture
vessels following harvest with or without an exposure to hypotonic solution
was compared to that of a control population. Controls were obtained by
shaking monolayer cultures, centrifuging the detached cells, resuspending
in conditioned medium, and seeding to Leighton tubes. Cells accumulated
in Colcemid for 4 hours were also detached by shaking and inoculated to
tubes. Additional cultures were incubated in Colcemid and metaphase
cells harvested in hypotonic HBSS. The proportion of each population
which attached to a culture vessel was then compared by counting cells

TABLE II

PERCENT ATTACHMENT FOLLOWING HARVEST OF A244
CELLS[a,b]

Time (hours)	Control	Colcemid	Colcemid + hypotonic
1	80	75	91
2	87	87	94
3	89	85	95
4	88	85	95

[a]Maximum deviation ±3%.
[b]From Gaffney and McElwain (1973). Reprinted by permission of the Tissue Culture Association, Inc.

from 4 replicate tubes at 1 hour intervals. The results in Table II, expressed as percentage of cells attached to the glass, demonstrate that those cells exposed to the hypotonic solution attach more readily that the control or Colcemid-treated groups (Gaffney and McElwain, 1973). Twelve to 15% of these populations had not attached to the glass within 4 hours.

Synchronous progression through a cell cycle was followed in the next experiments. Mitotic cells obtained by hypotonic treatment were resuspended in medium and inoculated to Leighton tubes at a concentration

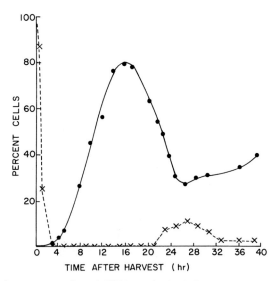

FIG. 3. Synchronous growth and DNA synthesis during the cell cycle of A244 cells. ●—●, Labeled cells; ×---×, mitotic cells. From Gaffney and McElwain (1973). Reprinted by permission of the Tissue Culture Association, Inc.

of 5×10^4 cells per tube. Duplicate tubes were stained at 1- or 2-hour intervals for 38 hours, and the average percentage of mitoses was determined. The results of the experiment plotted in Fig. 3 (Gaffney and McElwain, 1973) show that the proportion of the population in mitosis declined from 97% immediately after harvest to less than 1% by 3 hours. A second wave of mitoses involving 93% of the starting population was observed between 23 and 31 hours with a peak at 26.5 hours. The average cell generation time for a synchronous population was then compared to the average cell doubling time for asynchronous cultures. Random populations were established in Leighton tubes at cell densities of 1, 1.5, or 2.0×10^5 cells per tube and 4 replicate tubes were counted daily. The population doubling time during the exponential portion of the growth curves was found to be 25 hours. Thus, the average time for a synchronous population to replicate following exposure to colcemid and hypotonic solution was comparable to that of asynchronous controls.

Duplicate tubes of synchronized cells were maintained in the presence of tritiated thymidine for 30 minutes at intervals during the cell cycle and prepared for radioautography. The proportion of the population participating in DNA synthesis was then determined by counting 1000 cells in each sample. Each point on the curve plotted in Fig. 3 represents the average percent labeled cells from three separate experiments. DNA synthesis was initiated in some cells by 4 hours. This proportion of the population increased to a minimum of 80% by 15 hours and declined during the G_2 period. The lengths of the G_1, S, and G^2 periods of interphase can be estimated from these data according to the method of Quastler and Sherman (1959) as 9, 13, and 4 hours, respectively.

V.   Application to Other Cell Lines

The following experiments were performed to determine whether this modified approach to the selection of mitotic figures would be useful with other cell lines. The L-929 mouse fibroblast line, which has an average population doubling time equal to approximately 20 hours, has been frequently used in synchrony studies and was one of the first cell lines to be synchronized by the selective harvest technique. Mitotic cells in this population are characterized by a tenuous attachment to culture vessels during metaphase and are easily detached by agitation of supernatant fluids. In contrast, the CV-1 monkey cell line has a population doubling time of approximately 40 hours, and a small proportion of the cells are

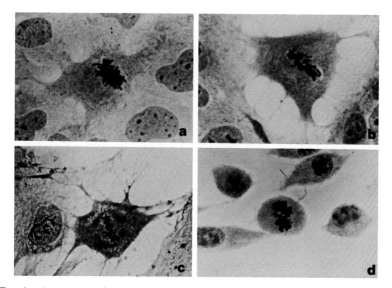

FIG. 4.   Appearance of metaphase figures from four cell types. (a) A244; (b) CV-1; (c) HEL; (d) L-929.

in mitosis at any one time. In addition, metaphase cells are firmly attached to the vessel surface and appear flattened throughout mitosis. The typical morphologies of cells during metaphase are shown in Fig. 4a–c.

Populations of each of these cell types were grown in Leighton tubes, incubated in Colcemid for 4 hours, and stained to determine the percentage of metaphase-arrested cells. Unlike cells of primate origin, cells from murine animals require higher concentrations of Colcemid (Sutbblefield and Klevecz, 1965; Romsdahl, 1968). Therefore, L-929 cells were treated with Colcemid at a concentration of 0.04 $\mu g/ml$. Cells of each line were also grown in Falcon plastic flasks, incubated in Colcemid and detached by selective harvest. The percentage of the total cell population recovered by detachment in Colcemid medium or hypotonic solution, the percentage of the mitotic population recovered, and the synchrony of the harvested cells were recorded (Table III) (Gaffney and McElwain, 1973). The modified selection procedure enhanced the detachment of A244 cells. In addition it appeared that a similar approach may be used with CV-1 cells since the hypotonic treatment resulted in a 3-fold increase in the number of mitotic cells harvested with the resulting population demonstrating an acceptable degree of synchrony (84%). However, the size of the mitotic population was not as dramatically increased by hypotonic treatment with

TABLE III

ESTABLISHMENT OF SYNCHRONOUS POPULATIONS[a]

Cell line	% Mitotic cells		% Cells recovered			% Mitotic cells recovered	% Synchrony
	Control	Colcemid	Colcemid	Hypotonic	Total		
A244	2.3	9.4	1.2	2.6	3.8	41	98
L-292	1.7	5.6	3.2	0.7	3.9	69	83
CV-1	1.8	6.8	1.3	3.6	4.9	72	84
HEL	2.3	8.3	3.8	2.3	6.1	73	96

[a]From Gaffney and McElwain. Reprinted by permission of the Tissue Culture Association, Inc.

strain L-929 cultures. In contrast to previous results (Gaffney and Nardone, 1968), it appeared that exposure to hypotonic salts decreased the synchrony of the harvested population.

VI.   Human Diploid Cells

Experiments were performed with diploid embryonic lung fibroblasts between passages 17 and 27 to determine the response of normal human cells to Colcemid and the potential enhancement of mitotic selection through exposure to hypotonic salts. Preliminary studies revealed that Colcemid accumulates in these cells in metaphase to a maximum of 12% with 8 hours of incubation, and that the proportion of mitotic cells was not markedly increased by concentrations of the inhibitor above 0.01 μg/ml. The reversibility of mitotic arrest was determined after incubation in Colcemid at concentrations of 0.01–0.04 for 6 hours. Triplicate cultures were fixed at 1-hour intervals, and the proportion of mitotic cells was observed. The results illustrated in Fig. 5 are similar to those with the A244 cell line. The mitotic population progressed more slowly into interphase after incubation at Colcemid concentrations above 0.02 μg/ml whereas at the lower concentration the number of metaphase cells returned to near control levels by 2 hours. Increasing concentrations of Colcemid led to increasing numbers of aberrant mitoses. This was apparent both at the end of the 6-hour incubation period and for 3 hours after removal of the inhibitor.

The efficiency with which mitotic cells can be recovered using selective harvest is shown in Table III. Cells removed following incubation in 0.01 μg of Colcemid per milliliter for 6 hours represent 3.8% of the total popula-

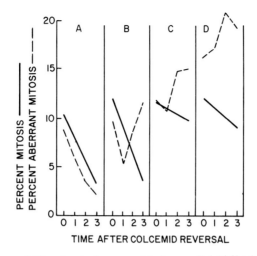

FIG. 5. The reversibility of metaphase arrest in human diploid fibroblasts incubated for 6 hours at 4 concentrations of Colcemid (μg/ml). A, 0.01; B, 0.02; C, 0.03; D, 0.04. Solid lines indicate the decline in number of mitoses 3 hours after removal of the inhibitor. Dashed lines represent the percentage of the mitotic cells which appear aberrant.

tion. Based on the results of 6 experiments, hypotonic treatment increased the size of the harvested synchronous population by 61%. The efficiency of hypotonic treatment, however, was not consistent. In one experiment short-term incubation in diluted salts enhanced the size of the harvested mitotic population by 150% over that obtained in medium alone. Although mitotic cells collected in medium supplemented with Colcemid traverse the cell cycle normally, cells exposed to hypotonic salts are partially delayed from entering interphase.

VII. Discussion

The data obtained from the previous experiments suggest the following: (1) Mitotic inhibitors such as Colcemid can be used to increase the proportion of cells in mitosis without adverse effects on morphology and cell cycle timing provided preliminary studies are conducted to determine the minimum concentration and incubation period necessary to arrest division. (2) The exposure of monolayer cultures of certain cell types to a hypotonic salt solution can selectively detach mitotic cells resulting in large synchronous populations that will progress through a subsequent cell cycle.

A report by Romsdahl (1968) suggested that a sufficiently low concentration of Colcemid might be found to just block mitosis and yet be readily reversible and free of toxic effects. This agreed with Taylor's hypothesis (1965) that only 3–5% of those sites which bind colchicine need to be complexed to inhibit mitosis.

Studies in this laboratory to determine a concentration of Colcemid which would inhibit division in A224 cells were followed by experiments to measure that rate at which normal growth would proceed subsequent to removal of the drug. Although cells were accumulated at a maximum rate by incubation in a concentration of 0.02 μg of Colcemid per milliliter of culture fluid, the reversibility of mitotic arrest occurred more rapidly in populations treated with a lower concentration. Counts of dividing cells at hourly intervals after reversal of the block revealed that the mitotic index of cells treated at 0.01 μg/ml fell from 0.1 to approximately the control value 0.03–0.04 within 2 hours.

Reports from other laboratories have also shown that Colcemid does not induce aberrant mitoses if the concentration or the blocking time is reduced (Stubblefield and Klevecz, 1965; Stubblefied and Deaven, 1966). In our experiments aberrant mitoses were defined as either multipolar divisions or unequal distributions of material between daughter cells. While absolute values obtained from observations of stained cells are not possible, the qualitative trend was obvious. Reducing Colcemid concentration to 0.01 μg/ml blocks mitosis in A244 cells, but maintains the number of aberrant division figures near the control level of 4%.

The 4-hour time of incubation in Colcemid was chosen on the basis of two findings. First, linear accumulation of mitotic figures in 0.01 μg of Colcemid per milliliter ends at 5–6 hours in cultures of A244 cells. Approximately 12% of the population was in mitosis at 4 hours and less than 15% was in mitosis at 6 hours. Second, the efficiency of metaphase arrest decreased after 4 hours and an increase in the proportion of altered cells appeared with further increase in the mitotic fraction in 0.02 μg of Colcemid per milliliter. According to Robbins and Gonatas (1964), ultrastructural alterations occurred in HeLa cells after 5 hours of incubation. The period of 4 hours is equal to one-sixth of the cell cycle, the same proportion suggested by Stubblefield and Deaven (1966) to avoid aberrant division figures.

The basis for the selective detachment of mitotic cells in hypotonic solutions is unclear. Other methods of selection, such as incubation in calcium- and magnesium-free medium, reduction in serum content or treatment with EDTA or trypsin act directly on the adhesive forces binding cells to the substrate. Hypotonicity does not act on the adhesive forces directly, but apparently breaks attachments by increasing cell surface area. The degree of attachment during mitosis differs slightly with each cell line.

Cells such as those of the A244 or CV-1 line seen in Fig. 4 maintain cytoplasmic adhesion to the culture vessel during division. The uniqueness of the hypotonic treatment is its ability to loosen and remove the less firmly attached mitotic cells, before affecting the tightly bound interphase cells. There is increasing evidence that the initiation of mitosis and the progression of cells from prophase into metaphase are directly related to localized ion concentration. The treatment of HeLa cells during interphase with hypertonic salts induced changes in nuclear morphology resembling those seen during prophase. This response was accompanied by markedly and reversibly depressed rates of macromolecular synthesis (Robbins *et al.*, 1970). A generalized increase in ion concentration during mitosis seems unlikely since this would induce a subsequent influx of water. A more acceptable explanation is the intracellular compartmentalization of small ions.

Thus, it appears that the degree to which ions are bound at different stages of the cell cycle has a profound influence on the state of aggregation of nuclear structures. This ion accumulation may reflect an increased membrane permeability and active transport mechanism during division and could account for the selective swelling and subsequent detachment of mitotic cells induced by hypotonicity. Compelling evidence for this theory stems from a study of surface to volume ratios during the division cycle of L-929 cells (Cone, 1969). A decrease in cellular surface/volume ratio was observed through time lapse cinematography during rounding of the cell in preparation for mitosis. Prophase initiation was brought about by an osmotically induced volume increase which resulted in an increase in the Na^+ concentration. In addition, experimental results showed that hypotonic pulsation with Na^+ has a pronounced stimulatory effect on mitosis initiation.

When the system for mitotic selection was established, the resulting populations were tested to determine normalcy. Growth curves on A244 cells showed that the population doubling time was 24–26 hours. Reasonably good agreement was obtained from mitotic counts of synchronized cells which showed a wave of mitosis between 23 and 31 hours, with the peak at 26.5 hours. When the 1- to 2-hour delay seen in the Colcemid reversal curve is considered, the cycle times and population doubling times are approximately equal. Although synchrony decay obscures more accurate interpretation, obvious effects such as the shortening of the $S + G_2$ period by one-third reported for other synchrony procedures (Bostock *et al.* 1971) are not present. Since plating efficiencies of Colcemid-arrested cells were said to decline after treatment, the efficiency of attachment of A244 cells dislodged by agitation was measured. The proportion of unattached cells at 4 hours after inoculation to tubes obtained in the control and Colcemid-treated cultures is in marked contrast to the 5% found with hypotonic treatment. The DNA synthesis in synchronized cultures was shown to follow the pattern observed in HeLa cells. However, Terasima and Tolmach (1963) showed a peak of 90% cells synthesizing DNA during S phase and a G_2 low of only 10%. The initial mitotic index of synchronized populations of A244 cells was as high as reported by these authors for HeLa cells; however, the corresponding values of 80 and 27% synthesizing DNA in this system indicate a high rate of synchrony decay. This may result from several sources. Colcemid incubation may introduce an initial degree of distortion into the cycle. It is also possible that the wide range of chromosome numbers contributed significantly to the heterogeneity of the harvested population (Gaffney *et al.*, 1970). These factors accompanied by the relatively long cycle time would induce the smearing of the synchronous band during the subsequent mitosis.

It has been shown for the A244 and CV-1 cell lines that flattened mitotic cells can be selected from random cultures. Both lines attach and begin to grow within 2 hours. The lower mitotic index seen in harvested populations of L-929 cells, a line for which mitotic selection is commonly used, suggests that the effects of hypotonic solutions is mainly to swell cells and that this treatment serves only to dislodge a greater number of interphase cells in a population where mitotic cells are already rounded and tenuously attached.

REFERENCES

Bostock, K., Prescott, D. and Kirkpatrick, J. (1971). *Exp. Cell Res.* **68**, 163–168.
Cone, C. D. (1969). *Trans. N.Y. Acad. Sci.* [2] **31**, 404–427.
Gaffney, E. V., and McElwain, E. G. (1973). *In Vitro* **9**, 56–63.
Gaffney, E. V., and Nardone, R. M. (1968). *Exp. Cell Res.* **53**, 410–416.
Gaffney, E. V., Fogh, J., Ramos, L., Loveless, J. D., Fogh, H., and Dowling, A. M. (1970). *Cancer Res.* **30**, 1668–1676.
Jensen, F., Girardi, A., and Gilden, R. (1964). *Proc. Nat. Acad. Sci. U.S.* **52**, 53–59.
Quastler, H., and Sherman, F. (1959). *Exp. Cell Res.* **17**, 420–438.
Reuckert, R., and Mueller, G. (1960). *Cancer Res.* **20**, 1584–1591.
Robbins, E., and Gonatas, N. (1964). *J. Histochem. Cytochem.* **12**, 704–711.
Robbins, E., Pederson, T., and Klein, P. (1970). *J. Cell Biol.* **44**, 400–416.
Romsdahl, M. (1968). *Exp. Cell Res.* **50**, 463–467.
Rosenberg, H., and Gregg, E. (1969). *Biophys. J.* **9**, 592–606.
Sanford, K. K., Earle, W. R., and Likely, G. D. (1948). *J. Nat. Cancer Inst.* **9**, 229–246.
Stubblefied, E., and Deaven, L. (1966). *J. Cell Biol.* **31**, 114A.
Stubblefield, E., and Klevecz, R. (1965). *Exp. Cell Res.* **40**, 660–664.
Studzinski, G. P., and Lambert, W. (1969). *J. Cell. Physiol.* **73**, 109–117.
Taylor, E. W. (1965). *J. Cell Biol.* **25**, 145–160.
Terasima, T., and Tolmach, L. (1963). *Exp. Cell Res.* **30**, 344–362.

Chapter 6

Use of the Mitotic Selection Procedure for Cell Cycle Analysis: Emphasis on Radiation-Induced Mitotic Delay[1]

D. P. HIGHFIELD[2] AND W. C. DEWEY

Department of Radiology and Radiation Biology,
Colorado State University,
Fort Collins, Colorado

I.   Introduction

Mammalian cells cultured as monolayers usually round up and become loosely attached as they enter mitosis (Axelrod and McCulloch, 1958; Terasima and Tolmach, 1963; Tobey *et al.*, 1967). By agitating the medium on

[1] This work was supported in part by Public Health Service Grant CA-08618.

[2] *Present address*: Department of Human Biological Chemistry and Genetics, Cell Biology Section, University of Texas Medical Branch, Galveston, Texas 71550.

the cells, usually accomplished by shaking the culture flask, the mitotic cells can be removed from the culturing surface. The medium poured from the culture vessel will contain the mitotic cells, while the interphase cells will be left behind on the surface. More medium can be added to the culture vessel, and the shaking procedure can be repeated after an incubation period (about 10 minutes) sufficient for more cells to move into but not through mitosis. This procedure can be repeated many times over a period of several hours, and the cells shaken loose can be collected and held in mitosis by keeping them at 4°C. The cells collected can be deposited on Millipore filters (Schneiderman *et al.*, 1972) and processed for analysis by light microscopy to determine the percentage of cells in mitosis (mitotic index) and, in certain cases, the percentage of cells labeled with tritiated thymidine (^{3}HTdR). Thus, after a correction (usually less than 5%) for the few interphase cells which are shaken loose, the number of cells collected in mitosis from each shake indicates the rate at which cells are leaving G_2 and entering mitosis.

Information on the position of cells relative to particular transition points (TPs) where specific biochemical events are terminated also can be obtained by using this mitotic selection technique. For example, if at $t = 0$, a monolayer of cells is pulse-labeled for 10 minutes with ^{3}HTdR, then at t_1, when t_1 is less than about 6 hours, unlabeled mitotic cells would have been in G_2 at $t = 0$, and labeled mitotic cells would have been in S at $t = 0$. Positions of cells at $t = 0$ relative to other transition points can be similarly determined. For example, if cycloheximide (CH) is added to the culture at $t = 0$, any cells collected in mitosis at later times would have been beyond the CH transition point at $t = 0$, i.e., beyond the point in G_2 where protein synthesis is no longer required for entering mitosis. Similarly, transition points can be obtained for the actions of heat or X-irradiation in causing division delay and for the action of actinomycin D (AMD) in blocking cells in G_2. The AMD transition point is usually identified as the point in G_2 beyond which RNA synthesis is no longer required for entering mitosis.

In this chapter, determinations of transition points and movements of cells between transition points will be illustrated. As an example, if at $t = 0$, the cells are pulse-labeled with ^{3}HTdR and then treated with AMD at a later time, t_1, any labeled cells reaching mitosis would have progressed during the t_1-t_0 interval from S phase to beyond the AMD transition point. This type of analysis revealed that cells X-irradiated in S phase sustain part of their cycle delay (division delay) in G_2 beyond an AMD transition point (G_2 block).

II. Materials and Methods

A. Cell line, Medium, and Growth Conditions

Chinese hamster ovary (CHO) cells (Tjio and Puck, 1958), 95% near diploid, were carried as a monolayer either on glass or Falcon plastic at 37°C in a humidified atmosphere of 6% CO_2 in air (Dewey and Miller, 1969). The growth medium, modified McCoy's 5a (GIBCo., Grand Island, New York) contained fetal calf serum (5%) and calf serum (10%) and was supplemented with the following antibiotics: neomycin sulfate (0.1 g/liter), polymixin B sulfate (0.05 g/liter), potassium penicillin G (0.05 g/liter), and streptomycin sulfate (0.05 g/liter). Fresh cultures were removed from a liquid nitrogen freezer every 2–3 months.

B. Collecting Mitotic Cells

This procedure is best accomplished in a walk-in incubator where the cells and incubation medium can be maintained at 37°C. Mitotic cells in exponentially growing cultures were collected by the mitotic selection procedure of Terasima and Tolmach (1963) as modified by Tobey et al. (1967). The 75 cm² Falcon flasks (as many as 12 in one experiment) were seeded with 2.2×10^6 cells, and 24 hours later they were shaken at 260 cycles/sec in an Eberbach reciprocating shaker for 10 seconds (Schneiderman et al., 1972). The medium containing the mitotic cells was poured off and rapidly cooled to 4°C to prevent cell division. Conditioned medium was returned to the flasks, pH was maintained at about 7.0 by flushing with air and CO_2, and 10 minutes later, the procedure was repeated. The number of cells from each flask in each collection (shake) was determined with a Coulter counter, and the mitotic index could be estimated from a plot of cell size distribution. Usually 5–8 "clean-up" shakes were required to get a fairly constant cell number and a mitotic index in excess of 90–95%. When the "clean-up" was completed, the cells were kept in test tubes placed in ice water until they could be collected on Millipore filters. At 4°C, CHO cells can be held in mitosis for as long as 24 hours (Nagasawa and Dewey, 1972).

As shown previously (Schneiderman et al., 1972), the mitotic cells are removed primarily in metaphase and anaphase, with the mean time of selection 16 minutes prior to completion of division. The width of the selection window, however, extends from 4 to 22 minutes prior to completion of division (Fig. 3).

C.   Collecting Cells on Millipore Filters for Analysis of Mitotic and Labeling Indices

Because of the low cell numbers obtained from each flask per shake, sometimes as low as 2000 cells in 8 ml of medium, standard cytological techniques where the cells are squashed or spread on slides could not be used. Instead, the cells were concentrated for observation by collecting them on Millipore filters. The basic technique described by Millipore Corporation (1963, 1966) was used, although several modifications were introduced. Details of the procedure follow. For best results, the cells should be stored in the cold no longer than 6 hours before they are filtered.

1. Number the Millipore filters (0.45 μm MF-2.5 cm) with a fine ball-point pen, and then divide them into groups with about 12 filters in each group.

2. Place a group of filters into 95% ethyl alcohol for about 10 seconds until they expand.

3. Gently place the filters on the filtering apparatus, which has been wet with Hank's Balanced Salt Solution (BSS). With the No. 3025 Sampling Manifold apparatus from Millipore Corporation, all 12 samples from a group can be processed at one time.

4. Place the top on the filtering apparatus, and draw a vacuum of about 5 mm Hg until any excess liquid is gone.

5. Turn off the vacuum, mix the cold cell samples well, preferably with a vortex mixer, and pour them onto their respective filters.

6. Turn on the vacuum as before, and just as the liquid is gone, but before there is any drying of the filter, quickly add 5–10 ml of 50% acetic acid at 4°C.

7. Before all of the acetic acid is gone, turn off the vacuum and place the filters in a tray with 50% acetic acid (room temperature). In 3–60 minutes remove the filters and rinse them thoroughly in water. Dry them on a paper towel and store them until they are to be stained or prepared for autoradiography.

D.   Staining Cells before Autoradiography

1. Use a 2% solution of aceto-orcein stain in 50% acetic acid. Filter the stain twice through a Whatman No. 1 filter just before using.

2. Wet the filters in 50% acetic acid and place them in stain for 0.5–1 minute. Then, to remove the excess strain, transfer the filters with agitation through three baths of n-propanol (about 2 minutes in each bath of 75 ml, which is sufficient for about 25 filters).

3. Clear the filters by placing them in xylene for at least 30–60 seconds, but for no longer than 1 day. If autoradiographs are to be prepared, remove the filters, place them on absorbent paper, and let them thoroughly air dry. If autoradiographs are not to be prepared, proceed with the next step.

4. To mount and fit a filter on a microscope slide, remove it from xylene, blot it on absorbent paper, and cut it to size. Then, dip the filter into a solution of 3 parts of Permount to 1 part of xylene. Drain the excess from the filter, and place it cell-side-up on a microscope slide. With a cover glass, press on the filter to flatten it and to remove excess Permount and air bubbles. Add more Permount to the edges of the cover glass; otherwise, the filter may dry and turn white. If this happens, add more Permount or xylene and let it flow under the cover glass to again clear the filter.

5. The slide can now be observed and/or set aside until the Permount hardens.

E. Preparing Filters for Autoradiography

1. If filters are to be stained after autoradiography with Wright's Stain, proceed with the dry filters from step 7, Section C. If filters have been stained with aceto-orcein, proceed from step 3, Section D. With flat tip (Millipore) forceps, place the *dry* filter into a 0.5% solution of Formvar in ethylene dichloride.

2. After the filter has expanded completely (in about 5–10 seconds), place it cell-side-up on a clean microscope slide. To avoid getting air bubbles under the filter, lay it down smoothly beginning with one edge of the filter. Smooth out the edges with the flat surface of the forceps.

3. Within 1 minute, place the slide with the filter down over a 250-ml beaker half-full of acetone at about 55°C. Immediately, the filter will expand slightly and smooth out onto the slide. After 2–2.5 minutes, when the filter becomes completely transparent, remove the slide and let the filter dry. If the filter remains over the acetone vapors too long, it will start to dissolve. Also, do not store the slides for more than one day; otherwise, the filter may detach.

4. Apply AR-10 stripping film by the standard procedure where the film is floated on water and picked up with the slide.

5. Develop the film at about 20°C following standard procedures, but do not agitate the slides in the developing solution; otherwise, the film may slip. Then, air dry the film keeping the slides cell-side-up. If the cells have been stained previously with aceto-orcein, prepare a permanent preparation by using Permount and a 22-mm diameter cover glass.

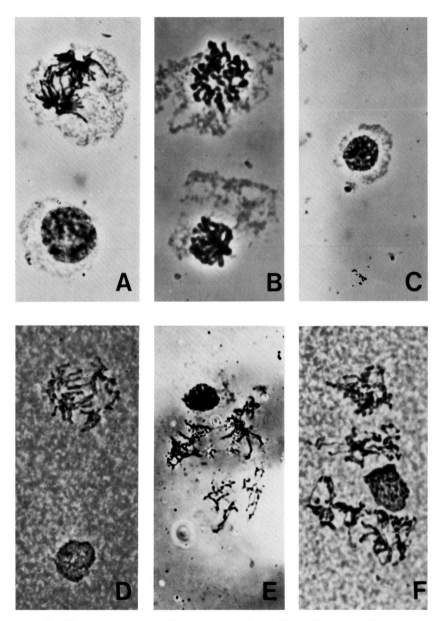

FIG. 1. Photomicrographs of cells squashed (A, B, and C) or collected on Millipore filters (D, E, and F). A and D each illustrate a mitotic and interphase cell from control samples. B, C, and F illustrate mitotic and interphase cells from a sample treated with actinomycin

F.   Staining Cells after Autoradiography

1. If the cells have not been stained previously, they can be stained through the film with Wright's Stain. With this procedure, special care must be taken to prevent the film from slipping.

2. Gently dip the slides with the filters into the following solutions at 20–25°C: 0.17% Wright's Stain in methyl alcohol for about 30 seconds, the above diluted 1:1 with phosphate buffer at pH 6.6 for about 45 seconds, phosphate buffer for about 45 seconds, and two washes of deionized water for about 15 seconds each.

3. After the last wash, the film should be dried rapidly, using a fan, while the slides are kept cell-side-up.

G.   Analysis of Slides

The slides were used to correct for the few interphase cells shaken loose (interphase background), when such corrections were deemed necessary. Examples of the cell preparations are illustrated in Fig. 1. For each sample (shake), 200 cells were selected at random, observed under phase contrast optics (500×), and classified as mitotic or interphase. When the cells had been labeled with ^{3}HTdR, they were also classified as labeled or unlabeled. From these data, the percentage of cells in mitosis (mitotic index) and the percent of cells in S phase at the time of labeling (labeling index) were determined. In all experiments, the mitotic indices were at least 90–95% for control untreated cultures. As the number of cells entering mitosis decreased because of a specific treatment, the mitotic index decreased and approached zero when only interphase background remained.

The reliability of the determinations made from slides prepared by the filtering technique was tested by comparing the results with determinations made for the same pool of cells from slides prepared by the standard squash procedure (Hsu and Kellogg, 1960). For the two methods, respectively, the mitotic indices were 88% and 92% for controls and 90% and 95% for cells treated for 60 min with AMD. The AMD analysis was most critical because an AMD treatment of about an hour caused the chromatin in interphase cells to condense somewhat (Fig. 1). Then, after about 4 hours of treatment with AMD, the interphase background began to increase.

D (AMD) for 60 minutes. The condensed chromatin in the interphase cells (C and F) is typical following AMD treatment. E is an autoradiograph of cells labeled with ^{3}HTdR; the mitotic cell in the center with the white grains is labeled. The cells were stained by the aceto-orcein method. Photographs were taken with phase-contrast optics at 200 × magnification on Kodak High Contrast Copy film. Total magnification of prints is 500 ×.

III. Results

An experiment in which the transition points for actions of X-irradiation, CH, and AMD are compared is illustrated in Fig. 2. Note that the differences of approximately 15 minutes between the transition points (TPs) are readily apparent in this type of analysis.

Results of similar experiments are summarized in Fig. 3. The transition from S into G_2 (S/G_2) was determined by pulse labeling with ^{3}HTdR and then plotting the number of unlabeled mitotic cells versus time (number of shakes). The TPs, which are all significantly different from one another at the 99% confidence level (each point based on 5–10 experiments), indicate several interesting facts. The TP for heat (46.5°C for 7 or 9 minutes) is located during prophase, while the TP for ionizing radiation (100 rads) is

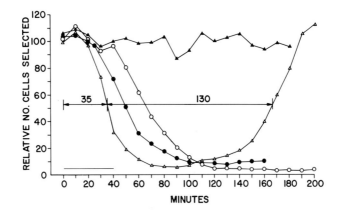

FIG 2. Asynchronous monolayer cultures growing in 4 replicate 75 cm^2 flasks were shaken every 10 minutes, and the number of cells shaken loose as they entered mitosis is plotted as a function of time. The actual numbers at $t = 0$, about 10,000/ml in a volume of 8 ml, were normalized to 100. At $t = 0$, one flask (X-105), was X-irradiated with 105 rads ($\triangle$–$\triangle$), and two other flasks (CH and AMD$_5$) were treated with 20 µg (per milliliter) of cyclo-heximide ($\bullet$——$\bullet$) or 5 µg/ml of actinomycin D ($\circ$——$\circ$), respectively. For all subsequent shakes, the drugs were left in the shake media. The transition points were determined from the times when the curves decreased to 50, i.e., 35 minutes for X-ray; however, 16 minutes were added to give the transition points relative to the end of mitosis (see Fig. 3). The average mitotic delay following irradiation was 130 minutes in this experiment. The solid line between 0 and 40 and slightly above the abscissa indicates the number of interphase cells being removed (interphase background).

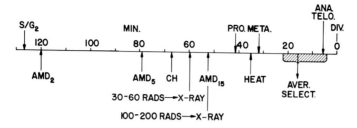

FIG. 3.   The average transition points (TPs) are plotted relative to completion of division. All the points are significantly different from one another at the 99% confidence level. See Fig. 2 for the method of determining the TPs. See Schneiderman *et al.* (1972) for the determination of the locations of anaphase-telophase (ANA. TELO.), metaphase (META.), prophase (PRO.), and the interval over which the cells are shaken loose (4–22 minutes). For the heat TP, the flasks were placed in a water bath at 46.5°C for either 7 or 9 minutes and then returned to 37°C. The location of the CH TP was the same for concentrations from 5 to 50 μg/ml; however, the location of the AMD TP shifted as the concentration was increased from 2 μg/ml (AMD$_2$) to 15 μg/ml (AMD$_{15}$).

located 10 minutes prior to prophase. For doses above 100 rads, the TP does not shift, but for lower doses of 30–60 rads, the TP shifts to about 20 minutes prior to prophase. The TP for CH is 15 minutes prior to the TP for 100–200 rads and is the same for concentrations varying from 5 to 50 μg/ml. Also, based on four determinations, the TP for puromycin (50 μg/ml) was only 3.5 minutes prior to the TP for CH (data not shown). For AMD, however, the TP is dependent upon concentration. At 5 μg/ml of AMD, the TP is located 12 minutes prior to the TP for CH, but at 2 μg/ml, the TP is located near the S/G$_2$ transition. At higher concentrations of 10–15 μg/ml AMD, the TP is the same as for 100–200 rads of X-irradiation.

An experiment investigating the movement of irradiated cells across the S/G$_2$ and AMD (5 μg/ml) TPs is illustrated in Fig. 4. In panels A, B, and C, the cells were pulse-labeled with ^{3}HTdR for 10 minutes either immediately before irradiation (L$_0$) or at 50 min (L$_{50}$) or 100 min (L$_{100}$) after irradiation to study movement of the cells from S into G$_2$. Data in panel A or B show that the total delay for S phase cells was about 95 minutes, and data in panel C show that the cells were moving from S into G$_2$ by 50 minutes after irradiation. (Note the increase in the UL$_{50}$ curve above the UL$_0$ curve.) Thus, the delay in S (S delay) was less than 50 minutes [about 45 minutes reported (Leeper *et al.* 1973)]. However, the total delay for S cells was about 95 minutes which means that the S cells sustained an additional delay of about 45 minutes in G$_2$.

Data in panel F indicate where these S cells were delayed during G$_2$; for this panel, the cells were pulse-labeled with ^{3}HTdR immediately before

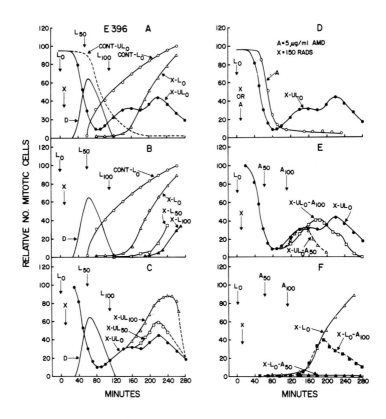

FIG. 4. In this experiment (396) involving 7 replicate flasks, the movement of irradiated cells from S into G_2 (panels A, B, and C) and then on past the AMD_5 TP (panels D, E, and F) was studied. See the legend of Fig. 2 for a general description of the mitotic selection procedure. For panels A, B, and C, the cells were pulse-labeled with ^{3}HTdR for 10 minutes either immediately before irradiation (L_0) or at 50 minutes (L_{50}) or 100 minutes (L_{100}) after irradiation (X). The mitotic cells shaken loose were identified by autoradiography as labeled (X-L_0, X-L_{50}, or X-L_{100}) or unlabeled (X-UL_0, X-UL_{50}, or X-UL_{100}). CONT represents unirradiated control cells. In panel A, the curve labeled D is the difference between the curve for unlabeled control cells (CONT-UL_0) and unlabeled irradiated cells (X-UL_0) and represents the division of unirradiated cells located between the X TP and the S/G_2 TP; the delay of irradiated X-S/G_2 cells can be seen by comparing the D curve with the X-UL_0 curve. The delay of the S cells can be seen by comparing the CONT-L_0 curve with the X-L_0 curve. For panels D, E, and F, the cells were pulse-labeled with ^{3}HTdR immediately before irradiation and then were treated with AMD (5 μg/ml) at 50 minutes (A_{50}) or 100 minutes (A_{100}) after irradiation. The mitotic cells shaken loose were identified as labeled (X-L_0-A_{50} or X-L_0-A_{100}) or unlabeled (X-UL_0-A_{50} or X-UL_0-A_{100}); i.e., X-L_0-A_{100} represents pulse labeled 10 minutes, irradiated, and then treated with AMD 100 minutes later. Once the AMD was added it remained in the shake medium. The curve designated A represents cells treated with AMD only.

irradiation and then were treated with AMD (5 μg/ml) at 50 or 100 minutes after irradiation. By this procedure, the movement of cells in S or G_2 at the time of irradiation to a point beyond the TP for AMD could be studied. The X-L_0-A_{100} curve shows that several S cells had progressed beyond the TP for AMD by 100 minutes. Since these S cells had been delayed in S for about 40–50 minutes and would require about 45 minutes to traverse from S/G_2 to the TP for AMD (Fig. 3), they must have sustained about 40 minutes of the total delay of 95 minutes between the AMD and X TPs.

Data in panels D and E indicate that the delay of the G_2 cells was quite varied; i.e., the unirradiated X–G_2 cells which normally divided over an interval of 80 minutes (curve D in panel A) divided over an interval of 170 minutes when they were irradiated (X-UL_0 in panels A and D). The early G_2 cells were delayed 110–160 minutes and the late G_2 cells located between the AMD and XTPs (X-UL_0-A_{50} curve in panel E) were delayed only about 60 minutes, a delay less than for late S cells and much less than for early G_2 cells. Also note in panel E (X-UL_0-A_{100}), that by 100 minutes after irradiation, a significant number of G_2 cells had moved beyond the TP for AMD. Therefore, the early G_2 cells, which had a total delay of about 120 minutes, must have sustained an additional delay beyond the TP for AMD.

A more quantitative way of analyzing data of this type is to measure the areas under the curves in Fig. 4 and to equate these areas to the number of cells entering the selection window (Fig. 3). The number of cells is expressed in terms of minutes required (min of cells) for an equal number of control unirradiated cells to enter the selection window. Such an analysis is shown in Figs. 5 and 6.

In Fig. 5, the delay of cells in late S at the time of irradiation (100 rads) can be analyzed. These cells which sustained a total delay of about 70 minutes were delayed about 45 minutes in leaving S. Then, they crossed between S/G_2 and TP for AMD without a delay before they sustained another delay of 36 min between the TPs for AMD and X. As shown by the curves XL_{40} and XL_{80}, the cells still in S at 40 or 80 minutes after irradiation, respectively, had part of the damage responsible for delay repaired by 40 or 80 minutes because the total delays were only 53 and 40 minutes, respectively. Although movement across the AMD TP was not studied for these cells still in S at 40 or 80 minutes after irradiation, it is quite likely that most of the damage responsible for S delay had been repaired by 80 minutes and that the 40-minute delay remaining occurred in G_2 past the AMD TP.

In Fig. 6, the delay of cells in G_2 at the time of irradiation can be analyzed. As shown in Fig. 4, the delay of about 75 minutes for those cells in late G_2 (between AMD and X) was much less than the delay of about 150 minutes

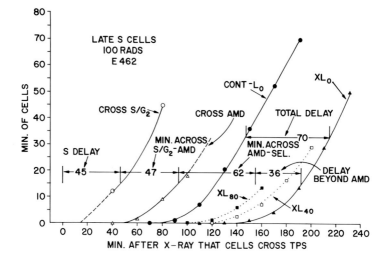

FIG. 5.   Data derived from curves like those plotted in Fig. 4 show the delay and movement of cells irradiated in late S phase. The areas under the curves (like those in Fig. 4) were equated to the times required for an equal number of control cells to enter the selection window. For example, the point with 80 on the abscissa and 45 on the ordinate of the curve, "cross S/G_2," was derived from the area under a curve plotting the number of unlabeled cells collected in mitosis when the cells were pulse-labeled with ^{3}HTdR 80 minutes after irradiation; this area (X-UL$_{80}$) minus the area for unlabeled cells when they were pulse-labeled immediately prior to irradiation (X-UL$_0$) was equal to the area of a 45-minute interval under the control curve, i.e., 45 × 100. Thus, by 80 minutes after irradiation, a number of cells equivalent to 45-minutes of control progression had moved from S into G_2. From the areas under curves for labeled cells when the cells were pulse-labeled with ^{3}HTdR immediately before irradiation and then treated with AMD (5 μg/ml) at later times, e.g., X-L$_0$-A$_{100}$ in Fig. 4, panel F at 100 minutes, the movement of labeled S cells across the AMD TP could be determined (cross AMD curve). The movement of the labeled S cells into the selection window (Fig. 2) was determined from the area under an X-L$_0$ curve for irradiated cells and under the area of a CONT-L$_0$ curve for unirradiated cells (as in Fig. 4, panel A). The total delay for late S cells is 70 minutes, the difference between the CONT-L$_0$ and X-L$_0$ curves. As seen from the "cross S/G_2" curve, the cells were delayed about 45 minutes in entering G_2, and then they moved on past the AMD TP without a delay because in this experiment the difference between the S/G_2 TP and the AMD TP in the controls was 55 minutes. Since the AMD TP was 62 minutes before selection (SEL), the cells had to sustain an additional delay of 36 minutes beyond the AMD TP. The curves, XL$_{40}$ and XL$_{80}$, show the movement of labeled cells into the selection window when they were pulse-labeled at 40 or 80 minutes after irradiation, respectively.

for those in early G_2. For these early G_2 cells having a total delay of 150 minutes, about 78 minutes of the delay was sustained in the AMD-X window. The data for the two separate experiments are summarized in Table I.

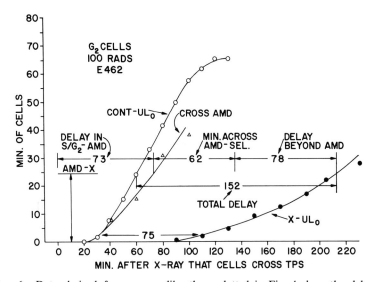

FIG. 6. Data derived from curves like those plotted in Fig. 4 show the delay and movement of cells irradiated in G_2. This is the same experiment illustrated in Fig. 5 for late S cells, and the minute equivalents of cells on the ordinate were determined from areas under curves. For example, the point with 100 on the abscissa and 38 on the ordinate of the curve "cross AMD" was derived from the area under a curve plotting the number of unlabeled cells collected in mitosis when the cells were pulse-labeled with ^{3}HTdR immediately before irradiation and then treated with AMD (5 μg/ml) 100 minutes after irradiation; this area (e.g., like $X\text{-}UL_0\text{-}A_{100}$ in Fig. 4) was equal to the area of a 38-minute interval under the control curve. Thus, when AMD was added 100 minutes after irradiation, the cells normally in the AMD-X window (25-minute equivalents) plus another 13 minute equivalents of cells which had crossed the AMD TP from G_2 were able to move toward the selection window. The movement of the unlabeled G_2 cells (located between X and S/G_2) into the selection window was determined from the areas under an $X\text{-}UL_0$ curve (Fig. 4) for irradiated cells and under a D curve (Fig. 4) for control cells (CONT.-UL_0). The total delay for G_2 cells increased from about 75 minutes for those in late G_2 to over 152 minutes for those in early G_2. Since the AMD TP was located 62 minutes prior to selection in this experiment, 78 minutes of the total delay of 152 minutes occurred beyond the AMD TP.

Note in Fig. 6 that not all of the AMD-X cells were able to repair their damage and enter division if AMD (5 μg/ml) was added sooner than 70 minutes after irradiation. This fact is illustrated by the recovery of fewer cells than those trapped in the AMD-X window, i.e., about 25 minutes in duration in this experiment. When AMD was added subsequent to 70 minutes after irradiation, however, all the AMD-X cells were able to enter mitosis. In fact, the curve rises above the 25-minute level for the A-X cells because, subsequent to 70 minutes, cells that were in early G_2 at the time of

TABLE I

X-Ray-Induced Delay (Minutes) in Relation to Transition Points[a]

	Late S cells		Early G$_2$ cells		Late G$_2$ cells	
Crossing S/G$_2$	45	48	—	—	—	—
Crossing between						
S/G$_2$ and AMD[b]	0	0	73	58	—	—
Beyond AMD[b]	36	50	78	59	75	60
Sum of delays	81	98	151	117	75	60
Total delay[c]	70	83	152	103	75	60

[a]See Figs. 4, 5, and 6 for method of analysis. Late S cells, early G$_2$ cells, and late G$_2$ cells were located, respectively, in late S, between the S/G$_2$ and actinomycin D (AMD) TPs, or between the AMD and X TPs at the time of irradiation. The first values are for experiment 462 (100 rads), and the second values are for experiment 396 (150 rads). The reduction in delay for experiment 396 compared with experiment 462 is probably associated with the reduction in length of G$_2$ (Dewey et al., 1965), i.e., S/G$_2$ TP at 117 and 135 minutes and periods between S/G$_2$ and AMD TPs of 39 and 55 minutes, respectively.

[b]AMD at 5 μg/ml was added at various times after irradiation.

[c] Total delay was determined from the difference between the control (CONT-L$_0$, CONT-UL$_0$) and irradiated (XL$_0$, X-UL$_0$) curves (Figs. 5 and 6).

irradiation crossed the AMD TP. When a concentration of 2 μg/ml AMD was added at the time of irradiation, however, the AMD$_2$-X cells were able to enter mitosis after a delay period which was actually 40 minutes shorter than for untreated cells (data not shown). Therefore, RNA synthesis apparently was not required for repair of the damage responsible for delay, and the reason for the different results for 2 μg/ml and 5 μg/ml remains to be resolved.

Figure 7 illustrates the results of a similar experiment in which the movement of irradiated cells across the TPs for AMD and CH was determined. These results were confirmed in another experiment not illustrated. The first important point is that the number of cells moving across the CH TP is only 5 minute equivalents greater than the number moving across the X TP. This means that when recovery from delay occurred, the TP for CH was located about 5 minutes earlier in G$_2$ than the TP for X-rays. In other words, CH acted within 5 minutes, and any cells located between X and CH (see Fig. 3) needed to synthesize protein to recover from the X-ray-induced delay. In other experiments (data not shown) cells located prior to the CH TP also had to synthesize protein to recover from their X-ray-induced delay. The second important point is that at 70 minutes, the cells started moving across the AMD TP much faster than they started moving across the CH or X TP. This clearly shows, as seen in Figs. 5 and 6, that during

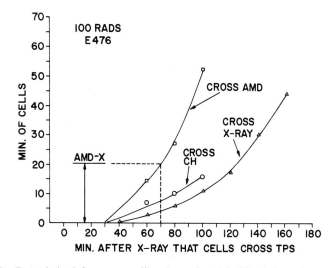

FIG. 7. Data derived from curves like those plotted in Fig. 4 show the movement of irradiated cells across the X, CH (20 μg/ml) and AMD (5 μg/ml) TPs. In this experiment, the cells were not labeled with ³HTdR, but CH or AMD was added to replicate flasks at 40, 60, 80, or 100 minutes after irradiation. Once a drug was added, it was kept in the shake medium. The minute equivalents of cells were determined from the total areas under curves plotting the numbers of mitotic cells entering the selection window. For example, when CH was added 100 minutes after irradiation, the cells entered the selection window at the same time as the irradiated cells not treated with CH, but only 16-minute equivalents of cells were able to enter the selection window; the remainder of the cells were trapped prior to the CH TP. For the curve "cross X-ray," the time of 37.5 minutes between X and the average point of selection (like the 35 minutes illustrated in Fig. 2) was subtracted in order to show when movement across the X TP occurred. Note that by 70 minutes after irradiation, the AMD-X cells could be recovered, and then subsequent to 70 minutes the cells were crossing the AMD TP faster than they were crossing the CH TP.

the delay period, cells were accumulating in the AMD-X window. Thus, a large part of the G_2 block is associated with cells piling up between the TP for AMD (5 μg/ml) and the TP for CH.

IV.   Discussion

The technique of monitoring the number of cells entering mitosis following various treatments provides a precise method for locating the points in the cycle where certain events are terminated (Fig. 2). For example, the completion of events which renders the cells resistant to radiation-induced mitotic delay is located 8 minutes earlier in G_2 for 30–60 rads than for doses exceeding 100 rads (Fig. 3). Furthermore, the transition point for cyclo-heximide (5–50 $\mu g/ml$) does not coincide exactly with the TP for X-rays, as has been reported (Walters and Petersen, 1968b), but instead precedes it (Schneiderman *et al.*, 1972) by 8–15 minutes depending on the radiation dose. The TP for heat, however, which probably damages spindle structures (Inove, 1952, 1964; Bergan, 1960; Ikeda, 1965; Sisken *et al.*, 1965; Dewey *et al.*, 1971; Westra and Dewey 1971) is located in prophase, and thus about 20 minutes later than the TP for X-rays. Also, the duration of delay for heat is about 12 hours (Dewey *et al.*, 1971; Westra and Dewey, 1971) com-pared with only 1–4 hours for a dose of X-rays which produces the same amount of cell killing as the heat shock. Finally, the TP for AMD is dose dependent. Since a concentration of 2 $\mu g/ml$ has been reported to inhibit at least 90% of all RNA synthesis (Tobey *et al.*, 1966) within a few minutes, the TP of 120 min for 2 $\mu g/ml$ may indeed be the point where RNA synthesis necessary for division is completed [114 minutes reported (Tobey *et al.*, 1966) for CHO cells for 2 $\mu g/ml$]. Then, at higher concentrations, the increase in amount of AMD bound to DNA may be having some direct effect on DNA and associated chromosomal proteins. The fact that the TP for AMD plotted against concentration approaches asymptotically the TP for X-rays (data not shown) is probably not coincidental and may imply that X-irradiation and AMD block the cells in G_2 because of damage to the same critical structures.

Another important aspect of the technique is being able to study the movement of cells between the different transition points (Figs. 4–7). For example, the finding that cells irradiated in late S or early G_2 sustain a large part of their delay late in G_2 in the AMD-X window (for 5 $\mu g/ml$) is par-ticularly significant (Fig. 7 and Table I), especially since cells irradiated in the AMD-X window sustain less delay than those irradiated in early G_2. Although formulation of a model for mitotic delay may be premature, our data are consistent with the hypothesis that ionizing radiation damages structures having some relationship with chromatin and that these structures are in their final stages of assembly late in G_2, i.e. in the AMD_5-X window. Thus, the cells should sustain a large part of their delay, as observed, during the final period when the assembly is being completed. Furthermore, the sensitivity to radiation, as measured by mitotic delay, would decrease, as observed, during the final period because once the structures are assembled, they would be refractory to radiation damage. The difference between the TP for 30–60 rads compared with the TP for 100 rads or more [see Walters and Petersen (1968a) for high doses] implies that the probability of getting a certain threshold of damage in these structures (a) decreases as the dose is reduced below 100 rads and (b) decreases as the final assembly process reduces the number of structures. However, if the structures are damaged before their assembly is completed, protein synthesis but not RNA synthesis is required for their repair [Walters and Petersen (1968a), Fig. 7, and data not shown]. In fact, the synthesis of proteins for these structures would be required until the CH TP is reached; then, during the 8–15 minutes period between the CH and X TPs, the very last stages of assembly would be completed.

REFERENCES

Axelrod, A. A., and McCulloch, F. (1958). *Stain Technol.* **33**, 344.

Bergan, P. (1960). *Nature (London)* **186**, 905.

Dewey, W. C., and Miller, H. H. (1969). *Exp. Cell Res.* **57**, 63.

Dewey, W. C., Humphrey, R. M., and Jones, B. A. (1965). *Int. J. Radiat. Biol.* **8**, 605.

Dewey, W. C., Westra, A., Miller, H. H., and Nagasawa, H. (1971). *Int. J. Radiat. Biol.* **20**, 505.

Hsu, T. C., and Kellogg, D. S. (1960). *J. Nat. Cancer Inst.* **24**, 1067.

Ikeda, M. (1965). *Exp. Cell Res.* **40**, 282.

Inoue, S. (1952). *Biol. Bull.* **103**, 316.

Inoue, S. (1964). *In* "Primitive Motile Systems in Cell Biology" (R. D. Allen and N. Kamiya, eds.), pp. 549–594. Academic Press, New York.

Leeper, D. B., Schneiderman, M. H., and Dewey, W. C. (1973). *Radiat. Res.* **53**, 326.

Millipore Corporation (1966). "Millipore Application Data Manual-40." Techniques for Microbiological Analysis, p. 15. Millipore Corp., Bedford, Massachusetts 01730.

Millipore Corporation (1963). "Millipore Application Data Manual-70," Microchemical and Instrumental Analysis, pp. 45–47. Millipore Corp., Bedford, Massachusetts 01730.

Nagasawa, H., and Dewey, W. C. (1972). *J. Cell. Physiol.* **80**, 89.

Schneiderman, M. H., Dewey, W. C., Leeper, D. B., and Nagasawa, H. (1972). *Exp. Cell Res.* **74**, 430.

Sisken, J. E., Morasca, L., and Kibby, S. (1965). *Exp. Cell Res.* **39**, 103.

Terasima, T., and Tolmach, L. J. (1963). *Exp. Cell Res.* **30**, 344.

Tjio, J. H., and Puck, T. T. (1958). *J. Exp. Med.* **108**, 259.

Tobey, R. A., Petersen, D. F., Anderson, E. C., and Puck, T. T. (1966). *Biophys. J.* **6**, 567.

Tobey, R. A., Anderson, E. C., and Petersen, D. C. (1967). *J. Cell. Physiol.* **70**, 63.

Walters, R. A., and Petersen, D. F. (1968a), *Biophys. J.* **8**, 1475.

Walters, R. A. and Petersen, D. F. (1968b). *Biophys. J.* **8**, 1487.

Westra, A., and Dewey, W. C. (1971). *Int. J. Radiat. Biol.* **19**, 467.

Chapter 7

Evaluation of S Phase Synchronization by Analysis of DNA Replication in 5-Bromodeoxyuridine

RAYMOND E. MEYN, ROGER R. HEWITT, AND
RONALD M. HUMPHREY

*Departments of Physics and Biology,
The University of Texas System Cancer Center,
M. D. Anderson Hospital and Tumor Institute,
Houston, Texas*

I. Introduction

Various techniques have been employed to monitor progression after synchronization of cells at the beginning of their DNA synthetic (S) phase. Each of these techniques, which include determination of (1) the fraction of labeled nuclei by autoradiography (Bootsma *et al.*, 1964; Galavazi and Bootsma, 1966; Sinclair, 1965); (2) the rate of incorporation of thymidine-^{3}H (TdR) into trichloroacetic acid-insoluble material (Taylor *et al.*, 1971); (3) DNA mass increase (Rueckert and Mueller, 1960; Mueller *et al.*, 1962); and (4) DNA content by flow microfluoremetric analysis (Tobey *et al.*, 1972), has its advantages and disadvantages. However, none of these techniques allows one to measure directly the proportion of DNA semiconservatively replicated as a function of time after release into S phase.

The only known approach to this type of measurement in mammalian cells is to density label the newly synthesized DNA with 5-bromodeoxyuridine (BUdR) and then determine the proportion of total DNA replicated by measuring the amount of DNA acquiring an increased buoyant density in CsCl gradients. The first demonstration that mammalian cells replicated DNA in a semiconservative manner utilized this method (Djordjevic and Szybalski, 1960; Simon, 1961). Although this technique has been applied to the study of a variety of problems associated with DNA replication it has not previously been used to evaluate methods of S phase synchronization.

In this chapter we will describe a method for evaluating S phase synchronization by analysis of DNA replication in BUdR and discuss the advantages that this technique has over the others mentioned above. As a practical demonstration of the technique the results of a comparative evaluation of excess TdR and hydroxyurea (HU) for synchronizing cells at the beginning of S phase will be presented.

II.   Methods

A.   Cell Cultures

Stock cultures of Chinese hamster ovary (CHO) cells were maintained as monolayers in McCoy's 5A medium (Grand Island Biological Co.) supplemented with 20% fetal calf serum (Grand Island Biological Co.). These stocks were subdivided each day by trypsinization. A Model ZBI Coulter counter was used to determine cell number. All incubations were carried out in a humidified, 37°C, 5% CO_2 atmosphere. Under these growth conditions, the CHO cells had a 12–14 hour doubling time and a generation time of 11.5 hours consisting of a 2.5-hour G_1, 6.75-hour S, 1.75-hour G_2, and 0.5-hour M. Procedures used for autoradiography have been described elsewhere (Barranco and Humphrey, 1971).

B.   Cell Synchronization

Asynchronous cells (1×10^7) were seeded into 32-ounce prescription bottles with 35 ml of medium containing TdR-^{14}C (0.25 μCi/ml, 33 mCi/mmole, Schwarz/Mann) and incubated for 18 hours. Nonradioactive medium containing 7.5 mM TdR was then substituted and incubation continued for 9 hours. This treatment produced a parasynchronous culture (Xeros, 1962; Bootsma et al., 1964) in S phase. The excess TdR medium was then replaced by normal medium and cells were incubated for 5 hours. During this period the parasynchronous S phase culture progressed through S and G_2. After 5 hours a greatly enhanced mitotic index (MI) was observed, which approached 15% during the sixth hour. Mitotic cells were selectively detached from the monolayer by agitation and decanting the medium (Tobey et al., 1967; Peterson et al., 1968). Fresh medium was replaced and mitotic selection was repeated at 15-minute intervals during the peak mitotic period. Mitotic cells were held in an ice bath.

Unfortunately, mitotic cells do not enter the subsequent S phase uniformly, and thus must be resynchronized at the G_1–S boundary. Aliquots of mitotic cells (5×10^5 cells, MI = 93 to 98%) were dispensed into petri dishes and the synchronizing agent to be tested was added. Cultures were then incubated at 37°C for 9 hours. During this period mitotic cells progress through G_1. The extent to which they progress into S phase is dependent on the effectiveness of a synchronizing agent in blocking DNA synthesis. Absolute blocking of S phase progression should provide a pure population at the G_1–S boundary. Removal of the block should then allow the synchronous progression of all cells through S phase with the result that 100% of the total DNA should replicate during this period. Thus, the kinetics and extent of DNA replication may be used to determine the effectiveness of the synchronizing treatment.

C. Analysis of DNA Replication

After removal of the synchronizing agents cells were incubated in medium containing 50 μg/ml BUdR and 0.1 μg/ml 5-fluorodeoxyuridine (FUdR). Inclusion of FUdR inhibits de novo biosynthesis of thymidine and allows complete replacement of BUdR for TdR during DNA synthesis (Taylor, 1968). Replicated parental DNA, about 50% of which is labeled with TdR-^{14}C, becomes heavier due to the BUdR and may be resolved from unreplicated DNA in CsCl gradients. Thus the density distribution of ^{14}C radioactivity in CsCl gradients may be used to determine the kinetics and extent of DNA replication as presented in Results.

Samples were taken as a function of the time of density labeling by removing a plate from the incubator and trypsinizing the cells. The cells were counted, washed, and centrifuged. The cell pellets were resuspended in buffer (0.15 M NaCl, 0.01 M EDTA, 0.01 M TRIS, pH 9.5) at a concentration of 1 $\times$ 10^6 per milliliter. At this point the samples were usually stored frozen until the next day when all samples were lysed at the same time. Storage for much longer than 24 hours prior to lysis resulted in DNA degradation. Cells were lysed by the addition of Sarkosyl (Geigy) to 0.1% and incubated for 30 minutes at 60°C in a constant-temperature water bath. The cell lysate was deproteinized by mixing with an equal volume of chloroform–isoamyl alcohol (24:1). The mixture was centrifuged and the aqueous layer was removed and saved. Deproteinized lysates could be stored frozen for many months without resultant DNA degradation.

CsCl gradients were prepared by mixing together 4.0 ml of CsCl solution (63.8% w/w) and 0.7 ml of cell lysate in a Teflon-pestle hand homogenizer. The homogenizer causes the DNA to be sheared to a uniform size which results in good separation of replicated DNA from unreplicated DNA in CsCl. The gradients were centrifuged for 45 hours (33,000 rpm at 21°C) in an SW50.1 rotor in a Spinco Model L-2 ultracentrifuge. Gradients were collected from the bottom of the tube directly into scintillation vials containing 1 ml of water. Fluor [1 part Triton X-100 (Rohm and Haas), 2 parts toluene, 4 gm/liter 2,5-diphenyloxazole, and 50 mg/liter 1,4-bis[2-(5-phenyloxazolyl)] benzene] was added to the vials, which were then placed in a liquid scintillation counter for determination of ^{14}C radioactivity.

During gradient collection, refractive index readings were made on one drop from each fifth fraction. These readings were used to determine the density of the CsCl solution as a function of fraction number.

D.   Labeling with ^{32}P

The relative amounts of DNA synthesized during different treatments were estimated by determining the ratio of $^{32}P:^{14}C$ in DNA. ^{32}P-Labeled DNA was separated from ^{32}P-labeled RNA as follows. After labeling, the cells were lysed and the DNA banded in preparative CsCl gradients as previously described. Fractions were collected into 1 ml of water in scintillation vials and counted without the addition of scintillation fluor. The fractions containing ^{32}P-DNA were identified by counting pulses produced by Cerenkov radiation from the ^{32}P and by their position in the gradient. These fractions were pooled, treated with a mixture of 25 units of T1 ribonuclease (CalBiochem) and 50 μg of pancreatic ribonuclease (Worthington) per milliliter, and dialyzed in buffer overnight. After dialysis, the DNA was banded a second time in CsCl. The amount of ^{14}C and ^{32}P radioactivity in each fraction was measured by liquid scintillation spectroscopy using scintillation fluor.

III.   Results

A.   DNA Replication Kinetics of TdR- and HU-Synchronized Cells

The kinetics of DNA replication in cells synchronized with TdR and HU are shown in Fig. 1. The two CsCl gradient profiles used to calculate the data points at 9 hours in Fig. 1 are shown in Fig. 2. The percent total DNA

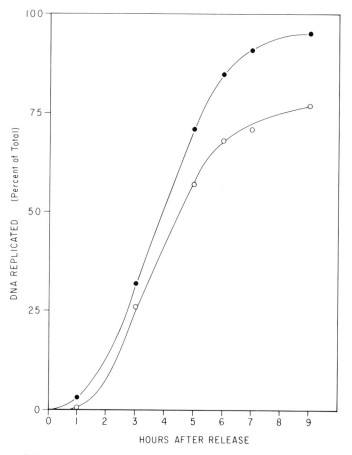

FIG. 1.   DNA replication in cell populations synchronized with thymidine (TdR) or hydroxyurea (HU). Cells were labeled with TdR-^{14}C, selected in mitosis, and treated for 9 hours with either TdR or HU. The cells were then released into replication test medium and periodically samples were prepared for analysis of the percentage of total DNA replicated. (●——●) Cells synchronized with HU; (○——○) cells synchronized with TdR. Reprinted with permission from Meyn et al. (1973).

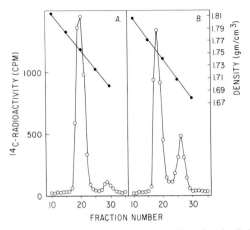

FIG. 2.   Distributions of ^{14}C-labeled DNA after centrifugation in CsCl density gradients for the 9-hour points in Fig. 1. (A) Hydroxyurea treated. (B) Thymidine treated. The density of the CsCl solution was determined in several fractions of each gradient from refractive index readings (●——●). Reprinted with permission from Meyn *et al.* (1973).

replicated is determined by summing the ^{14}C radioactivity (cpm per fraction) in each of the two peaks and making the following calculation:

$$\% \text{ DNA replicated} = \frac{\text{total cpm in hybrid peak}}{\text{total cpm in both peaks}} \times 100$$

The hybrid peak containing the replicated DNA has a density of 1.75 gm/cm^3 and the peak of unreplicated DNA has a density of 1.70 gm/cm^3 (Fig. 2). A visual comparison of the two profiles in Fig. 2 indicates that substantially more DNA was replicated during the 9 hours after release into S phase in HU-synchronized cells than in TdR-synchronized cells.

The DNA replication kinetics curves shown in Fig. 1 are generated by analyzing the CsCl gradients at each time point in the experiment by the method above and plotting the percentage of DNA replicated as a function of the time of density labeling. The main difference in DNA replication kinetics between TdR and HU synchronized cells, as seen in Fig. 1, is that after 5 hours into S phase the curve for the TdR-treated cells plateaus to about 75% total DNA replicated, while the HU curve continues to about 95%.

B.   Measurement of DNA Synthesis during TdR and HU Treatment

Our initial interpretation of the fact that both of the curves in Fig. 1 did not proceed to 100% was that 25% and 5% of the DNA had replicated during

TABLE I

RELATIVE DNA SYNTHESIS DURING THYMIDINE (TdR)
AND HYDROXYUREA (HU) TREATMENT[a]

Sample	$^{32}P/^{14}C$	$^{32}P/^{14}C$ as % of control
Control	0.301	100
Thymidine treated	0.092	31
HU treated	0.017	6

[a]From Meyn et al. (1973).

the TdR or HU treatments, respectively. This DNA would not replicate again during the same S phase. This idea was confirmed by measuring the relative amounts of DNA synthesized during the 9-hour treatments with these synchronizing agents. TdR-^{14}C-labeled mitotic cells were plated into normal medium or medium containing either 7.5 mM TdR or 2 mM HU. After the addition of 10 μCi of carrier-free ^{32}PO$_4$ per milliliter, the cells were incubated for 9 hours. The DNA was then extracted and purified, and the ^{32}P/^{14}C ratio was determined as described in Methods. The ratios for each sample are presented in Table I. These ratios are proportional to the

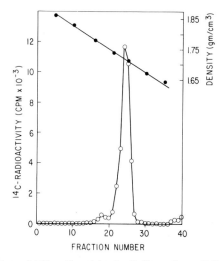

FIG. 3. Distribution of ^{14}C radioactivity in CsCl gradient. Cells were labeled for 18 hours with thymidine-^{14}C and selected in mitosis. Mitotic cells were plated into medium containing 2 mM hydroxyurea and 50 μg of bromodeoxyuridine and 0.1 μg of fluorodeoxyuridine per milliliter and incubated for 9 hours. The density of the CsCl solution was determined in several fractions from refractive index readings (●———●). Reprinted with permission from Meyn et al. (1973).

specific activity of the newly synthesized DNA and, hence, are proportional to the relative amounts of DNA synthesized. Treatment with TdR limited DNA synthesis to 31% of that in control cells, while HU limited synthesis to 6% of control.

The actual amount of DNA replicated during an HU treatment was determined by plating TdR-^{14}C-labeled mitotic cells into medium containing 2 mM HU and 50 μg of BUdR and 0.1 μg of FUdR per milliliter. The rationale of this experiment was that any DNA replicated during treatment with HU would contain BUdR instead of TdR. At the end of treatment the cells were harvested and analyzed for percentage of total DNA replicated. It can be seen from the CsCl gradient profile shown in Fig. 3 that about 4% of the total DNA replicated during 9 hours in HU. A comparable experiment with TdR-blocked cells is not possible.

C.   Comparison with Conventional Progression Analysis

Experiments in which the cells were synchronized in exactly the same way as in Fig. 1 were performed in order to compare the conventional method of progression analysis with that proposed here. In these experiments, S phase progression was analyzed by determining the TdR-^{3}H-labeling index by autoradiography. The results presented in Fig. 4 indicate that the length of S phase, as determined from labeling index, differs after the two types of synchronization by only 30 minutes. S phase progression in medium containing BUdR and FUdR was also studied in cells treated with HU and was found to be nearly the same as progression in normal medium (Fig. 4).

IV.   Discussion

The results presented in Fig. 1 indicate that a significant amount of DNA can be synthesized during treatment with 7.5 mM TdR. Thus, cells synchronized by this agent are not blocked at the G_1–S boundary, but accumulate in early S phase. These findings confirm the earlier investigations into this problem which utilized other methods of analysis [labeling index (Bootsma et al., 1964); DNA mass increase (Studzinski and Lambert, 1969); incorporation of ^{32}P into DNA (Bostock et al., 1971); and DNA content by flow microfluorometry (Tobey et al., 1972)]. Comparison of the DNA replication kinetics (Fig. 1) with the labeling index (Fig. 4) clearly illustrates the advantage of the BUdR technique over the conventional method. While the labeling index patterns (Fig. 4) show that S phase is somewhat shorter in

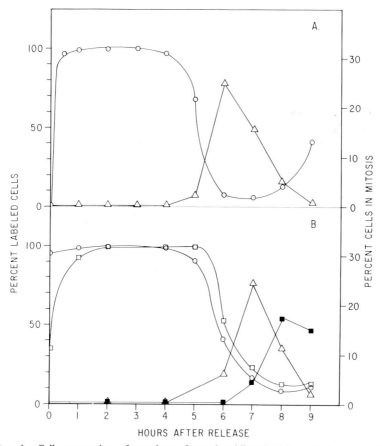

FIG. 4. Cell progression after release from thymidine (TdR) or hydroxyurea (HU) treatment. Cells were selected in mitosis and treated for 9 hours with TdR (A) or HU (B). After synchronization, the cellular progression was studied by determining the percentage of labeled cells (O——O) and percent cells in mitosis (△——△) as a function of time. The percentage of labeled cells was measured by periodic pulse-labeling with TdR-³H and autoradiography. Progression of cells in medium containing 50 μg of bromodeoxyuridine (BUdR) and 0.1 μg of fluorodeoxyuridine per milliliter was also studied after synchronization with HU (B). The percentage labeled cells was determined using BUdR-³H in this case (□——□). The percentage of cells in mitosis was also scored (■——■). Reprinted with permission from Meyn *et al.* (1973).

TdR-treated cells compared to HU-treated cells, the DNA replication kinetics (Fig. 1) give a better description of the S phase in terms of rates and amounts of replication, and also suggest the reason for its shortening after TdR treatment. In contrast to TdR, HU seems an excellent agent for synchronization of DNA replication, since only 4% of the total cellular DNA replicates during 9 hours in HU (Fig. 3).

It is important to distinguish between DNA replication and DNA synthesis in studies of cell cycle progression, synchronization, and treatments affecting DNA. DNA replication, as defined by the BUdR density-labeling method described here, requires duplication of parental DNA segments of at least 10×10^6 daltons, which is the approximate size of the segments analyzed in the CsCl gradients. DNA replication always requires DNA synthesis, whereas some types of synthesis do not yield the expected amounts of replicated DNA. For example, the limited amounts of replacement synthesis associated with "repair replication" in damaged DNA, or reiterative duplication of a limited number of genes associated with gene amplification would not yield the amounts of replicated DNA segments that would be expected from the amount of DNA precursor incorporated. Thus the density labeling method allows a clear distinction to be made between DNA replication during progression through S phase and DNA synthesis associated with other types of DNA metabolism.

We propose that by measuring the rate and maximum extent of DNA replication with BUdR during the subsequent S phase, the relative effectiveness of different synchronizing treatments may be evaluated. Once the proper synchronizing technique has been chosen, this method can then also be used to measure the kinetics of DNA replication in many types of investigations which include: (1) studies of the temporal regulation of DNA replication; (2) determination of the time of replication of RNA cistrons; and (3) determination of the influence of radiation or chemicals on the rate and sequence of DNA replication.

ACKNOWLEDGMENTS

This investigation was supported by Public Health Service Research Grant No. CA-04484 from the National Cancer Institute. We wish to acknowledge the excellent technical assistance of Mrs. Beverly A. Sedita and Mrs. J. M. Winston.

REFERENCES

Barranco, S. C., and Humphrey, R. M. (1971). *Mutat. Res.* **11**, 421.
Bootsma, D., Budke, L., and Vos, O. (1964). *Exp. Cell Res.* **33**, 301.
Bostock, C. J., Prescott, D. M., and Kirkpatrick, J. B. (1971). *Exp. Cell Res.* **68**, 163.
Djordjevic, B., and Szybalski, W. J. (1960). *J. Exp. Med.* **112**, 509.
Galavazi, G., and Bootsma, D. (1966). *Exp. Cell Res.* **41**, 438.
Meyn, R. E., Hewitt, R. R., and Humphrey, R. M. (1973). *Exp. Cell Res.* **82**, 137.
Mueller, G. C., Kajiwara, K., Stubblefield, E., and Rueckert, R. R. (1962). *Cancer Res.* **22**, 1084.
Peterson, D. F., Anderson, E. C., and Tobey, R. A. (1968). *Methods Cell Physiol.* **3**, 347–370.
Rueckert, R. R., and Mueller, G. C. (1960). *Cancer Res.* **20**, 1584.
Simon, E. H. (1961). *J. Mol. Biol.* **3**, 101.

Sinclair, W. K. (1965). *Science* **150**, 1729.
Studzinski, G. P., and Lambert R. C. (1969). *J. Cell Physiol.* **73**, 109.
Taylor, J. H. (1968). *J. Mol. Biol.* **31**, 579.
Taylor, J. H., Myers, T. L., and Cunningham, H. L. (1971). *In Vitro* **6**, 309.
Tobey, R. A., Anderson, E. C., and Peterson, D. F. (1967). *J. Cell. Physiol.* **70**, 63.
Tobey, R. A., Crissman, H. A., and Kraemer, P. M. (1972). *J. Cell Biol.* **54**, 638.
Xeros, N. (1962). *Nature (London)* **194**, 682.

Chapter 8

Application of Precursors Adsorbed on Activated Charcoal for Labeling of Mammalian DNA in Vivo

GEORGE RUSSEV AND ROUMEN TSANEV

Institute of Biochemistry, Bulgarian Academy of Sciencies, Sofia, Bulgaria

I. Introduction

Continuous labeling of DNA for a long period of time is often required in biochemical experiments. Such is the case when cell cycle kinetics or proliferative pools are studied and when a high degree of incorporation of precursors such as 5-iodo- and 5-bromodeoxyuridine (BUdR) in DNA is the aim. Unfortunately, a few minutes after administration the concentration of exogenous thymidine in the organism begins rapidly to decrease (Bresciani, 1965). For this reason, such experiments are preferably carried out on cell cultures that can be grown on media containing the desired precursor for practically an indefinite time. For animals, a technique of repeated injections is used. However, together with the technical inconveniences, this treatment has the serious shortcoming that each injection affects the mitotic activity of the cells (Markov and Raitcheva, 1967).

Here we describe a technique which avoids the above-mentioned disadvantages by administration of the precursor as a "depot." In this way a reliable and constant concentration of the precursor is maintained in the organism for several days.

II.   Preparation of Adsorbed Precursors

We have performed several attempts to obtain thymidine and BUdR in a "depot" form. The precursors were included in stabilized oil emulsions, embedded in Sephadex, or adsorbed on ion exchangers. However, the results were far from satisfactory until we used activated charcoal as adsorbent. The procedure we finally adopted is as follows:

Activated charcoal (pulver form, Merck) is washed several times with 5% NaOH and 5% HCl, then with distilled water to neutral pH, and dried at 105°C. The charcoal is added to the aqueous solution of thymidine or BUdR to make a 1–10% suspension and is occasionally stirred at room temperature for 2 hours. This results in a 98–99.5% adsorption of the precursor on the charcoal. The value of 98–99.5% was obtained when charcoal was added to a solution containing as little as 5×10^{-4} mg/ml of the radioactive precursor, on increasing the amount of the precursor in the solution up to 25% of the weight of the charcoal sample, the result was the same. Further increase of the thymidine-to-charcoal ratio did not lead to additional adsorption, the result showing that the limit of the adsorption capacity of the charcoal was 25% of its weight. Thus, it appears that charcoal quantitatively adsorbs thymidine and its analogs even from very diluted aqueous solutions, provided that the thymidine to charcoal ratio is up to 0.25 (w/w).

In order to check the stability of the charcoal-precursor complex, charcoal with adsorbed radioactive precursor was extensively washed with water or suspended in water to make a 0.001% suspension and stirred at room temperature for several hours. In either case less than 0.5% of the radioactivity were desorbed.

III.   Kinetics of Labeling of Mammalian DNA after a Single Injection of Adsorbed Thymidine-[14]C

Because of the high and constant proliferative rate of the intestine epithelium, mouse ileum was chosen as a model for studying the kinetics of labeling of mammalian DNA *in vivo*. Two sets of experiments were carried out. In the first one, referred to as "continuous labeling," the animals received 2×10^{-3} mg of thymidine-^{14}C per gram body weight; in the second set of experiments, designated as "heavy substitution," the amount of thymidine was 0.5 mg per gram body weight.

A. Continuous Labeling

The animals received a single intraperitoneal injection of 5 μCi of thymidine-^{14}C adsorbed on 5–6 mg of activated charcoal (Russev and Tsanev, 1973). The control animals were injected with the same amount of free thymidine-^{14}C, and in both cases the kinetics of incorporation of the radioactive precursor was followed by determination of the specific radioactivity of DNA at different intervals after administration of the precursor. In the case of free thymidine-^{14}C, the specific radioactivity of DNA rapidly increased for an hour, remained constant during the next 40 hours, and then began to decrease (Fig. 1B). The shortness of the period during which the specific radioactivity of DNA increased is due to the rapid elimination of the exogenous thymidine (Bresciani, 1965). The observed decrease in the specific radioactivity of DNA after about 40 hours reflects the loss of cells labeled during the first hour. The length of this interval is in agreement with data indicating a 40-hour mean lifetime for the intestine epithelium cells (Hagemann *et al.*, 1970).

Quite a different curve was obtained after a single intraperitoneal injection of the same amount of adsorbed thymidine-^{14}C (Fig. 1A). The specific radioactivity of DNA in this case constantly increased showing that the labeled thymidine was available for incorporation into DNA during the whole period studied. It can be concluded that thymidine, when applied as a "depot," is gradually desorbed and maintained at a constant concentration in the mammalian organism. Judging from the quantity of thymidine incorporated into DNA, this concentration is about 1% of the total amount administered.

It should be noted that in these experiments the specific radioactivity of DNA from "depot"-treated animals is always lower than the specific radioactivity of DNA isolated after administration of the same amount of radioactive thymidine without charcoal (Fig. 1).

B. Heavy Substitution

In the second set of experiments the control animals received intraperitoneally 0.5–1.0 ml of an aqueous solution containing 10 mg of "cold" thymidine (Fluka) and 5 μCi of thymidine-^{14}C (0.1 mCi/mg, ČSSR). The experimental animals received the same amount of "cold" and labeled thymidine adsorbed on 50 mg of activated charcoal suspended in 0.5–1.0 ml

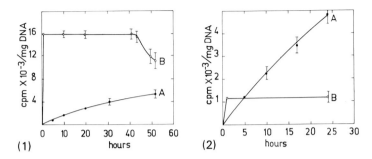

Fig. 1.   Specific radioactivity of mouse ileum DNA isolated at different intervals after a single intraperitoneal injection of 5 μCi/animal of thymidine-^{14}C adsorbed on 6 mg of activated charcoal (A), or of the same amount of free thymidine-^{14}C (B). Each point represents the mean of 3–5 different experiments. From Russev and Tsanev (1973).

Fig. 2.   Specific radioactivity of mouse ileum DNA isolated at different intervals after a single intraperitoneal injection of 5 μCi of thymidine-^{14}C (0.1 mCi/mg, ČSSR) together with 10 mg of "cold" thymidine (B), or of the same mixture adsorbed on 50 mg of activated charcoal (A). Each point represents the mean of 3–5 different experiments.

of distilled water (Section II). The animals were killed at different intervals, and the specific radioactivity of ileum DNA was determined (Fig. 2). The character of the curves obtained was the same as in Fig. 1, except that the specific radioactivity of DNA 5–6 hours after injection of adsorbed thymidine-^{14}C reached the specific radioactivity of DNA isolated after administration of "free" thymidine-^{14}C and continued to increase. At the 24th hour, it showed 4- to 6-fold higher values (Fig. 2).

This result is in agreement with the finding that the utilization of exogenous thymidine by mouse ileum DNA is proportional to the amount of administered thymidine up to about 0.05–0.1 mg per gram body weight, and any further increase in the amount of the precursor administered by a single injection does not cause a higher degree of incorporation (Lang *et al.*, 1968). From Fig. 2 is clear that administration of a rather large amount of the precursor as a "depot" makes it possible to achieve, after a certain period of time, a higher degree of substitution in DNA than after a single injection of the same amount of "free" precursor.

IV.   Applications of the Technique

Recently we have applied the technique of adsorbed precursors for several purposes, using both the procedures of continuous labeling and of heavy substitution.

A.   Continuous Labeling

As a demonstration of the technique of continuous labeling, the proliferative pool of mouse ileum crypts was determined, and in recent experiments the technique was applied in prescreening tests for anticancer drugs.

1.   PROLIFERATIVE POOL OF MOUSE ILEUM CRYPTS

One hundred microcuries per animal of thymidine-^{3}H (49 mCi/mg, DDR), adsorbed on activated charcoal were injected intraperitoneally to albino mice strain Agnes-Blum. At different intervals the animals were killed in groups of 2–3, paraffin sections of the ileum, fixed in Lillie fixative were covered with Ilford 2K emulsion and exposed at 4°C for 40 days (Russev and Tsanev, 1973). In Fig. 3 the percentage of labeled nuclei for three different areas of the villi as a function of the time of labeling is shown. It is clearly seen that in the first hours after injection only the cells in the crypts are labeled, then labeled cells appear in the middle area of the villi and finally, after a period of about 30 hours, labeled cells appear at the top of the villi. The new cells produced during this period were also labeled, which confirmed that thymidine-^{3}H adsorbed on activated charcoal was permanently

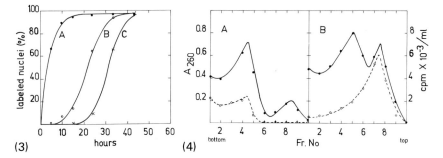

FIG. 3. Percentage of labeled nuclei in mouse epithelium at different intervals after a single intraperitoneal injection of 100 μCi thymidine-^{3}H per animal, adsorbed on 6 mg of activated charcoal. The labeling was followed by radioautography. For each point thousand cells from 2–3 different animals were counted. A, Crypts; B, middle area of the villi; C, top of the villi. From Russev and Tsanev (1973).

FIG. 4. Separation of the old from the newly synthesized DNA strands together with the corresponding histones. ——, A_{260}; ----, ^{3}H-radioactivity. Newly synthesized DNA in regenerating rat liver was labeled with thymidine-^{3}H and heavily substituted with 5-bromodeoxyuridine (BUdR) by the technique of adsorbed precursors; rat liver chromatin was isolated, irradiated at 312 nm in order to introduce breaks at BUdR-containing points of the newly synthesized DNA strand, denatured at pH 12.2, and centrifuged in 5 to 30% alkaline sucrose density gradients. (A) Sedimentation profile prior to irradiation; (B) after irradiation. From Tsanev and Russev (1974).

available for incorporation into DNA. At the 40–44th hour, practically all epithelial cells were labeled, and the proliferative pool in the crypts was estimated to be 98%. This finding is in agreement with the value of 96% found by Hagemann et al. (1970) for the crypts of mouse ileum and of 93% found by Frankfurt (1967) for the basal layer of mouse gastric epithelium. In all these experiments the authors used the technique of repeated (up to 14) injections.

2. PRESCREENING FOR CYTOSTATIC EFFECT

In the search for cytostatic drugs, the inhibition of DNA synthesis is used as a criterion for testing. However, here again it is difficult to overcome the inconveniences connected with the rapid elimination of the exogenous thymidine from the organism. The technique of continuous labeling can be used as a simple and reliable method for prescreening of potential cytostatic drugs (Russev et al., 1974).

Albino mice strain Agnes-Blum were injected intraperitoneally with 0.2–0.3 ml of undiluted ascites fluid of the Ehrlich-Lettré hyperdiploid ascites tumor strain. At day 7 after inoculation, the animals were treated with different cytostatic agents and in an hour with 5 μCi of thymidine-^{3}H (49 mCi/mg, DDR) adsorbed on 3–5 mg of activated charcoal. Ten hours later the animals were killed, ascites cells were collected, and the specific

TABLE I

INHIBITION OF DNA SYNTHESIS IN EHRLICH ASCITES TUMOR CELLS
TREATED WITH CYTOSTATIC AGENTS *IN VIVO*[a]

Cytostatic agent	Dose (mg/animal)	DNA specific radioactivity (%)	N[b]
None	—	100	10
Hydroxyurea	50	3	5
Actinomycin D	0.5	31	3
Mitomycin C	0.2	9	3
6-Mercaptopurine	6	67	3
5-Fluorouracil	20	64	3
Endoxan	40	45	6
Sarcolysin	4	24	3

[a]At day 7 after inoculation with 0.2–0.3 ml of undiluted ascites liquid, albino mice were treated with different cytostatic agents and in an hour with 100 μCi of thymidine-^{3}H as a "depot." Ten hours later, the animals were killed and the specific radioactivity of the ascites DNA was determined (Russev et al., 1974).

[b]Number of animals.

radioactivity of DNA was determined. The results summarized in Table I clearly show the usefulness of the method.

B.   Heavy Substitution with 5-Bromodeoxyuridine

Using the technique of heavy substitution we were able to study the effect of BUdR on the synthesis of rat parotid α-amylase *in vivo* (Russev and Tsanev, 1974), and in quite a different biochemical experiment—the distribution of newly synthesized histones between the two DNA strands during DNA replication (Tsanev and Russev, 1974).

1.   INCORPORATION OF BUdR INTO RAT PAROTID DNA

The parotid glands of albino rats 3–4 months old weighing 180–200 g were stimulated to proliferate with isoproterenol (IPR), and 24 hours after the treatment the animals received intraperitoneally 100 mg of BUdR adsorbed together with 50 μCi of BUdR-^{3}H (3 mCi/mg, Amersham) on 500 mg of activated charcoal. Four days later, the animals were killed, the glands were removed, and the specific radioactivity of BUdR-^{3}H-labeled DNA was determined. From the specific radioactivity of DNA and the specific radioactivity of BUdR-^{3}H measured under the same experimental conditions, it was calculated that by this treatment up to 14% of thymidine in the total parotid DNA was substituted with BUdR.

2.   INCORPORATION OF BUdR INTO RAT LIVER DNA

Albino rats, 3–4 months old, were partially hepatectomized; 16 hours after operation they were injected with 100 mg of BUdR adsorbed together with 100 μCi of thymidine-^{3}H (49 mCi/mg, DDR) on 500 mg of activated charcoal. After another 24–48 hours the animals were killed, liver DNA isolated, hydrolyzed and analyzed chromatographically. It was found that 18–22% of thymidine in the total DNA were substituted with BUdR (Russev and Tsanev, 1973). This high degree of substitution made possible the separation of the old DNA–protein complex from the newly synthesized complex due to the sensitivity of the latter to UV irradiation (Fig. 4).

V.   Concluding Remarks

The technique of adsorbed precursors can be successfully used for continuous labeling of mammalian DNA, and it has been shown that when applied as a "depot," a constant concentration of the precursor is maintained in the organism for 4–5 days. However, as mentioned above (Section

III,A), in the case of continuous labeling the specific radioactivity of DNA is always lower, and it never reaches that of the controls. Therefore, if high specific radioactivity of DNA rather than constant and continuous labeling for a certain, relatively long period of time is the aim, the use of adsorbed precursors is not recommended.

In several cases the described technique of adsorbed precursors has demonstrated its ability to meet problems hardly to be solved in any other way. It should be specially emphasized that for heavy substitution of DNA *in vivo* it is unique, and in no other way could such a high degree of incorporation of exogenous precursor be obtained. However, it possesses several disadvantages, the most disturbing of which is the nature of charcoal itself. In the case of continuous labeling, when $0.1–0.3$ mg/gm body weight of activated charcoal were administered intraperitoneally, the animals were practically not affected. However, in "heavy substitution" experiments, when a rather great amount of charcoal ($2–3$ mg/g body weight) was introduced intraperitoneally, they suffered acute sterile peritonitis, and some died $3–5$ days after the treatment. We have tried to inject the charcoal suspension subcutaneously, but unfortunately, in this case the incorporation of the precursor into DNA was $20–50$ times less effective. If a different adsorbent could be found, which is less irritating than charcoal and could be degraded and metabolized in mammalian organism, the application of the method would be considerably extended.

REFERENCES

Bresciani, F. (1965). *Exp. Cell Res.* **38**, 13–32.

Frankfurt, O. (1967). *Cytologia* **9**, 175–184.

Hagemann, R. F., Sigdestad, C. P., and Lesher, S. (1970). *Amer. J. Anat.* **129**, 41–52.

Lang, W., Müller, D., and Maurer, W. (1968). *Exp. Cell Res.* **49**, 558–571.

Markov, G. G., and Raitcheva, E. (1967). *C. R. Acad. Bulg. Sci.* **20**, 587–590.

Russev, G., and Tsanev, R. (1973). *Anal. Biochem.* **54**, 115–119.

Russev, G., and Tsanev, R. (1974). *Cell Differentiation* **3**, 127–133.

Russev, G., Emanuiloff, E., Angelof, I., and Golovinsky, E. (1974). *C. R. Acad. Bulg. Sci.* **27**, 26–29.

Tsanev, R., and Russev, G. (1974). *Eur. J. Biochem.* **43**, 257–263.

Chapter 9

Growth of Functional Glial Cells in a Serumless Medium

SAM T. DONTA

Department of Medicine, University of Iowa, and Veterans Administration Hospitals, Iowa City, Iowa

I. Background Leading to Current Studies

One of the major advances in the study of mammalian cell biology was the development of techniques to clone and propagate cells that express differentiated functions in tissue culture (Yasumura, 1968). The two most important factors responsible for successful adaptation of functional cells to tissue culture are (1) the use of transplantable animal tumor cells and (2) the technique of alternate passage of these cells from tissue culture to animal in order to select for those cells that would be capable of sustained *in vitro* growth and propagation (Sato and Buonassisi, 1964). This goal having been achieved with many different cell types, the laboratories of Dr. Gordon Sato, then at Brandeis University, turned its attention in part to the growth of some of these cell lines in a chemically defined, totally synthetic, and serum-free medium. Of the many different cell lines tested, the one that showed the greatest promise from the outset was a rat glial cell line, designated as C_6, which is capable of producing the brain-specific protein S-100. Most of the discussion in this chapter is centered around the adaptation of this clonal cell line to a serumless medium and a comparison of some of the characteristics of the adapted cell line with the parent strain.

II. Composition, Preparation, and Storage of Synthetic Medium

The medium that was chosen for use in these studies was that formulated by Waymouth as MAP 954/1 and used by Haggerty and Sato (1969) to prove the biotin dependence of mouse L cells. This medium was chosen primarily because it was generally richer in synthetic nutrients than were other available media and because of the possibility that certain nutrients, reducing agents, trace metal salts, and folinic acid present in this medium, but not in others, would be significant factors in the ability of cells to grow without serum.

Table I lists the ingredients of the medium, the amounts that were used in the preparation of stock solutions, and the final concentrations of each ingredient. Table II describes the procedure that was employed in preparing the medium. It was generally found advisable to add the calcium and magnesium salts last to avoid possible precipitation of some of the salts.

Most of the stock solutions could be stored frozen for long periods of time (more than 1 year), and some were best stored at 4°C to avoid precipitation of the ingredients. The medium itself, once prepared, could be stored at 4°C for at least 2–3 months, and probably for periods of time up to 6 months. Without added serum, if the medium were frozen, some precipitation of salts and the "insoluble" amino acids would occur.

III. Adaptation and Care of Cells in Serumless Medium

Originally, the C_6 cells were in a Ham's F-10 medium supplemented with 15% horse serum and 2.5% fetal calf serum. Transfer of the cells into the Charity Waymouth (CW) medium supplemented with 5% or 10% fetal calf serum was accomplished without any special adaptative procedures. Attempts to subculture the cells directly into serum-free medium resulted

TABLE I
Composition of Synthetic Medium[a]

Component	Stock solution (gm/liter)	Final concentration in medium (mg/liter)
I. Soluble amino acids		
Phenylalanine	1.0	50
Lysine-HCl	4.8	240
Methionine	1.6	80
Threonine	1.5	75
Valine	2.0	100
Isoleucine	1.5	75
Tryptophan	0.8	40
Arginine-HCl	1.5	75
Histidine-HCl	3.0	150
Glycine	1.0	50
Glutamic acid	3.0	150
Proline	1.0	50
Serine	1.28	64
Leucine	2.0	100
(Phenol Red)	0.20	10
II. Insoluble amino acids		
Tyrosine	1.0	40
Cystine	0.375	15
III. Vitamins		
Thiamine-HCl	0.02	10
Calcium pantothenate	0.02	1.0
Riboflavin	0.02	1.0
Pyridoxine-HCl	0.02	1.0
Folic Acid	0.01	0.5
i-Inositol	0.02	1.0
Nicotinamide	0.02	1.0
B_{12}	0.004	0.2
IV. Reducing agents and basal salts		
NaCl	60	6000
KCl	1.5	150
$Na_2HPO_4 \cdot 7H_2O$	3.0	300
KH_2PO_4	0.8	80
Ascorbic acid	0.175	17.5
$Choline\text{-}HCl \cdot H_2O$	2.4	240
Cysteine-HCl	0.9	90
Glutathione	0.15	15
Glucose	50	5000
V. Folinic acid	0.008	0.4
VI. Biotin	0.004	0.02
VII. Glutamine	7.0	350
VIII. Hypoxanthine	1.25	25
IX. Trace salts		
$CuSO_4 \cdot 5H_2O$	0.05	0.05
$ZnSO_4 \cdot 7H_2O$	0.03	0.03
$FeSO_4$	0.260	0.26

(*continued*)

Table I (*continued*)

Component	Stock solution (gm/liter)	Final concentration in medium (mg/liter)
$(NH_4)_6MO_7O_{24} \cdot 4H_2O$	0.025	0.025
$MnSO_4 \cdot 4H_2O$	0.016	0.016
$CoCl_2 \cdot 6H_2O$	0.022	0.022
X. Calcium and magnesium salts		
$CaCl_2 \cdot 2H_2O$	2.4	120
$MgCl_2 \cdot 6H_2O$	4.8	240
$MgSO_4 \cdot 7H_2O$	2.0	100
XI. $NaHCO_3$	—	2240
XII. Antibiotics		
Penicillin	100,000 units/ml	100,000 units
Streptomycin	100 mg/ml	100 mg/liter

[a]Adapted from Charity Waymouth (CW) MAP 954/1.

TABLE II

PREPARATION OF CW MEDIUM[a]

Stock solution (concentration)	Storage	Amount/liter of medium
I[b] (20x)	Frozen	50
II[c] (25x)	Refrigerated	40
III (20x)	Frozen	50
IV (10x)	Refrigerated	100
V (20x)	Refrigerated	50
VI (200x)	Frozen	5
VII[d] (20x)	Frozen	50
VIII[e] (50x)	Frozen	20
IX (1000x)	Refrigerated	1
X (20x)	Refrigerated	50

[a]Procedure:
1. Add stocks I, II, III, IV, and VII and adjust pH to 7.0 with 10 N NaOH.
2. Add remaining stocks, leaving X for last after volume made up to ~900 ml and H_2O.
3. Add $NaHCO_3$ and antibiotics, then H_2O up to 1.0 liter.
4. Filter through 0.22 μm filters.

[b]Adjust to ~pH 7.4 with 10 N NaOH.
[c]Dissolve by boiling in 250 ml H_2O + 1 ml of concentrated HCl, then bring up to volume with H_2O.
[d]Preferably should be prepared fresh; should not be repeatedly frozen and thawed.
[e]Dissolve in 125 ml H_2O + 5–10 drops of concentrated NH_4OH, then bring up to volume.

TABLE III

ADAPTATION OF CELLS TO SERUMLESS MEDIUM

Medium	Serum concentration (%)	Time in culture[a] (weeks)	Number of subcultures[a]
F10	17.5	1	1
CW	5	2	2–3
CW	2	2–3	3–4
CW	1	2–3	3–4
CW	0.5	3	3–4
CW	0.25	3	3–4
CW	0		

[a]Minimum time or number of subcultures in medium before going to next lower serum concentration.

in initial attachment of the cells to the plates, but no sustained growth. It was found necessary to slowly decrease the amount of fetal calf serum supplementing the CW medium in order to eventually achieve successful growth in serum-free medium. Table III schematically depicts the minimum periods of time necessary before decreases in serum concentrations could be tolerated by the cells at the various serum concentrations. Other factors that were found to be important were (1) the use of large numbers of cells (not splitting the cells more than 1:3 on subculture); (2) not using trypsin for dislodging cells at time of subculture after the cells had been adapted to a 2% fetal calf serum concentration (the cells could be easily dislodged with the use of a Pasteur pipette); and (3) the use of conditioned medium, i.e. the medium in which the cells had been growing for 4–7 days prior to the subculture. Practically speaking, a 20–40% concentration of conditioned medium was used for the freshly subcultured cells. It was also found that keeping the cells in the tissue culture vessel for longer than 10–14 days was not helpful, and may have been detrimental to the cells.

During the process of adaptation, the cells demonstrated some typical morphological signs of stress; i.e., development of long axonlike processes which tended to form cytoplasmic bridges with other cells, and increased numbers of intracellular refractile bodies. These signs were especially prominent when too few cells were used for plating of new cultures and can still be seen if too few cells are used for seeding, even with fully adapted cells.

Once the cells are fully adapted, their care and propagation are fairly simple. Stock cultures are maintained for only 7–10 days, and subcultures are made (without trypsin) at a ratio of stock to new plates of no greater than 1:5 or 1:6. The medium can be changed completely at 3–4-day intervals, although care should be exercised on addition of new medium onto the monolayer cultures, as this addition may result in some cells being dislodged. The medium is best added by directing the stream onto the sides of the plates, or onto the opposite sides of flasks.

Following subculture, the cells attach to new plates within the first few minutes, then begin to grow after a lag phase of 1–4 days, depending on the number of cells used for plating, and the use of conditioned medium. The cells remain firmly attached to the vessel surface until they reach near confluence in their stationary phase of growth, usually at a density of 2–3 × 10⁶ cells/60 mm plate. After reaching this state, cells or clumps of cells can occasionally spontaneously detach from the monolayer surface and can often be found at the air–liquid interphase. These cells are viable as proved by their ability to form monolayer cultures when placed into new culture vessels. Several attempts to grow fully adapted cells (C_6s) in suspension culture were unsuccessful.

IV. Functional Characteristics of Adapted Cells

A. Growth Curve and Plating Efficiency

While in medium supplemented with 5% or more of serum, C_6 cells proliferate with a doubling time of approximately 19 hours. Without any serum, even after a period of up to 2 years after adaptation to serum-free medium, the generation time of C_6s cells is not shorter than 40 hours, and is usually closer to 48–60 hours. Figure 1 demonstrates a typical growth curve of C_6s cells (Donta, 1973).

Ordinarily, no fewer than 10⁵ cells are required for proliferation to occur on 60 mm plates. Use of fewer numbers results in attachment of the cells and occasional formation of clusters of cells, but no sustained growth. Smaller (35 mm) plates and fewer numbers have not been tried.

B. Effect of Conditioned Medium

C_6s cells must produce substances during the course of their growth that are beneficial to their own growth and plating efficiency. With 10% conditioned medium, C_6s cells can grow at faster rates and from lower numbers than cells growing without the conditioned medium (Fig. 2). Similarly, this amount of conditioned medium added to fresh medium increases the plating efficiency by 50% (Table IV).

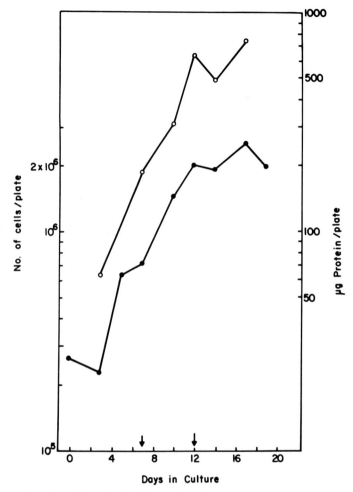

FIG. 1. Growth of C_6s cells in CW medium MAP 954/1. Cells were grown on 60 × 15 mm plates, and the medium was changed on the days indicated by arrows. ●——●, Number of cells/plate (average of duplicate counts of duplicate plates); ○——○, μg protein/plate (average of duplicate determinations on duplicate plates). From Donta (1973).

C.   Storage of Cells

C_6s cells can be stored under liquid N_2 in CW medium with added 20% glycerol after first being kept at $-70°C$ for 3 hours. Approximately 10^7 cells/ampoule in 1.0 ml are necessary to assure successful cell growth following storage in liquid N_2. After quick-thawing to $37°C$, the glycerol must be removed from the medium by centrifugation in order for the cells

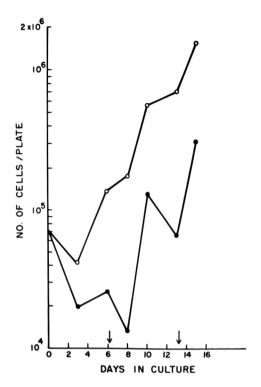

Fig. 2. Growth of C_6s cells with (O——O) or without (●——●) 10% added conditioned medium. Cells were grown on 60 × 15 mm plates, and the medium was changed on the days indicated by arrows.

TABLE IV

EFFECT OF CONDITIONED MEDIUM ON PLATING EFFICIENCY OF C_6s CELLS

No. of Cells/ 60 mm plate	Days post-plating[a]					
	2		6		14	
	+ CM[b]	− CM[b]	+ CM	− CM	+ CM	− CM
1.21×10^5	+	+	+ + +	+ +	+ + + +	+ + +
1.02×10^5	+	−	+ +	+	+ + +	+ +
0.40×10^5	+	−	+	−	+ +	−
0.15×10^5	−	−	−	−	−	−
0.10×10^5	−	−	−	−	−	−

[a]Results expressed on scale from 1+ to 4+ and represent degree of confluence on plates; − represents no areas of cellular proliferation.
[b]+ CM = with 10% conditioned medium;
− CM = without conditioned medium.

to grow successfully. Different concentrations of glycerol or the use of dimethyl sulfoxide as a protective agent during the freezing process have not been tried.

D. Production of S-100, Prostaglandins, and Fatty Acids

Similar to the parent C_6 cell line, C_6s cells produce the brain-specific protein, S-100, and in similar amounts; i.e., 0.8–1.8 μg of S-100/mg of cell protein. Quantitation of the amounts of intracellular S-100 was done by using quantitative microcomplement fixation methods (Donta, 1973). Also similar to findings with the parent clone (Pfeiffer et al., 1970), the production of S-100 by C_6s cells did not become readily apparent until near the end of their logarithmic phase of growth (Fig. 3). The amounts of S-100 produced by C_6s cells did not seem to vary over a 2-year period of time in which the cells were grown without serum.

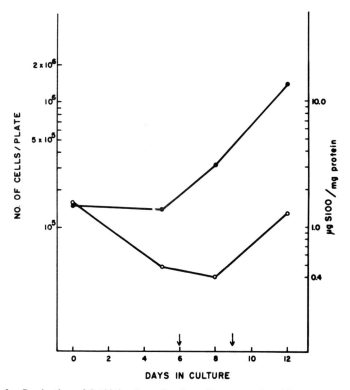

FIG. 3. Production of S-100 by C_6s cells. C_6s cells were analyzed for content of S-100 (O——O) at various time points during their growth (●——●). Arrows indicate the days on which medium was changed.

Prostaglandins are elaborated by many tissue culture cells, and the amounts of prostaglandins E_2 and $F_2\alpha$ produced by C_6s cells were found to be equal to or slightly greater than those produced by the parent C_6 line. When a comparison of the types of fatty acids produced by C_6 and C_6s cells is made, using gas chromatographic analyses of the methylated fatty acids extractable from the two cell lines, there is an increased production of the monounsaturated fatty acids, oleic and palmitoleic, by C_6s cells and a concomitant decrease in the respective saturated fatty acids (Fig. 4) (Donta, 1973). The significance of these changes is not yet known, but similar changes have been noted when other cell types have been grown in serum-free medium.

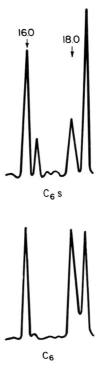

Fig. 4. Gas chromatographic elution patterns of the methylated fatty acid esters of C_6s and C_6 cells. Portions of the patterns are shown along with arrows indicating the elution points of a standard solution of mixed saturated fatty acids. From Donta (1973).

E.   Effect of Various Nutrients

Although a systematic study of all of the components of the CW 954/1 medium was not made, attention was focused on some of the components that were present in the CW medium but not usually present in many of the other commercial media.

1.   TRACE METAL SALTS

With the concentrations listed, none of the trace salts save for $FeSO_4$ could be shown to be important for cell growth and propagation. With Fe as the only trace metal, propagation of the cell line was sustained for more than 15 generations. Longer time periods were not tested. In the absence of Fe, no cell growth was observed. As no measurements of the exact amounts of the other trace metals present in the presumably metal-free media were not made, it cannot be stated that those metals were indeed absent and therefore not involved with cellular proliferation and function. Experiments in which varying amounts of Fe were used to assess rates of cell growth revealed that the amounts of this metal present in the original CW medium were probably suboptimal (Fig. 5), and that probably larger amounts (at least 2-fold more) should be incorporated into the synthetic medium.

The role of Fe in functions other than glial cell growth has not yet been tested. It is of interest that one of the major proteins produced, i.e., S-100, is an iron-containing protein.

2.   GLUCOSE

Glucose is the major source of energy present in most tissue culture media. Again, as with the Fe experiments, the addition of more glucose than that present in the original CW medium is beneficial for cellular growth rates (Fig. 5).

3.   GLUTAMINE

Many authors stress the importance of glutamine for cellular proliferation, and large amounts, usually in excess of that actually required, are added to media because of the instability of the amino acid. The results of experiments with C_6s cells show that the amount of glutamine required for optimal growth can be less than that prescribed in the original formula (Fig. 5).

4.   CHOLINE

Choline is a precursor for membrane lipids, and, especially for cells growing in a serum-free environment, may represent an important factor in the ability of cells to adapt to this type of environment. The amounts of

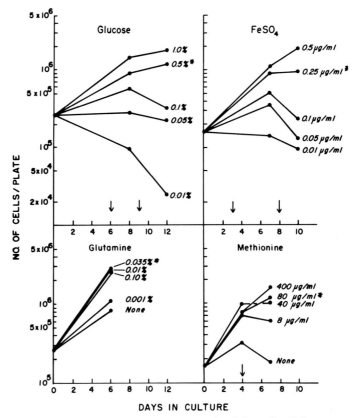

FIG. 5. Influence of various nutrients on growth of C_6s cells. Cells were grown on 60×15 mm plates, and the medium was changed on days indicated by arrows. * indicates the concentration present in the original formulation of CW MAP 954/1 medium.

choline present in the CW medium are much greater $(50–250\times)$ than those amounts found in other media. Although experiments have not been done to determine the optimal concentrations of choline necessary for maximal rates of cell growth, the absence of choline from the medium was incompatible with continued cellular proliferation; i.e., cells began to grow fairly normally in choline-free medium in the first 4–5 days, but soon developed many large refractile bodies and became detached from the plates.

5. METHIONINE

The only amino acid systematically evaluated was methionine, primarily because of a concurrent interest in the laboratory in the development of B_{12}-deficient cell lines. Experiments using different concentrations of methionine suggested that perhaps larger amounts than that present in the original medium might be slightly better, although the differences were not significant (Fig. 5). In the absence of methionine, cell proliferation ceased. Addition of homocysteine to medium lacking methionine seemed to be of some help in sustaining cell growth.

6. VITAMIN B_{12}

This vitamin can be eliminated from the medium without any effects on cell growth rates and functions in so far as elaboration of S-100 and prostaglandins are concerned. These cells, however, can be shown to be vitamin B_{12} deficient in that their ability to convert propionate into CO_2 is impaired (Table V). The pathway from propionate goes through methylmalonate into succinate and into the tricarboxylic acid cycle, the transformation from methylmalonate into succinate being a B_{12}-dependent step. Other differences between the cells grown in the presence or absence of the vitamin is in their fatty acid compositions (S. T. Donta, unpublished observations).

7. OTHER NUTRIENTS

No effects on C_6s growth rates have been noted with CW media free of folinic acid, cysteine, ascorbic acid, hypoxanthine, glutathione, and biotin for as long as 10 generations. Longer times were not tested. It could be, however, that trace amounts of biotin were present that were important to cellular growth rates. The effects of these agents on the various glial cell functions were also not evaluated.

TABLE V

EFFECT OF B_{12} DEFICIENCY ON METABOLISM OF
PROPIONATE

Cells[a]	Conversion of propionate to CO_2[b]
C_6s − B_{12}	53 ± 17
C_6s + B_{12}	231 ± 65

[a]Cells were grown with or without B_{12} for 2 months before use in assay.

[b]Results are expressed as cpm/10^6 cells per 120-minute incubation period; 0.03 μCi of propionate-1-^{14}C was incubated with cells in CW − B_{12} medium (total volume 2.1 ml), then the $^{14}CO_2$ released from the medium with 4 N sodium acetate, pH 5.5, and the $^{14}CO_2$ trapped in a well containing 0.1 ml of 1 N NaOH.

V.   Attempts to Grow Other Cells with Differentiated Characteristics in Serumless Medium

In addition to C_6 rat glial cells, successful adaptation to sustained growth in the CW medium was achieved with two other cell types; one is a human glial cell line, CHB_4, derived from a human astrocytoma (Lightbody et al., 1969) and similar in many ways (production of S-100, prostaglandins) to C_6 cells; the other cell type is a mouse interstitial cell line, I_{10}, derived from a Leydig cell tumor (Yasumura, 1968). Although the CHB_4 line was shown to retain its functional characteristics after adaptation to serum-free medium, no functional assessments were done with the I_{10} cells.

Other cell types that were used in the studies included melanoma cells (Yasumura et al., 1966b), neuroblastoma cells (NB_{42}) (Augusti-Tocco and Sato, 1969), adrenal cells (Y1) (Yasumura et al., 1966a), and a functional hepatoma (MHC) (Tashjian et al., 1970) cell line. None of these cell types could be grown in less than 2% serum, and the MHC cells in not less than 5% fetal calf serum. The reasons for these failures are not readily apparent, but it may be that the requirements for the in vitro growth of glial cells are simpler than those of hepatoma and adrenal cells, perhaps because they are "more fibroblastic" and therefore have less-complex functions and requirements than do the other cell types.

VI.   Significance of Studies and Future Directions

The development of clonal cell lines that express differentiated functions in tissue culture has simplified the study of the complex biochemical events surrounding these functions. These studies ideally, at times, would be best conducted in a totally synthetic medium, devoid of serum, in which the contents and concentrations of all the components are known and can be identified. Although successful adaptation of the other cell lines tested was not possible, the success achieved with the glial cell lines may result in the development of information that may be helpful in understanding the specific requirements for the in vitro cultivation of even more complex cell types. The successful adaptation of glial cells to a serum-free medium may also simplify studies designed to assess the role of these cells in various neurobiological functions. The studies with the glial cells have demonstrated that for ideal growth rates, and probably cellular function, the culture medium has to be tailored to the specific needs of each individual cell line. These studies have also demonstrated that the use of parameters such as cell growth may be too crude for assessing the effects of various nutrients or conditions, and that measurement of specific cellular biological functions would be more sensitive assay systems.

Appropriate future directions would include a study of cellular lipid metabolism, especially as related to membrane lipids, and their role in the ability of cells to survive and propagate in a serum-free environment. Isolation, characterization, and an understanding of the controlling factors surrounding the growth-promoting substances produced by these cells and elaborated into the medium would also seem to be of value in the study of normal, and perhaps of abnormal, cellular biology.

ACKNOWLEDGMENTS

These studies were carried out in the laboratories of Drs. Gordon Sato and Lawrence Levine at Brandeis University, and I am grateful to both of them for their encouragement and advice. Support for these studies was made possible by training grant CA-05174 from the National Cancer Institute.

REFERENCES

Augusti-Tocco, G., and Sato, G. (1969). *Proc. Nat. Acad. Sci. U.S.* **64**, 311.
Donta, S. T. (1973). *Exp. Cell Res.* **82**, 119.
Haggerty, D. F., and Sato, G. H. (1969). *Biochem. Biophys. Res. Commun.* **34**, 812.
Lightbody, J., Pfeiffer, S. E., Kornblith, P. L., and Herschman, H. R. (1969). *J. Neurobiol.* **1**, 411.
Pfeiffer, S. E., Herschman, H. R., Lightbody, J., and Sato, G. (1970). *J. Cell. Physiol.* **75**, 329.
Sato, G. H., and Buonassisi, V. (1964). *Nat. Cancer Inst., Monogr.* **13**, 81.
Tashian, A. H. Jr., Bancroft, F. C., Richardson, U. I., Goldlust, M. B., Rommel, F. A., and Ofner, P. (1970). *In Vitro* **6**, 32.
Yasumura, Y. (1968). *Amer. Zool.* **8**, 285.
Yasumura, Y., Buonassisi, V., and Sato, G. (1966a). *Cancer Res.* **26**, 529.
Yasumura, Y., Tashjian, A. H. Jr., and Sato, G. H. (1966b). *Science* **154**, 1186.

Chapter 10

Miniature Tissue Culture Technique with a Modified (Bottomless) Plastic Microplate

ELISEO MANUEL HERNÁNDEZ-BAUMGARTEN

Research Project on Bovine Paralytic Rabies (Derriengue), Instituto Nacional de Investigaciones Pecuarias (National Institute of Livestock Research), México-Toluca, Mexico D.F.

I. Introduction

The search for miniature assay procedures has lead many researchers to develop miniature microplates for various uses, ranging from microhemagglutination tests to bacteriology. The 1973 Microtiter Bibliography (Conrath, 1973) quotes over 1200 references on the various uses of these techniques. For tissues culture work, Rosembaum *et al.* (1963) used the round-bottom microplate, but this plate is cumbersome to read in the microscope, since one has to focus constantly on the receding walls of the wells. A welcome improvement was introduced by Fuccillo *et al.* (1969) when these authors developed a flat-bottom microplate. These microplates are available from a number of commercial sources, e.g., Microtiter, Linbro, Microtest II.

The flat-bottom microplate has been very useful for a number of tissue culture and virological applications. When critical optical (i.e., fluorescent microscopy or phase contrast microscopy) or other close-range examination (i.e., autoradiography, radioisotope work) are needed, however, the flat-bottom microplate is of little value.

A chamber slide has been marketed by Lab-Tech in which up to eight square chambers are mounted on a regular glass slide. Although this chamber slide obviates some of the problems mentioned in the case of the flat-bottom microplate, it cannot be used with the automatic equipment available for the latter.

In order to take advantage of the convenience of the flat-bottom microplate and the automated systems now available, which enable both convenience and reproducibility, the microplates were modified to accommodate four standard microscope slides, which could be removed from the microplate in due time for examination. Although this work has been briefly reported elsewhere (Hernández-Baumgarten, 1972), it is presented and expanded here for convenience.

II. Preparation of Microplates

Although the plate can be made entirely from thick plastic by an expert machinist, the procedure may be difficult and expensive. The commercially available rigid styrene flat-bottom microplates (Microtiter, Linbro, etc.) containing 96 wells of 6 mm diameter each, can be conveniently modified to accommodate four standard microscope slides. For convenience, two methods of modifying the microplates are described: (A) by the use of a milling machine, and (B) by hand, using sand paper.

A. The Milling-Machine Procedure

The microplates are fixed in inverted position on the table of the milling machine with double adhesive tape (tape covered by adhesive on both sides). The milling is performed at low speed with a fine toothed flat steel milling tool. The bottoms must be removed gradually, advancing a few mils (thousands of an inch) each time. It will be noted that the bottom of the wells do not lie in exactly the same plane. When the milling has opened most of the wells and a few remain closed by a thin film of plastic, the milling process should stop. The milling should be done on the bottom of the wells leaving the rim intact to serve as a support for the plates to stand or stack on top of each other. After milling is completed, the few wells not yet opened should be opened by gentle tapping with a tapered steel reamer and each well should then be reamed to remove plastic edges that partially occlude the openings of the now bottomless wells. In order to obtain a smooth finish, the milled area can be lapped with a fine-grain waterproof sand paper, taking care to wet both the paper and the plate. The sand paper should be wrapped around a small wooden block for best results.

B. The Manual Procedure

When a milling machine and machinist to go with it are not available, the microplates can be prepared manually. The manual procedure is slightly less satisfactory.

On a flat surface, such as a thick glass plate or a marble-top table, a coarse-grade waterproof sand paper should be fixed. The sand paper should be kept wet by constantly dribbling water over it. The water acts as a coolant to the grinding and prevents the plastic in contact with the grit from melting, thus ruining the plate.

Start the grinding by applying a light pressure on the plate with a smooth rotating motion. The rim of the plate will be removed first. Then as the grinding proceeds the bottom of the wells will come in contact with the sand paper. The grip on the microplate should be changed often to prevent warping or groove formation. After the bottoms of the wells are removed, the sand paper should be changed in decreasing coarseness to the very fine grade (increasing grit numbers indicate finer paper; a 180, 220, 400, and 700 grit sequence gives good results). When the finest grade paper is used, the precaution should be taken to wash and scrub the plate to avoid carrying coarser-grit carbide grains on it, for this would ruin the finish of the plate. Although the procedure seems long and time consuming, a good plate can be prepared in this manner in about half an hour. Sanding with paper a given grit can be ended when the ground area of the plate is not further smoothed by the paper. An alternate procedure for sanding out the bottoms of the wells can be carried out with a belt sander, such as those available for lapidary work. The wide choice of woodworking paraphernalia has given very poor results for this purpose because water cannot be used as a coolant; this results in blades sticking to the plastic, sanding disks smudging plastic on the plate, and similar disasters.

The disadvantage of the manual procedure is that in removing the bottoms of the wells, the rim of the plates is also removed. Thus microplates cannot be stacked on top of one another, and if the plates with slides in place are put on a wet surface, the slides will be dislodged when the microplate is picked up. This disadvantage can be overcome by mounting the microplate thus prepared on top of a table as illustrated in Fig. 1. Alternatively, four glass-head pins, such as the ones used in maps, can be driven into the corners of the microplate by holding each with tongs in the flame until the pin is red hot and then quickly pushing it in all the way, leaving the glass head to function as a little leg on a corner. Stacking the plates thus prepared may still be difficult.

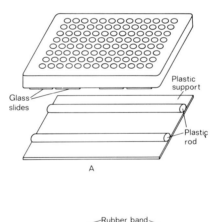

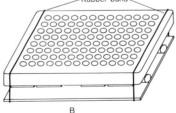

FIG. 1. Carrying support to mount the microplate as prepared by the manual method. (A) Exploded view of assembly. (B) Assembled plate and support.

C.   Washing the Plates

Regardless of the procedure used to prepare the plates, when they are ready, they should be throughly washed with a suitable detergent, such as Microsolve or a similar detergent appropriate for tissue culture work. The scrubbing should be done first with a flat brush in the bottom area to remove oils, grit, and dirt and then each well should be cleaned with a fine test tube brush. The plates should be thoroughly rinsed first in running tap water, then in several changes of distilled water, and finally soaked overnight in 95% ethanol. Next morning the plates are removed from the ethanol bath and left to dry at room temperature. Prior to assembly, each plate should be carefully examined. If the rim of any well is nicked or cracked, the plate should be discarded, for such wells will leak when the plate is in use. If only one well is damaged, the plate may be used provided that this well is marked on the top of the plate so that one can avoid filling it with liquid.

The microplates are now ready for assembly.

D.   Choice of Glass Slides

The plate can accommodate four standard (25 × 75 mm) microscope glass slides by following one of the two patterns illustrated in Fig. 2. Pattern 2A uses all 96 wells, but it requires extreme care in positioning the slides on the plate, for the standard microscope slide is a little too narrow to cover the three rows of wells to the edge of the rim. If the slide is a little off position, an entire row of wells may leak. Pattern 2B, on the other hand, uses the microplate less efficiently, for only 56 of the wells can be used, but the positioning of the slides is less critical and there is little danger of leakage.

Several manufacturers of glass slides and coverslips offer special custom-made slides, cut to your specifications; Table 1 may serve as a guide to special sizes that may be useful for a given purpose. The 8.3 mm round cover-slip is included for special purposes even though the assembly of a plate with 96 coverslips would be a little cumbersome (see Section V, F).

III.   Assembly of the Microplate for Use

After the plates have been prepared either manually or with a milling machine as outlined in Section I, A and I, B and washed as explained in I, C, they are ready to be assembled with clean slides.

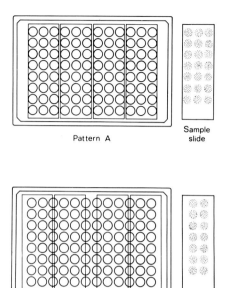

FIG. 2. Two possible patterns to accommodate four standard microscope slides. (A) Each side is used to cover three rows of wells. (B) Each slide is used to cover two rows of wells. Fig. 2A is redrawn from Hernández-Baumgarten (1972) with permission.

TABLE I

SPECIAL-SIZE GLASS SLIDES AND COVERSLIPS FOR THE MODIFIED PLASTIC MICROPLATE[a]

Width (mm)	Length (mm)	Number of wells covered	Number of slides per plate	Position on plate
26.5	78.0[b]	Three rows of 8 wells each	4	Short axis
26.5	72.0[c]	Three rows of 8 wells each	4	Short axis
35.3	78.0	Four rows of 8 wells each	3	Short axis
35.3	72.0	Four rows of 8 wells each	3	Short axis
35.2	120	Four rows of 12 wells each	2	Long axis
35.2	108	Four rows of 12 wells each	2	Long axis
8.3	Round	Single well	96	—

[a] The sizes indicated can be used either for microscope slides or coverslips. All of them make use of all 96 wells in the plate.

[b] Maximum length to be used.

[c] Minimum length to cover all wells.

A.   Cleaning the Slides

The slides can be cleaned by soaking overnight in 96% ethanol and then wiping with a clean lint-free cloth just prior to assembly of the microplate. Always use new slides to start with.

Certain types of cell lines are difficult to grow in slides cleaned with this simple procedure; this can be determined after a few trials, and a more elaborate washing may be needed.

Using a stainless steel slide basket, such as those used for staining procedures, place the slides in a hot tissue culture detergent in an ultrasonic cleaner. Clean the slides in the ultrasonic bath for half an hour; rinse thoroughly in hot running tap water, then in three succesive trays of twice-distilled or deionized water; and finally set to dry in an oven. Let the dry slides cool to room temperature, then proceed to assembly of the plate. If the slides are clean, water should not form beads on the surface of the slides in the last rinse.

B.   Assembly of the Plates

There are two ways of assembling a plate.

1.   SILICONE GREASE METHOD

Working on a clean lintless cloth, place the plate in an inverted position with the machined area facing you, cover the rims of the wells with a uniformly thick coat of silicone stopcock grease. Use a patting motion of your finger since rubbing will tend to fill the wells, thus reducing the area of the culture (Fig. 3). Set the slides carefully in place (Fig. 4) by following either pattern indicated in Fig. 1. Care should be taken to avoid moving or relocating the slides once in contact with the silicone grease, for this will cause smearing of the slides, thus reducing the area of the culture. The cells will not grow in the siliconized area. If smearing occurs, remove the smeared slide, discard it, apply more silicone grease and set a clean slide. Press the slides firmly against the plate with a rubber stopper (Fig. 5) to ensure a good seal of every well.

The assembled plate should now be examined with the reflection of a lamp. The outline of the silicon grease should be seen on the slide, thus indicating a good seal. Some practice is required before the plate can be successfully assembled. If breaks of the silicone seal are evident, disassemble the plate, and start over.

2.   SILICONE SEALANT METHOD

The microplate assembled with silicone grease has been termed "precarious." The danger of the slides coming off the plate while it is in use is

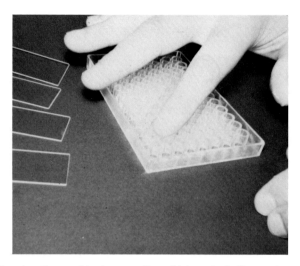

FIG. 3.   Covering the machined area with a thick coat of silicone stopcock grease. A patting motion should be used.

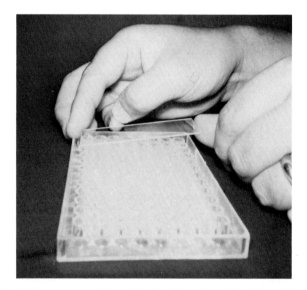

FIG. 4.   The slides are set carefully on the plate. Reproduced from Hernández-Baumgarten (1972) with permission.

FIG. 5. The slides are pressed firmly against the plate to ensure a good seal. A rubber stopper is useful in this step.

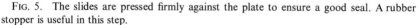

real, and when one is dealing with infectious material this can be an important hazard. The plate can also be assembled with silicone sealant (Dow Corning 781 building sealant) in order to have a more sturdy assembly. The silicone sealant is sold in cartridges, and a special gun is needed to push the bottom of the cartridge toward the tip, thus allowing extraction of the sealant. This particular silicone sealant releases acetic acid on contact with air. The procedure recommended for assembling the plate with silicone grease is not appropriate for silicone sealant because skin irritation may result.

Work in a well ventilated area or fume hood. Squeeze a small blob of the sealant on a glass plate (8 × 10 inches) and then extend it to a flat film with a roller, such as the one used for fixing the sealing tape to the microplates. Using the roller, now coated by a thin coat of silicone sealant, coat the rims of the wells with a thin, uniform coat of the sealant. Assemble the plate as discussed in Section II,B,1 and check for leaks.

It is important that the film of silicone sealant be thin with no blobs of the product entering either the wells or the crevices between them. The sealant will stick to the glass slide, and when it is later disassembled for staining and observation, the blobs of silicone rubber, produced when the sealant "cured," will protrude and make the use of coverslips very difficult.

To seal the small gaps left between the slides, squeeze a small amount of the sealant into a disposable syringe and then use the syringe without the needle, to squirt a thin "worm" of the sealant. The glass plate and the roller should be cleaned before the sealant hardens to rubber. The assembled plates should be left inverted until the sealant has had time to harden, about 1 hour. The entire process of assembly should be carried out in about 10 minutes, that is, before the sealant has had time to start hardening. If it takes longer to place the slides on the plate, an improper seal may result.

The sealant releases acetic acid during the process of hardening and for sometime afterward. It is convenient to dry the assembled plates in a forced-draft oven at 40°C overnight. Prior to sterilization, the plates should be soaked in a sodium bicarbonate solution for a few hours to neutralize as much of the acetic acid as possible. Then the plates should be rinsed twice in distilled water and dried before proceeding to sterilization.

C. Sterilization of the Plates

The plates may be sterilized either by ultraviolet light or in an ethylene oxide gas cabinet.

1. If UV light is used, a tissue culture hood equipped with a UV lamp is very useful. Place a table about 10 inches directly under the UV lamp and cover it with a white lintless cloth. Place the assembled plates right side up beneath the lamp and place a lid for each plate upside-down. Leave the plates under UV light with the door of the cabinet closed for 3 hours or overnight. When sterilization is complete, the plates should be covered with the lids and stored in a plastic bag. The plates remain sterile in the closed plastic bag for a few days. Repeated UV sterilization will turn the plastic yellow and brittle, and it will eventually crack.

2 a. An alternative sterilization procedure is offered by the ethylene oxide gas sterilization systems now commercially available. It has the advantage that the plates can be sterilized in sealed cellophane bags and stored for long periods of time without losing sterility. The sterilization cycle should vary according to the equipment available and the manufacturer's indications. Extreme care should be taken to aerate the plates properly after sterilization in order to eliminate all residual ethylene oxide. The plates are now ready for use.

b. A new development was recently brought to my attention. This is the Anprolene[1] brand gaseous sterilant for room temperature atmospheric pressure sterilization. I have tested this ethylene oxide sterilization system for microplates with excellent results. So far this seems to be the most con-

---

[1] Manufactured by H. W. Andersen Products, Inc., 45 East Main Street, Oyster Bay, New York 11771.

venient form of sterilizing the plates as repeated sterilization has no detrimental effect on the plates (as does UV light). The sterility obtained is perfect and does not require the large investment of traditional ethylene oxide sterilization systems.

IV.   Tissue Culture Procedures

A.   Growing Monolayers

The bottomless microplate has been used in our laboratory to grow a variety of primary mammalian cells as well as several cell lines. The procedure outlined here is used for BHK/21 and BHK/21-13S Cl 3. The first was derived by Stoker and MacPherson (1964), and the second is a clone of the BHK/21-13S derived in our laboratory (Bijlenga and Hernández-Baumgarten, 1974).

Each well is seeded with 0.1 ml of the cell suspension containing 4×10^5 cells/ml (4×10^4 cells per well) in growth medium. The growth medium consists of 80% BHK-21 medium (Gibco), 10% tryptose phosphate broth (TPB, Gibco), and 10% heat-inactivated fetal bovine serum (FBS, Flow). The growth medium is supplemented with 50 IU of penicillin, 50 μg of streptomicin, and 20 μg of mycostatin.

After the entire plate is seeded, it is sealed with sealing tape and incubated in a CO_2 incubator. The plates are examined daily in an inverted microscope for cell growth. The cells grow to a confluent monolayer in 2 or 3 days. For infection, the medium is withdrawn by aspiration with a fine-tipped Pasteur pipette, with the plate slanted, the pipette is placed on the lower corner of each well, taking care not to touch the monolayer. Each cell layer is rinsed with 0.1 ml of phosphate-buffered saline (PBS), and the PBS is withdrawn. The cells are ready for infection.

B.   The "Instant Monolayer"

Alternatively, the monolayer can be prepared with some cell lines by adding enough cells to form a complete monolayer upon settling of the cells.

When infecting a previously grown cell monolayer, there is always a fluid film left on the surface of the culture. When large vessels are used, i.e., Roux bottles and the like, the fluid layer still covering the cells will be small.

This amount of fluid will be composed of the water of hydration of the monolayer (the amount of tissue culture medium or PBS that clings to the cells upon drainage), and the small amount of fluid not aspirated by the pipette. As the container where the cells are grown becomes smaller, the proportion of fluid layering the monolayer will represent a higher proportion of the cell volume used in the test. When receptacles up to the size of a test tube (10 $\times$ 70 mm) are used in a specific test, this amount of fluid may still be considered negligible. In the microassay procedures, however, this amount of fluid will represent several times the cell volume contained in the well. This fluid, besides acting as a diluent of the virus-containing fluid, might prevent the random contact of virus to cell membrane, necessary for successful infection. This is especially true when no specific receptor sites are involved in the virus cell interaction. This "prevention of contact effect" would be caused by the layers of liquid tightly bound to the cells, which would tend to act as a shield on the cells (Rein and Rubin, 1968). Virological research on rabies virus was greatly hindered by lack of penetration of the cells by the virus. It was not until the use of polyions (Kaplan et al., 1967) was introduced as an enhancer of viral penetration that reproducible results could be obtained with rabies virus. Rabies virus causes an irregular cytocidal effect (CPE) on infected monolayers. An agarose suspension plaquing system was developed (Sedwick and Wiktor, 1967) to produce plaques with rabies virus, lymphocytic choriomeningitis, and other noncytocidal RNA viruses. The main advantage of this technique probably lays in the fact that the virus diffuses into the agarose, thus impinging upon the cells suspended in the agarose. This effect causes more successful virus–cell interactions to take place than would be the case if the cells were grown on a solid substrate and the virus sloshed on top of them

It is in these special cases that the "instant monolayer" concept will give best results. The virus-containing fluid is stationary while the cells settle through it, thus the cells have a higher probability of a virus–cell interaction.

The technique has the limitation that only a few established cell lines will tolerate this type of work. Most primary and some established cell lines will clump at high concentrations.

V.   Use of the Plates

The following are but a few examples of the uses of the bottomless plate. It makes no attempt to be a complete list of uses, but is only illustrative of the possibilities of the technique.

A. Decimal Dilutions for Virus Titrations

The manufacturers of microculture equipment do not recommend the use of their equipment for decimal dilutions.

In our laboratory the decimal dilutions of virus are first made in test tubes, and then the appropriate amounts of the virus dilution are pipetted with the calibrated micropipettes into the cell layer.

The cell layer can be washed prior to infection by aspirating the growth medium with a Pasteur pipette, then 0.1 ml of PBS is added and aspirated. The cell layer is then infected with the suitable virus dilution, incubated, and withdrawn as before. For convenience the plate is covered with a sterile lid during these procedures and only sealed after the cells are infected and replenished with maintenance medium.

B. Serum Neutralization Test

Double and 4-fold dilutions can be carried out very conveniently by the microtitration equipment directly on the plate.

To avoid scratching the monolayers, platinum dilutors can be used. A second possibility is the microtransfer plate as described by Catalano et al. (1969). This is a plate where the dilutions can be carried out separately, and when the incubation is complete, the virus–serum mixtures can be transferred to the infected monolayers. It is important that the slides lie on the same plane as the original bottoms of the well (hence the controlled depth milling); otherwise the microtransfer plate, when pushed all the way into the bottomless plate, may dislodge the slides. This is particularly true if the microplate has been assembled with silicone grease. A third possibility is to prepare the serum–virus mixtures on a clean empty plate and then form an "instant monolayer" into it. The dilution factor produced by the medium in which the cells are suspended should be taken into consideration for calculations of serum neutralization titers (SN titers).

C. The "Instant Monolayer" Procedure

The "instant monolayer" concept has been introduced above (Section IV,B) as well as our reasons for using it.

In our laboratory we use a fluorescent microfoci technique for rabies virus titrations as well as for serum neutralization tests. This technique has been presented (Hernández-Baumgarten and Bijlenga, 1970, 1973), and a combination of the two techniques (instant monolayer and fluorescent microfoci) is being used in our laboratory to conduct routine on-the-line checking of vaccine production and serum neutralization tests (Hernández-Baumgarten *et al.*, 1974b). A brief description follows as an illustration of the technique.

Decimal dilutions of virus are carried out in test tubes and serological pipettes as usual. The diluent used is growth medium supplemented with 50 μg of DEAE dextran per milliliter. With 50-μl calibrated pipettes, the diluted virus is placed on two repeat wells per dilution starting with the highest dilution and working down to the lowest dilution. The plates are assembled following pattern 2B (Fig. 2) so that only two rows of wells are accommodated on each slide. Only 7 of the 8 wells of each row are used to provide a small space on the glass slide for identification. The plates are always handled in large petri dishes to prevent accidental spillage of rabies virus.

After the wells are filled with the appropriate dilution of virus, the cells are added in a monodispersed suspension to the virus-containing wells. Since the virus and the cell suspension will dilute each other, the amount of cells used is 1.6×10^6 cells/ml. We use the BHK/21-13S Cl 3 cells for this test; 50 μl of the cell suspension is added to each well. The plates are sealed, the petri dish is closed with its lid, and the contents of the wells are mixed by tipping the dish. The plates are incubated in a CO_2 incubator with manual mixing every 10–15 minutes, and then the plates are left undisturbed to allow the cells to settle and attach to the glass slide. Twelve hours after infection, the cells have settled and attached to the glass slide. At this time the medium is changed to remove unadsorbed virus and the DEAE-dextran. Thirty hours after infection (this has been found to be the optimal time for growth of the microfoci, while secondary foci formation has not yet occurred), the plates are examined with an inverted microscope to determine the condition of the monolayer. The medium is removed and the plate is disassembled. The individual slides are washed in PBS, fixed in acetone at room temperature for 15 min and stained by the fluorescent antibody technique (FA) (Dean, 1966). The monolayers are examined under a fluorescent microscope, and the number of fluorescent microfoci per monolayer is determined. A fluorescent microfoci may be from a single, clearly infected cell to a group of cells closely associated. The titer is expressed as number of fluorescent microfoci-forming units per milliliter (FMFU/ml) of undiluted virus. All material coming in contact with infectious material is either autoclaved or disinfected with an appropriate disinfectant, and stringent laboratory safety procedures are followed throughout.

For the SN test, a constant amount of virus, capable of producing 50 microfoci per well under the conditions of the test, is used. Double dilutions of serum, 25 μl per well, are prepared on the plate, to which is added the appropriate amount of virus contained in 25 μl. The serum–virus mixture is incubated at 36°C for 1 hour and then 50 μl of the cell suspension is added. The infection of the cells is carried out as outlined, and the titer of the serum is expressed as the dilution of serum capable of neutralizing 50% of the microfoci found in the control monolayers.

D.   Centrifugation of Plates

Only plates assembled with silicone rubber can be centrifuged. The plate is centrifuged in the centrifuge carriers designed for this purpose. To prevent the glass slides from coming off the plate during centrifugation a rubber mat should be cut to cover the area of the slides. The mat should be over 3 mm thick to ensure that the entire plate is supported by it; i.e., the rim of the plate should not touch the carrier. This is only for low speed centrifugation. No high speed centrifugation has been attempted.

E.   Radioisotope Work

The plate is ideally suited for repeat autoradiography work. Once the infected plates are removed, the slide can be coated with photographic material, incubated, and developed as usual. The removal of the isotope tracers from the plate may be difficult, and in this case it is probably better to eliminate the plate after use.

F.   Scanning Electron Microscopy

The scanning electron microscope (SEM) is increasingly being used to examine cultured cells and infected tissue cultures (Hernández-Baumgarten et al., 1974a). The small size of the critical point driers makes the use of small samples necessary. The plate is assembled with 8.3-mm round coverslips, and used as necessary. For examination, the coverslips are removed, fixed in the coils of a thin steel spring, and critical-point dried as usual (Anderson, 1951). Each stub will accommodate one coverslip for coating and examination. Wherever coverslips are used on the plate it must be assembled with silicone grease. The silicone rubber will hold the coverslips too firmly and when they are removed from the plate, many will break. For SEM work the silicone grease will offer no special problem because most of it will be removed during dehydration and amyl acetate treatment.

VI.   Removal of Slides

A.   When Fixed with Silicone Grease

The media contained in the plates may be removed by aspiration or simply by turning the plates upside-down over a tray covered with a cloth soaked with disinfectant. The slides may be removed by lifting each one with a pair of tweezers from one end. Care should be taken to avoid smearing and peeling off the cultures against the plate.

B. When Fixed with Silicone Rubber

Remove the medium as before (Section VI,A). Using a scalpel cut the rubber used to seal the spaces between the slides; using the handle of the scalpel or other sturdy instrument, carefully lift the slide. If too much force is applied at one time, the slide may break. Lift alternately the slide from each end and from the side. The air allowed to enter between the rubber and the plate can be observed as one lifts the slide. Work slowly until the slide gently comes off the plate. The silicone rubber stays firmly attached to the slide. If the plate was assembled sloppily, the rubber will now protrude from the slide making the use of a coverslip very difficult or impossible.

The slides are now ready for staining.

VII. Staining of Slides

A. As Individual Monolayers

For fluorescent antibody work, it is necessary to use the conjugate mixed either with normal mouse brain or with challenge virus standard (CVS) infected brain (Dean, 1966). Each slide should have a duplicate monolayer for each dilution tested. One duplicate will become the test, and the other the control. The advantages of this technique being the ease of handling multiple samples on one slide, thus avoiding handling too many coverslips. Another advantage is that each monolayer has a known area (6 mm diameter) so that a similar area of infected cells is counted each time. Also, each miniculture will be rimmed by either silicone grease or silicone rubber, preventing the conjugate from spilling or mixing with that of neighboring cultures.

B. As a Slide

For certain purposes all the monolayers need to be stained in the same manner, i.e., hematoxylin–eosin to stain the cells and detect inclusion bodies, or to prepare demonstration slides for teaching. This procedure offers no special problem, and the slides can be stained in a basket as usual.

VIII. Recycling the Plates

A. Washing after Use

When the plates have been used with infectious material it is important that they be disinfected with a product tested to inactivate the particular agent involved. The plates cannot be autoclaved if one wishes to use them again.

For rabies virus we use a 1:100 solution of sodium hypochlorite (bleach) and Microsolve detergent. The plates are completely immersed in this solution for 24 hours before washing.

After disinfection, the plates are scrubbed under running hot tap water to remove the silicone grease and all dirt left after use of the plate.

B. Recycling the Plates

The plates are washed in three changes of distilled water, soaked overnight in 95% ethanol, air dried, and assembled for use as outlined in Section III.

REFERENCES

Anderson, T. F. (1951). *Trans. N.Y. Acad. Sci.* [2] **13**, 130–136.
Bijlenga, G., and Hernández-Baumgarten, E. M. (1974). *Amer. J. Vet. Res.* (In press).
Catalano, L. W., Jr., Fuccillo, D. A., and Sever, J. L. (1969). *Appl. Microbiol.* **18**, 1094–1095.
Conrath, T. B., ed. (1973). "Microtiter Bibliography." Dynatech Laboratories, Cooke Engineering Company. 735 North Saint Asaph Street, Alexandria, Virginia 22314.
Dean, D. (1966). *World Health Organ., Monogr. Ser.* **23**, 59–68.
Fuccillo, D. A., Catalano, L. W., Jr., Moder, F. L., Debus, D. A., and Sever, J. L. (1969). *Appl. Microbiol.* **17**, 619–622.
Hernández-Baumgarten, E. M. (1972). *Appl. Microbiol.* **24**, 999–1000.
Hernández-Baumgarten, E. M., and Bijlenga, G. (1970). *7th Annu. Meet. Inst. Nac. Invest. Pecuar. Mex.*
Hernández-Baumgarten, E. M., and Bijlenga, G. (1973). *Tec. Pecuar. (Mex.)* **24** (in press).
Hernández-Baumgarten, E. M., Nowell, J., and Tyler, W. (1974a). *Science* (in press).
Hernández-Baumgarten, E. M., Jiménez, J. L., and Hernández-Baumgarten, O. J. (1974b). To be published.
Kaplan, M. M., Wikton, T. J., Maes, R. F., Campbell, J. B., and Koprowski, H. (1967). *J. Virol.* **1**, 145–151.
Rein, A., and Rubin, H. (1968). *Exp. Cell Res.* **49**, 666–678.
Rosembaum, M. H., Phillips, I. A., Sullivan, E. J., Edwards, E. A., and Miller, L. F. (1963). *Proc. Soc. Exp. Biol. Med.* **113**, 224–229.
Sedwick, W. D., and Wiktor, T. J. (1967). *J. Virol.* **1**, 1224–1226.
Stoker, M., and MacPherson, I. (1964). *Nature (London)* **203**, 1355–1357.

Chapter 11

Agar Plate Culture and Lederberg-Style Replica Plating of Mammalian Cells

TOSHIO KUROKI

Department of Cancer Cell Research, Institute of Medical Science, University of Tokyo, Tokyo, Japan

I. Introduction

The use of semisolid agar medium was first attempted as long ago as 1911. According to A. Fischer's textbook "Gewebezüchtung" published in 1930, Loeb (1911), Lewis and Lewis (1911), and Ingebrigtsen (1912) observed growth of various cells in agar. But until recently, agar medium was not widely used because of the toxicity of some preparations of agar. The procedure of agar overlay was introduced by Temin and Rubin (1958) for assay of Rous sarcoma virus and Rous sarcoma cells. Plaque assay of animal lytic viruses was achieved with cells suspended in a layer of soft agar (Cooper, 1955). Agar suspension culture, a technique in which cells form colonies suspended in soft agar (0.33%), was developed by Sanders and Burford in 1964. MacPherson and Montagnier (1964) made use of the ability of transformed cells to grow as suspensions in agar as the basis for a quantitative assay of transformation. Since then agar medium has become widely used in tissue culture.

Attempts to grow animal cells on the surface of semisolid agar were first reported by Wallace and Hanks (1958). They used agar slants, agar plates, and agar-impregnated filter papers for culture of HeLa and L cells, although cell growth seemed to be slow. Some details of agar slant culture have been reported in J. Paul's (1970) textbook, "Cell and Tissue Culture." With amphibian cells, Arthur and Balls (1971) reported that nine permanent cell lines of *Xenopus* could form colonies when seeded in liquid medium on plates containing a basal layer of hard agar. We have reported that most cell lines of mammalian cells readily grow to form colonies, in much the same way as bacteria, on the surface of solid agar (agar plates) in the absence of overlay liquid medium (Kuroki, 1973). This has resulted in the successful application of Lederberg's replica plating technique to mammalian cells. This chapter reports the fundamental conditions for agar plate culture and Lederberg-style replica plating and the availability of these techniques for studies on the somatic cell genetics of mammalian cells.

II.   Agar Plate Culture

Like bacterial cultures, most mammalian cell lines can grow and form colonies on the surface of solid agar medium (Fig. 1), and their plating efficiencies are comparable to those observed using conventional cloning procedures in liquid medium or in agar suspension culture. This section describes details of the preparation of agar plates and the characteristics of the colonies formed on agar plates. The cell lines and media used in these studies are summarized in Table I.

A.   Preparation of Agar Plates and Isolation of Clones

The following procedure has been used in our laboratory for clonal growth of mammalian cells on agar plates containing, for example, 0.5% agar and 10% serum.

1. Make a 1.1% solution of special agar-Noble (Difco Laboratories, Detroit, Michigan) and sterilize in an autoclave.

TABLE I

CELLS AND MEDIA USED IN THESE EXPERIMENTS[a]

Cells	Description	References
A. Cells that grow in suspension		
FM3A	A cultured cell line of C3H mouse mammary ascites tumor	Nakano (1966)
L5178Y	A cultured cell line of lymphocytic leukemia from a DBA/2 mouse	Fischer (1958)
YSC	A cultured cell line of Yoshida sarcoma	Kuroki *et al.* (1966)
Primary culture of Yoshida sarcoma	Primary culture of this tumor, as already described	Katsuta *et al.* (1959) Sato *et al.* (1967)
B. Cells that grow attached to a substrate		
HeLa S3	A clonal line derived from HeLa cells	Puck *et al.* (1956)
L-929	A cell line of C3H mouse fibroblasts	Sanford *et al.* (1948)
V79	A cell line obtained from male Chinese hamster lung cells	Ford and Yerganian (1958)
CHO-K1	A cell line obtained from Chinese hamster ovary	Kao and Puck (1967)
JTC-16	A cultured cell line of Yoshida ascites sarcoma AH-7974	Takaoka *et al.* (1969)
BHK-21/C13	A clonal cell line of BHK-21 cells (fibroblasts derived from baby hamster kidney)	MacPherson and Stoker (1962)
BHK-21/py 6	A polyoma virus transformant of BHK-21	Stoker and MacPherson (1964)
BHK-21/SR	A Rous sarcoma virus (Schmidt-Ruppin) transformant of BHK-21	MacPherson (1965)
BHK-21/SR reverted	A revertant from BHK-21/SR	MacPherson (1965)

[a]The sources of these cell lines were mentioned previously (Kuroki, 1973). The media used were as follows: L5178Y cells were cultured in Fischer's medium plus 10% fetal calf serum; CHO-K1 cells were grown in F12 medium plus 10% fetal calf serum; all other cells were cultured in Eagle MEM plus 10% fetal calf serum. Media were obtained from Nissui Seiyaku Co., Tokyo, and fetal calf serum was obtained from GIBCO, Grand Island, New York.

2. After cooling to 45–50°C in a water bath, mix with an equal volume of double-strength nutrient medium and 10% serum to give 0.5% agar medium. The medium and serum should be prewarmed to about the same temperature in a water bath.

3. Pour an aliquot (10 ml for a 90-mm dish and 4–5 ml for a 60-mm dish) into a petri dish and allow it to solidify on horizontal plane. Bacteriological plastic petri dishes can be used instead of the usual plastic petri dishes for tissue culture cells.

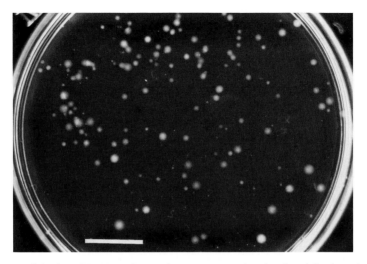

FIG. 1. Colonies of FM3A cells growing on an agar plate incubated for 2 weeks. Not fixed or stained. The bar indicates 1 cm.

FIG. 2. Conradi rod and turning table used for spreading cells evenly over the surface of agar. Spreading can also be done by tilting the plates without using these devices.

4. Allow the agar plates to stand overnight at room temperature to reduce the amount of excess moisture on the agar surface. Sealed plates can be stored for several days in a refrigerator before use. It is not usually necessary to adjust the pH of the agar plate by incubation in a CO_2-incubator before seeding the cells.

5. Prepare a suspension of cells containing more than 95% single cells (cf. Section II,B,2). Plate an aliquot (0.2 ml for a 90-mm petri dish and 0.1 ml for a 60-mm petri dish) on the agar plate and spread evenly over the surface, either by tilting the plate or with a Conradi rod (Fig. 2).

6. Incubate the agar plate in a CO_2-incubator for 2 or 3 weeks. The pH, temperature, and humidity of the incubator should be suitably controlled (cf. Section II,B,3).

7. Count the number of colonies formed by naked eye or under a dissecting microscope without fixing and staining (Fig. 1).

8. If necessary, individual clones can be isolated easily and quickly with a Pasteur pipette as follows. Draw a well isolated colony into a Pasteur pipette. Place the colony in a test tube, and disperse the cells either with a small amount of protease (trypsin or Pronase) or by mechanical agitation. Add the resulting cell suspension to growth medium in a 35-mm petri dish or small culture flask. Isolation of "pure clones" is described below (Section II,B,4).

B. Special Considerations for Agar Plate Culture

The essential conditions for cloning of cells are a suitably rich medium, careful regulation of the pH and moisture content of the CO_2 incubator, and careful handling of cells, as reviewed by Ham (1972). In addition, the following considerations will result in colony formation by many established cell lines on agar plates with a good plating efficiency.

1. AGAR

Colony formation on agar plates is largely dependent on samples and concentration of agar. Agar is a polysaccharide extracted from *Gelidium* and closely related algae. Some specimens of agar contain impurities that are toxic to the cells. In our experience the most suitable agar can be obtained from Difco Laboratories, Detroit, Michigan. As shown in Table II, plating efficiencies as high as 60% were obtained with FM3A cells on plates of 0.5% agar prepared with Bacto-agar or special agar-Noble purchased from Difco. At concentrations above 1.0%, Bacto-agar was toxic to cells, and even with 1.5% special agar-Noble 23% of the FM3A cells formed colonies. Comparable results were obtained with L5178Y cells. The following experiments were therefore done using special agar-Noble, unless otherwise stated.

The toxicity of agar seems to result from the presence of sulfated polysaccharides (agaropectin). Takemori and Nomura (1960) and Takemoto and Liebhaber (1961) reported that certain strains of polio virus are inhibited by sulfated polysaccharides. Elimination of these inhibitors can be attained by (1) washing the agar thoroughly with cold water or alcohol, (2) addition of the polymers such as DEAE-dextran or dextran sulfate (molecular weight

TABLE II

Plating Efficiency of FM3A Cells on Plates Prepared with Various
Specimens of Agar and Agarose

Concentrations of agar and agarose (%)	Agar		Agarose	
	Bacto[a]	Noble[b]	Bio-Rad[c]	Sigma[d]
0.5	60.7 ± 5.0[e]	59.4 ± 3.3	42.0 ± 4.2	55.5 ± 6.1
0.75	26.3 ± 4.5	47.3 ± 7.9	21.3 ± 3.6	46.0 ± 7.5
1.0	1.0 ± 0.6	46.6 ± 6.5	0	38.8 ± 3.8
1.2	0	39.3 ± 8.6	0	37.0 ± 7.9
1.5	0	22.8 ± 6.6	0	20.5 ± 7.6

[a]Bacto-agar, Difco Laboratories, Detroit, Michigan.
[b]Special agar-Noble, Difco Laboratories, Detroit, Michigan.
[c]Product of Bio-Rad Lab., Richmond, California.
[d]Product of Sigma, St. Louis, Missouri.
[e]Plating efficiency (%) ± standard deviation; 100 or 200 cells were plated per 90-mm petri dish and incubated for 2 weeks at 37°C in a CO_2 incubator.

5×10^5 to 2×10^6) or (3) using the agarose fraction of agar, which is devoid of sulfated polysaccharides. Montagnier (1968, 1971) found that BHK-21/C13 cells are sensitive to sulfated polysaccharides; the cells could form colonies in agar suspension culture prepared with agarose or with agar containing 100 μg/ml DEAE-dextran, while no colonies were formed in agar medium in the absence of polycation. Liebhaber and Takemoto (1961) reported that the size of plaques of encephalomyocarditis virus increased when the agar-overlay contained 50 μg/ml of DEAE-dextran.

Formation of colonies varies with different samples of agarose. As shown in Table II, agarose obtained from Sigma, St. Louis, Missouri, was apparently more suitable than that obtained from Bio-Rad, Richmond, California; agarose from Bio-Rad was toxic at a concentration of 1.0%, whereas that from Sigma, like special agar-Noble, could sustain growth of cells even at a concentration of 1.5%.

Tables II and III indicate the dependencies of the plating efficiency and the size of colonies on the concentration of agar at concentrations of 0.5% to 1.5%. With increasing concentrations of agar, the plating efficiency and size of colonies decreased. This seems to be due partly to the toxicity of the agar. It is also probably due partly to the lower supply of nutrients at higher agar concentrations. In agar plate cultures, nutrients pass from the agar layer by capillarity, and the rate of this depends on the hardness, or concentration, of the agar layer.

TABLE III

SIZES OF COLONIES OF FM3A AND L5178Y CELLS
ON AGAR PLATES PREPARED WITH VARIOUS
CONCENTRATIONS OF SPECIAL AGAR-NOBLE

Concentration (%)	FM3A (mm)	L5178Y (mm)
0.5	2.7 ± 0.6[a]	2.6 ± 0.7
0.75	2.6 ± 0.7	2.3 ± 0.5
1.0	1.8 ± 0.6	1.8 ± 0.6
1.2	1.5 ± 0.5	1.5 ± 0.6
1.5	1.5 ± 0.4	1.2 ± 0.3

[a]Mean ± standard deviation.

2. PREPARATION OF VIABLE SINGLE CELLS

Cells for cloning experiments must be handled more gently than those for usual experiments, which are inoculated at a high density. The proportion of single cells in a cell suspension is particularly important in ensuring that colonies originate from single cells. For colony formation, more than 95% of the cells in a cell suspension should be single. This can be achieved as follows.

1. The mother cells for cloning experiments should be in the exponential growth phase. Cells in the confluence or postconfluence stage are hard to disperse and tend to yield cell clumps.

2. The procedure used to disperse cells is as follows; wash the cell sheet with Ca^{2+} and Mg^{2+}-free PBS. Disperse the cells by treatment with trypsin (0.1%) or Pronase (0.05%) for 3–7 minutes at 37°C, and then add medium containing serum to neutralize the action of protease. Then make the cell suspension by pipetting. After centrifugation of the cell suspension for 5 minutes at 500–1000 rpm, resuspend the cells in fresh medium and disperse them well by pipetting. Count the number of cells and the proportion of single cells in a hemacytometer. The essential factors for obtaining a high proportion of single cells are choice of a suitable cell-dispersing agent and of an adequate concentration and incubation time, and prompt neutralization of the action of the agent.

3. When the proportion of single cells in the cell suspension obtained is not satisfactory, allow the cell suspension to stand in a test tube for 20–30 minutes to allow cell clumps to settle, and then carefully take the upper one-third or half of the cell suspension. This procedure is useful for elimination of cell clumps.

4. If cell clumps composed of 2 or 3 cells are still found in the cell suspension, disperse the cell suspension 4 times with a tuberculin syringe and neddle (26 G). This results in complete separation of cells without impairment of their viability.

3. CO_2-INCUBATOR

In agar plate culture, cells grow on the surface of agar with the agar layer on one side, and the atmosphere of the incubator on the other. The optimum ranges of pH, temperature, and osmotic pressure for clonal growth are not wide. Moreover, most cloning experiments require a relatively long period of incubation of more than 2 weeks without addition or renewal of the medium. Thus, it is essential to control the pH at the optimum value and to keep the humidity close to saturation to avoid loss of water from the cells and agar layer. Special considerations for control of a CO_2 -incubator are described by Ham (1972).

4. ISOLATION OF PURE CLONES ON AGAR PLATES

There is a possible risk with the colonial clones mentioned above (Section II,A,8) that the clonal population may not have originated from a single cell but from 2 or more cells. To isolate a "pure clone" which is absolutely certain to have originated from a single cell, it is necessary to ensure that only one cell was present when plated and to place the cell in a separate container to avoid possible contamination with other cells during growth of the clonal population. This can be achieved easily on agar plates as follows.

1. Plate a single cell suspension, prepared as described above, on an agar plate. After allowing the cells to settle on the agar, but before the first division (usually after incubation for 3–5 hours), mark well isolated single cells under an inverted microscope.

2. Place a sterilized cylinder on the agar layer around each selected cell, using sterilized forceps. Confirm under an inverted microscope that there is only one cell inside the cylinder. Cylinders (5–10 mm in diameter and 10 mm high with a 1mm thick wall) can be made from autoclavable silicon rubber tubing. No silicon grease is needed on the bottom of the cylinder.

3. After an adequate period, reconfirm that only one colony is growing inside the cylinder. After incubation for 2 or 3 weeks, isolate the clonal population with a Pasteur pipette.

C. Colony Formation by Various Cell Lines on Agar Plates

Under the appropriate culture conditions described above, all the cell lines shown in Table I, except BHK-21/C13 and BHK-21/SR reverted, formed distinctive colonies on the surface of agar. The colonies were round

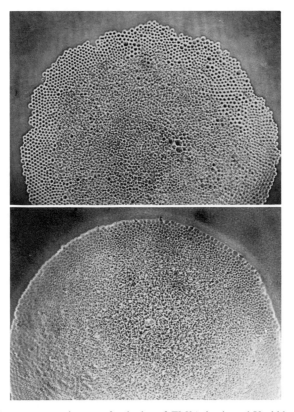

FIG. 3.   Phase contrast pictures of colonies of FM3A (top), and Yoshida sarcoma cells (bottom) after culture for 2 weeks.

with piled up cells in the center. Typical colonies of FM3A, Yoshida sarcoma cells, HeLa S3, V79, and CHO-K1 are shown in Figs. 3 and 4.

A comparison was made of the efficiencies of plating using three techniques for cloning, i.e., conventional cloning in liquid medium (Puck *et al.*, 1956), cloning by agar suspension culture (Sanders and Burford, 1964), and cloning on agar plates as described here. The results are summarized in Table IV. FM3A, YSC, and L5178Y cells that grow in suspension formed colonies with as high a plating efficiency as 60% or more both in agar suspension cultures and on agar plates. In primary cultures of Yoshida ascites sarcoma cells, the plating efficiency on agar plates was 8.9%, or approximately 4 times that of 2.1% of agar suspension culture. With HeLa S3, CHO-K1, V79, L-929, and JTC-16 cells that grow attached to a substrate, the plating efficiencies on agar plates were almost the same as those in liquid medium.

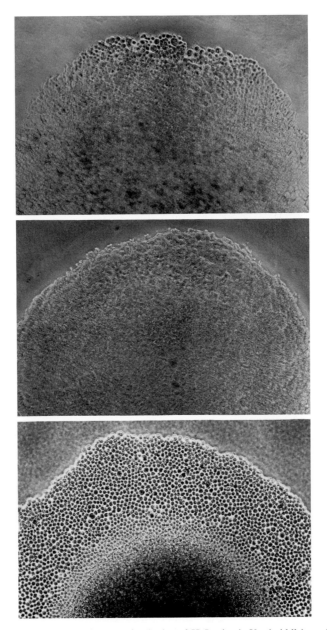

FIG. 4. Phase contrast pictures of colonies of HeLa (top), V_{79} (middle), and CHO-K1 (bottom) formed on culture for 2 weeks.

TABLE IV

COMPARISON IN PLATING EFFICIENCIES OF CELLS USING THREE PROCEDURES FOR
CLONING: CLONING IN LIQUID MEDIUM, CLONING BY AGAR SUSPENSION CULTURE,
AND CLONING ON AGAR PLATES[a]

Cells	Plating efficiency (%) ± standard deviation		
	Liquid medium	Agar suspension culture	Agar plate
A. Cells that grow in suspension			
FM3A	—	87.5 ± 4.3	59.4 ± 3.3
L5178Y	—	63.4 ± 0.7	69.8 ± 11.6
YSC	—	58.3 ± 6.0	64.6 ± 13.0
Primary culture of Yoshida sarcoma	—	2.1 ± 0.5	8.9 ± 0.6
B. Cells that grow attached to a substrate			
HeLa S3	51.5 ± 8.1	47.7 ± 7.7	44.6 ± 3.4
L-929	43.8 ± 3.6	14.2 ± 1.5	38.1 ± 4.3
V_{79}	52.0 ± 6.0	41.0 ± 1.3	38.7 ± 3.7
CHO-K1	48.3 ± 6.6	56.1 ± 5.6	49.7 ± 3.0
JTC-16	43.5 ± 3.4	19.8 ± 3.0	38.8 ± 4.1
BHK-21/C13	59.3 ± 6.7	0	0
BHK-21/PY6	66.0 ± 9.0	92.3 ± 11.1	82.0 ± 16.8
BHK-21/SR	95.0 ± 13.0	51.3 ± 4.4	51.5 ± 7.0
BHK-21/SR reverted	48.0 ± 3.9	0	0

[a]Special agar-Noble 0.5%.

The efficiencies of plating of L-929 and JTC-16 cells were about twice as great on agar plates as in agar suspension culture, while with other cells the plating efficiencies were similar on agar plates and in agar suspension cultures. Thus, these results clearly indicate that the efficiency of cloning on agar plates is comparable to those obtained using conventional cloning techniques in liquid medium or agar suspension culture.

Since the discovery by MacPherson and Montagnier (1964) that transformed BHK-21 cells can form colonies when suspended in an agar layer while untransformed cells cannot, agar suspension culture has been used widely as a selective method for assay of transformation. We investigated the possibility that agar plates allow selective growth of transformed cells, using a series of BHK-21 cells. As shown in Table IV, there was a close agreement between these two culture systems using agar; cells transformed with virus (polyoma and Rous sarcoma virus) could form colonies on agar plates as well as in agar suspension culture with high plating efficiency, while no colonies were formed when 10,000 cells of untransformed BHK-21/C13 or the revertant from the Rous sarcoma virus-transformed cells were plated.

This suggests that agar plate culture can be used, like agar suspension culture, as a selective system for assay of transformed cells.

Stoker and his associates (1968) studied selective growth of transformed cells in detail and termed it "anchorage dependency," because when untransformed BHK-21 cells become attached to fragments of glass in the agar layer, they can divide to form colonies. Montagnier (1968, 1971), however, found that BHK-21/C13 cells could grow in agar suspension culture when agar plus DEAE-dextran or agarose was used, suggesting that selective growth of transformed cells is due to their lower sensitivity to sulfated polysaccharides. We have also observed that BHK-21/C13 cells form colonies on agar plates prepared with Sigma-agarose with almost the same plating efficiency as in liquid medium.

III.   Lederberg-Style Replica Plating of Mammalian Cells

Lederberg and Lederberg (1952) first used pile fabrics for copying clones and since then replica plating has been widely used for isolation of mutant bacteria. This technique and variations on it have been applied to actinomycetes (Braendle and Szybalski, 1957), to unicellular algae (Eversole, 1956), and to a filamentous fungus (Roberts, 1959). A replica-plating technique has also recently been developed for mammalian cells by Goldsby and Zipser (1969), Robb (1970), and Suzuki et al. (1972; Suzuki and Horikawa, 1973), in which clones growing in the well of a microplate are transferred to a series of replica plates using a mechanical replicating device (Goldsby and Zipser, 1969) or simplified hand replicator (Suzuki et al., 1972; Suzuki and Horikawa, 1973). However, there is no report of application of Lederberg's replica plating technique to mammalian cells. One reason for this is that mammalian cells have been thought not to form colonies on the surface of solid agar in the absence of overlay liquid medium. As reported previously (Kuroki, 1973), the success of agar plate culture described above makes it possible to apply Lederberg's replica plating to mammalian cells. In this section, Lederberg style replica plating and its modification are described in detail.

A.   Method I: Replica Plating Using Pile Fabrics

In the original replica plating technique of Lederberg, colonies growing on the master plate are imprinted on pile fabrics and then transferred to a series of replica plates. In a similar way, colonies of mammalian cells on an

FIG. 5. Apparatus used for Method I of replica plating. This is essentially as described by Lederberg and Lederberg (1952). A sterilized 15-cm square of pile fabric (velveteen or velvet) is placed, nap upward, on a cylinder support and fixed firmly with a hoop. The master plate is inverted onto the fabric to imprint the colonies, and then used to imprint to a series of replica in the same way.

agar plate can be transferred to identical positions on replica plates. In our method, velveteen (cotton) or velvet (silk or synthetic fibric) are used for imprinting. A sterilized 15-cm square of fabric is placed, nap upward, on a cylindrical stainless steel support (8.5 cm in diameter for a 90-mm petri dish) and fixed firmly in place with a metal hoop (Fig. 5). Master plate carrying 30–40 well isolated colonies on its surface is inverted over the fabric with slight digital pressure to transfer the colonies. Then, the fabric is used for imprinting replica-inocula on a series of replica plates by impression in the same way. The replica plates are incubated in a CO_2-incubator for 1 week.

Table V shows the fidelity of replica plating of clones of FM3A, using two specimens of velvet and velveteen as imprinting fabrics and master and replica plates of 0.5% special agar-Noble. When velveteen A was employed for imprinting, about 94% of the colonies were transferred to a series of 4 replica plates, while specimen B of velveteen gave 76% fidelity. However, velvet did not seem adequate, because the fabric is less absorbent and accumulates moisture, so that colonies tended to be scattered over the plate. At an agar concentration of 0.75%, transfer of about 95% of the clones of the master plate was achieved with velvet. Thus, replica plating was influenced by both the fabric and the concentration of agar in the plates. The highest fidelities of replication were around 95% with a combination of velveteen and 0.5% agar or of velvet and 0.75% agar. This is comparable to that

TABLE V

Replica Plating Efficiencies of FM3A Cells Using Method I[a]

Fabric	No. of colonies on the master plate[b]	No. of colonies transferred[c]				Replica plating efficiency[d] (%)
		R1	R2	R3	R4	
A (velveteen)	35	33	30	34	35	94.3
	32	31	29	30	—	93.8
B (velveteen)	43	33	30	33	35	76.2
C (velvet)	38	Colonies were scattered over the plates				
D (velvet)	52	Colonies were scattered over the plates				

[a]Special Agar-Noble, 0.5%, was used for the Master and Replica Plates.
[b]Master plates were cultured for 2 weeks.
[c]Replica plates were cultured for 1 week.
[d]Average of values on 3 or 4 replica plates.

obtained by Goldsby and Zipser (1969) and by Suzuki *et al.* (1972; Suzuki and Horikawa, 1973).

In practice, however, we encountered the following difficulties: (1) adjacent colonies tended to fuse on replica plates unless colonies on the master plate were well separated from each other; (2) the efficiency depended on the nature of the pile fabric; and (3) the agar concentration was lower than for bacterial culture, so excess moisture often accumulated after 4 or 5 successive replica plates had been made. Because of these difficulties, we tried to modify Lederberg's replica plating technique as described below.

B. Method II: Manual Imprinting of Clones Using a Glass Rod

In this technique, colonies on the master plate are transferred manually, using a glass rod, to a series of replica plates. The positions of replica inocula are identified by a grid (Fig. 6). In practice, 200 clones can be transferred to 4 replica plates within 1 hour.

This manual imprinting gives more faithful reproduction than Method I. As seen in Table VI and Fig. 7, all the 75 clones of FM3A, L5178Y, and YSC cells, except for one clone of FM3A on the R1 plate, were successfully transferred to the identical positions on grids of 4 or 5 successive replica plates. The replica efficiencies of FM3A, L5178Y, and YSC cells were 99.7%, 100%, and 100%, respectively. More than 20 serial replicas can be carried out by this method.

The above-mentioned experiments were done with cells that grow in suspension. With cells that grow attached to the substrate, dispersion of cells in a colony *in situ* is necessary before replica plating. This can be done by

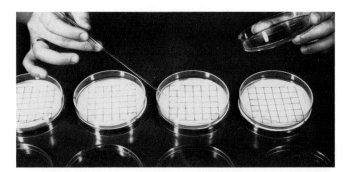

FIG. 6.   Method II of replica plating. Colonies on the master plate are transferred manually, using a glass rod, to a series of replica plates, and their positions are identified with a grid.

TABLE VI

REPLICA PLATING EFFICIENCIES OF FM3A, YSC, AND L5178Y CELLS USING METHOD II

Cells	No. of clones inoculated[a]	No. of clones transferred[b]					Replica plating efficiency[c] (%)
		R1	R2	R3	R4	R5	
FM3A	75	74	75	75	75	75	99.7
YSC	75	75	75	75	75	75	100
L5178Y	75	75	75	75	75	—	100

[a]Master plates were cultured for 2 weeks.
[b]Cultured for 1 week.
[c]Average of values on 4 or 5 replica plates.

spraying protease evenly onto the master plates with an atomizer and then incubating the plate at 37°C for 10–15 minutes. As seen in Table VII and Fig. 7 almost 100% efficiency of replica plating of CHO-K1 and V_{79} cells was obtained by pretreatment with 0.1% Pronase or 0.25% trypsin.

IV.   Application of Agar Plate Culture and Replica Plating to Mammalian Cell Genetics

Progress in mammalian cell genetics has been based on the isolation of mutants. At present a number of techniques are available for this: cloning procedures, as described above, cell hybridization (Barski *et al.*, 1961;

TABLE VII

REPLICA PLATING OF CELLS THAT GROW ATTACHED TO THE SUBSTRATE USING
PROTEASE TREATMENT AND THEN MANUAL IMPRINTING WITH A GLASS ROD

Cells	Protease pretreatment[a]	No. of clones inoculated	No. of clones transferred[c]				Replica plating efficiency (%)
			R1	R2	R3	R4	
CHO-K1	None	44	44	42	41	38	93.8
	Trypsin (0.25%)	44	44	44	44	43	99.4
	Pronase (0.1%)	44	44	44	44	42	98.9
V_{79}	None	44	43	43	42	43	97.2
	Trypsin (0.25%)	44	44	44	43	42	98.3
	Pronase (0.1%)	44	44	44	44	44	100

[a]The master plates were sprayed evenly with protease, using an atomizer, and then incubated at 37°C for 10–15 minutes. A glass rod was used for replica plating (Method II). Pretreatment with protease dispersed the cells in colonies *in situ*.

[b]The master plate was cultured for 2 weeks.

[c]Replica plates were cultured for 1 week.

Littlefield, 1964; Harris and Watkins, 1965), selection and isolation of drug-resistant clones (Chu and Malling, 1968), or nutritional auxotrophs (Kao and Puck, 1968, 1969), or temperature-sensitive mutants (Thompson *et al.*, 1970; Meiss and Basilico, 1972), and the replica plating techniques described above. In this section, the availability of agar plate culture and Lederberg's replica plating for isolation of mutant cells is discussed. Thompson and Barker (1973) have recently reviewed the isolation of mutant cells from mammalian cell cultures in Volume 6 of this series.

A. Isolation of Drug-Resistant Clones Using Agar Plates

In principle, the selection of drug-resistant clones on agar plates is straightforward and is the same as that in liquid medium; a mass culture of wild-type cells is plated on an agar plate containing a cytotoxic drug and after incubation for 2 or 3 weeks, surviving clones of the drug-resistant phenotype are isolated. In our laboratory, we have utilized the following procedure for assay and isolation of 8-azaguanine-resistant cells.

Selective agar plates are prepared by dissolving 20 μg/ml of 8-azaguanine in agar medium cooled to 45 to 50°C (step 2 in Section II,A: preparation of agar plates). 8-Azaguanine is dissolved in dimethyl sulfoxide at a concentration of 4 mg/ml so that the final concentration of the solvent in the agar medium is 0.5%, which does not affect the growth of cells significantly. The cells to be selected are then inoculated into the agar plates, usually at a

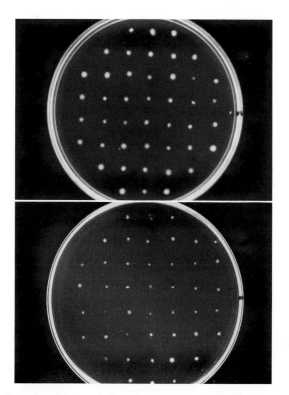

Fig. 7.   Replica plate (90-mm dish) with 44 clones of FM3A (top) and V_{79} (bottom) which had been transferred by manual imprinting using a glass rod (Method II). Master plates of V_{79} cells had been sprayed with 0.1% Pronase with an atomizer and then incubated at 37°C for 10–15 minutes before replica plating, in order to disperse cells in a colony *in situ*. Cultured for 1 week.

density of 10^5 cells per 90-mm dish. The observed frequency of appearance of resistant clones is corrected by the plating efficiency, which is obtained by plating an aliquot (usually 200 cells/60-mm dish) onto an agar plate containing no 8-azaguanine. In our experiments, 1–10 8-azaguanine-resistant colonies were obtained per 10^5 cells plated. Pretreatment of the cells with mutagens increased the frequency of mutation. These resistant clones were found to be stable during further cultivation.

Backward mutation from 8-azaguanine resitance, which is known to be closely associated with the activity of hypoxanthine–guanine phosphoribosyltransferase (HGPRTase), can be assayed using agar plates containing hypoxanthine, aminopterin, thymidine, and glycine (HAT medium). The frequencies of reversion of 8-azaguanine-resistant FM3A cells were less than 5×10^{-5}.

B. Isolation of Auxotrophic Mutants

Isolation of auxotrophic mutants, which can only grow in the presence of a particular nutrient, is facilitated by use of the replica plating technique. For example, Suzuki and Horikawa (1973) successfully isolated nutritional auxotrophs from Chinese hamster *hai* cells using a replica plating technique which they devised. In their experiments clones growing in the wells of microplates were transferred, by a hand replicator, to 17 replica plates containing various selection media, each of which lacked one of the ingredients, such as an amino acid or nucleoside. Nutritionally deficient mutants, which failed to grow in the selection media, were isolated from the identical well of the master plate. Backward mutation from a nutritionally deficient mutant to a requiring mutant can be assayed using a selection medium which lacks the ingredient. Agar plate culture and Lederberg style replica plating are also applicable for isolation of auxotrophic mutants in the similar way. However, agar plate culture does not seem suitable for the isolation of auxotrophs related to carbohydrate metabolism, since agar usually contains various species of carbohydrate even after it has been washed with cold water.

C. Isolation of UV-Sensitive Clones

Replica plating should be particularly useful for isolation of clones that are sensitive to ultraviolet light (UV), X-rays, antibiotics, or high temperature. We have recently tried to isolate UV-sensitive clones from mouse cell lines using agar plate culture and replica plating, and some of the results have been reported elsewhere (Kuroki, 1974).

The procedure for isolation of UV-sensitive clones consisted of the following 5 steps and took 6–7 weeks.

Step 1: Cells were treated with a mutagen such as N-methyl N'-nitro-N-nitrosoguanidine (MNNG) at a concentration which killed 70–90% of the population.

Step 2: Cells were cultured for 2 days to allow expression of mutagen-induced mutation.

Step 3: Master plates were made by plating 500–2000 cells (according to the toxicity of the mutagen) on a 90-mm agar plate, and the plates were incubated for 2 weeks in a CO_2-incubator.

Step 4: A series of 4 replica plates (R1 to R4 by order of plating) was made from the master plates by manual imprinting with a glass rod. Plates R1 and R3 were used as unirradiated controls, while R2 and R4 were irradiated at doses of 50 or 100 erg/mm². In agar plate cultures, culture medium was not

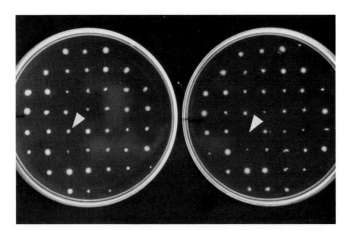

FIG. 8.   Control (left) and irradiated (right) replica plates at step 4. The arrows indicate
a possible UV-sensitive clone, which formed a distinctive colony on the control plate but
not on the irradiated replica plate.

present on the surface, so that no correction was made for absorption of
UV by the medium. After incubation in a CO_2-incubator, clones that formed
a distinctive colony on control plates (R1 and R3), but failed to grow on
irradiated replica plates (R2 and R4) were isolated from the control plates
as possible UV-sensitive clones. A typical example of control and irradiated
plates is shown in Fig. 8.

Step 5: Isolated clones were allowed to grow, and then their UV-sensitivity
was examined precisely by plating the cells on agar plates, irradiating them
with various doses of UV and scoring the survival fractions after incubation
for 2 weeks. To exclude possible shadowed areas, colonies in the periphery
of the dishes were not scored. The D_0 value was obtained from the exponential portion of the survival curve.

Using this procedure, we have isolated UV-sensitive clones from the
mouse cell lines FM3A and L5178Y. The frequency of appearance of UV-
sensitive clones after MNNG-treatment was 0.11% (1/886) with FM3A cells
and 1.43% (6/420) with L5178Y cells, while no such clones were found
among 552 control clones of FM3A and 111 control clones of L5178Y.
Figure 9 indicates the survival curves of a freshly isolated UV-sensitive
clone and the parental FM3A. The D_0 values of the UV-sensitive clones
ranged from 11 to 20 erg/mm², and are comparable to those of xeroderma
pigmentosum cells. The UV-sensitive clones were stable for some time (2–4
months) but then tended to lose their sensitivity, showing an increase in
their D_0 value or a shoulder in the survival curve. Details of the results will
be reported elsewhere.

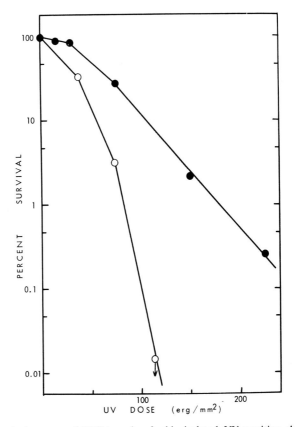

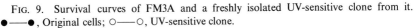

FIG. 9. Survival curves of FM3A and a freshly isolated UV-sensitive clone from it. ●——●, Original cells; ○——○, UV-sensitive clone.

V.   Concluding Remarks

Progress in tissue culture is closely dependent on development of suitable techniques. The availability of antibiotics and aseptic bench, plastic flasks, and petri dishes, and commercial supplies of sera and powdered media have permitted the use of tissue culture for medical and biological studies in many laboratories. Modern technology of tissue culture began with the introduction by Puck and his co-workers (1956) of the colony-forming technique to tissue culture. A number of techniques have been developed since then, and it is now possible to handle and recognize mammalian cells as individual cells, rather than as constituents of a multicellular organism. This chapter describes a technique for obtaining colonies of most established mammalian cells on the surface of agar plates in the absence of overlay liquid medium. This technique is closely analogous to that for culture of bacteria. The reason why such a simple technique has not been described before seems to be simply that mammalian cells, unlike bacteria, have been believed not to form colonies on the surface of agar. Agar plate culture has the following advantages: (1) It is easy to prepare agar plates and isolate clones. (2) The plating efficiency is comparable to that of already established cloning techniques. (3) It can be used for assay of transformation. (4) It is applicable to Lederberg-style replica plating.

Interest in genetics and mutagenesis of mammalian cells is increasing, and for studies on these problems methods are required for isolation of stable, well-defined mutants from cultured cells. At least three methods are available for isolating mutant cells. The first is selection of drug-resistant mutants by exposing a mass culture of wild-type cells to a cytotoxic drug. The second is selection of cells carrying a mutation from a wild-type population by killing the normally growing wild-type cells but allowing nondividing mutant cells to survive. A typical example is the experiment of Kao and Puck (1968, 1969) in which auxotrophic mutants were isolated using a procedure which involved 5-bromodeoxyuridine incorporation followed by irradiation with visible light. The third procedure is replica-plating techniques, analogous to those which have greatly facilitated the isolation of mutant bacteria. At present two replica-plating methods are available for mammalian cells; replica plating of cells growing in the wells of microplates using replicating devices (Goldsby and Zipser, 1969; Suzuki et al., 1972; Suzuki and Horikawa, 1973) and Lederberg-style replica plating as described here. Although these techhniques should be useful for isolation of mutant cells, they need improvement to increase their efficiency and to facilitate the procedures involved. The replica-plating method itself cannot be used for selection of mutants, but when used in combination with other methods of selection, it is valuable for isolation and purification of mutant cells.

REFERENCES

Arthur, E., and Balls, M. (1971). *Exp. Cell Res.* **64**, 113.
Barski, G., Sorieul, S., and Cornefert, F. J. (1961). *J. Nat. Cancer Inst.* **26**, 1269.
Braendle, D. H., and Szybalski, W. (1957). *Proc. Nat. Acad. Sci. U.S.* **43**, 947.
Chu, E. H. Y., and Malling, H. V. (1968). *Proc. Nat. Acad. Sci. U.S.* **61**, 1306.
Cooper, P. D. (1955). *Virology* **1**, 397.
Eversole, R. A. (1956). *Amer. J. Bot.* **43**, 404.
Fischer, A. (1930). "Gewebezüchtung, Handbuch der Biologie der Gewebezellen in Vitro," p. 15. Rudolph Müller & Steinicke, Munich.
Fischer, G. A. (1958). *Ann. N.Y. Acad. Sci.* **76**, 673.
Ford, K. D., and Yerganian, G. (1958). *J. Nat. Cancer Inst.* **21**, 393.
Goldsby, R. A., and Zipser, E. (1969). *Exp. Cell Res.* **54**, 271.

Ham, R. G. (1972). *In* "Methods in Cell Physiology" (D. M. Prescott, ed.), Vol. 5, pp. 37–74, Academic Press, New York.

Harris, H., and Watkins, J. F. (1965). *Nature (London)* **205**, 640.

Ingebrigtsen, R. (1912). *J. Exp. Med.* **16**, 421.

Kao, F. T., and Puck, T. T. (1967). *Genetics* **55**, 513.

Kao, F. T., and Puck, T. T. (1968). *Proc. Nat. Acad. Sci. U.S.* **60**, 1275.

Kao, F. T., and Puck, T. T. (1969). *J. Cell. Physiol.* **74**, 245.

Katsuta, H., Takaoka, T., Mitamura, K., Someya, Y., and Kawada, I. (1959). *Jap. J. Exp. Med.* **29**, 143.

Kuroki, T. (1973). *Exp. Cell Res.* **80**, 55.

Kuroki, T. (1974). *In* "Chemical Carcinogenesis Essays," (R. Montesano and L. Tomatis, eds.) IARC Sci. Pub. **10**, pp. 147–160, Int. Agency Res. Cancer, Lyon.

Kuroki, T., Isaka, H., and Sato, H. (1966). *Gann* **57**, 367.

Lederberg, J., and Lederberg, E. M. (1952). *J. Bacteriol.* **63**, 399.

Lewis, M. R., and Lewis, W. H. (1911). *Johns Hopkins Hosp. Bull.* **22**, 126.

Liebhaber, H., and Takemoto, K. K. (1961). *Virology* **14**, 502.

Littlefield, J. W. (1964). *Science* **145**, 709.

Loeb, L. (1911). *Science* **34**, 414.

MacPherson, I. (1965). *Science* **148**, 1731.

MacPherson, I., and Montagnier, L. (1964). *Virology* **23**, 291.

MacPherson, I., and Stoker, M. (1962). *Virology* **16**, 147.

Meiss, H. K., and Basilico, C. (1972). *Nature (London)* **239**, 66.

Montagnier, L. (1968). *C.R. Acad. Sci., Ser. D* **267**, 921.

Montagnier, L. (1971). *Growth Cont. Cell Cult., Ciba Found. Symp., 1970*, pp. 33–44.

Nakano, N. (1966). *Tohoku J. Exp. Med.* **88**, 69.

Paul, J. (1970). *In* "Cell and Tissue Culture," 4th ed., p. 223. Livingstone, Edinburgh.

Puck, T. T., Marcus, P. I., and Cieciura, S. J. (1956). *J. Exp. Med.* **103**, 273.

Robb, J. A. (1970). *Science* **170**, 857.

Roberts, C. F. (1959). *J. Gen. Microbiol.* **20**, 540.

Sanders, F. K., and Burford, B. A. (1964). *Nature (London)* **201**, 786.

Sanford, K. K., Earle, W. R., and Likely, G. D. (1948). *J. Nat. Cancer Inst.* **9**, 229.

Sato, H., Goto, M., and Kuroki, T. (1967). *In* "Gann Monograph Vol. 2, Cancer Chemotherapy, Endeavors to Break through the Barriers" (A. Goldin, *et al.*, eds.), pp. 127–140. Maruzen, Tokyo.

Stoker, M., and MacPherson, I. (1964). *Nature (London)* **203**, 1355.

Stoker, M., O'Neill, C., Berryman, S., and Waxman, V. (1968). *Int. J. Cancer* **3**, 683.

Suzuki, F., and Horikawa, M. (1973). *In* "Methods in Cell Biology" (D. M. Prescott, ed.), Vol. 6, pp. 127–142. Academic Press, New York.

Suzuki, F., Kashimoto, M., and Horikawa, M. (1972). *Exp. Cell Res.* **68**, 476.

Takaoka, T., Katsuta, H., Ohta, S., and Miyata, M. (1969). *Jap. J. Exp. Med.* **39**, 267.

Takemori, N., and Nomura, S. (1960). *Virology* **12**, 171.

Takemoto, K. K., and Liebhaber, H. (1961). *Virology* **27**, 434.

Temin, H. M., and Rúbin, H. (1958). *Virology* **6**, 669.

Thompson, L. H., and Barker, R. M. (1973). *In* "Methods in Cell Biology" (D. M. Prescott, ed.), Vol. 6, pp. 209–281. Academic Press, New York.

Thompson, L. H., Mankovitz, R., Barker, R. M., Till, J. E., Siminovitz, L., and Whitemore, G. F. (1970). *Proc. Nat. Acad. Sci. U.S.* **66**, 377.

Wallace, J. H., and Hanks, J. H. (1958). *Science* **128**, 658.

Chapter 12

Methods and Applications of Flow Systems for Analysis and Sorting of Mammalian Cells[1]

H. A. CRISSMAN, P. F. MULLANEY, AND J. A. STEINKAMP

Biophysics and Instrumentation Group,
Los Alamos Scientific Laboratory,
University of California,
Los Alamos, New Mexico

[1] Work performed under the auspices of the U.S. Atomic Energy Commission.

179

I. Introduction

The development and application of high-speed flow systems for auto-mated analysis and sorting of cell populations have added a new dimension to the field of cell biology. Rapid advances in instrument technology and cell preparative techniques now provide unique, quantitative methods for prob-ing specific cytological properties and gaining new insight into cellular mechanisms. Physical and biochemical characteristics of single cells pre-sently can be analyzed at rates of 10^5 cells per minute with the use of high-speed, cell-analysis systems. Such techniques permit rapid and simultaneous analysis of DNA, protein, cell volume, light scattering, and other properties of cells with the added option of electronically sorting from a given cell population particular cells fulfilling a set of preselected criteria. Thus, from a large cell population a relatively small fraction of inconspicuous cells can be numerically enriched and subjected to a visual examination and morpho-logical identification. Such investigative techniques are certain to have a profound impact on the field of biology and medicine, especially in the area of clinical diagnosis. However, rather than dwell on any one field of applica-tion, this chapter will be devoted to the methodology and application of flow systems that demonstrate the versatility of the technique for performing a variety of biological investigations. From the examples presented here, undoubtedly the reader will envision many potential applications relevant to his own research program.

A. Development of Flow Systems for Automated Cytology

Over a generation ago, Casperson (1936) made the pioneering steps in the then infant science of quantitative cytophotometry. Most of this early work involved absorption measurements on cellular nucleic acids. The instruments developed for these measurements consisted of photometers coupled to microscopes for analysis on cells mounted on microscope slides in the traditional fashion. Continued development of automated cytophotometry on single cells has led to such instruments as the Cytoanalyzer (Tolles and Bostrom, 1956), designed for automated screening for vaginal cancer, and CYDAC (Mendelsohn *et al.*, 1964), a variant of the Cytoanalyzer, which has been used for pattern recognition studies of individual cells as well as chromosome analysis (Mendelsohn *et al.*, 1974). A system called TICAS (Wied *et al.*, 1968) more recently has evolved which is capable of recognizing optical differences between cell types which the human observer cannot distinguish.

The image analysis techniques outlined above all attempt to mimic the pattern recognition capabilities of the human observer. Observations of this kind have high information content but are limited by the relatively small numbers of the sample population that can be analyzed by cell-scanning methods. Recognizing these limitations, several laboratories in the mid to late 1960s developed another approach by devising instrumentation to make photometric measurements on individual cells in a flow arrangement without requiring pattern recognition. Research groups at the International Business Machines–Watson (Kamentsky *et al.*, 1965), the Los Alamos Scientific Laboratory (Van Dilla *et al.*, 1967, 1969; Mullaney *et al.*, 1969), Stanford University (Hulett *et al.*, 1969), the University of Munster (Dittrich and Göhde, 1969), and the University of Freiburg (Sprenger *et al.*, 1971) pursued this flow-system methodology using a variety of instrumental designs. However, the basic similarity of all these approaches is that they require individual cells to flow in single file across a beam of exciting light. As the cells pass through the beam, photometric measurements of (1) fluorescence, (2) absorption, and (3) light scattering, or various combinations of these parameters can be made on each and every cell in a population. Many of the principles employed in aerosol counting and gamma-ray spectroscopy have been applied to the flow-systems approach. Various flow systems have been attempted: laminar flow cells (Mullaney *et al.*, 1969; Holm and Cram, 1973), variations of microscope slide chambers (Kamentsky *et al.*, 1965; Dittrich and Göhde, 1969), as well as injection of the sample stream into air followed by subsequent measurement (Hulett *et al.*, 1969; Van Dilla *et al.*, 1973; Jovin and Jovin, 1974).

There is one common point shared by all these methods: measurements made on cells are not always related to direct human observation. The various instruments detect cells and perform only those measurements as based on their design: quantitative photometry (fluorescence, light scattering, and absorption). To integrate the power of these devices into the present biological and medical experience, it is necessary to provide a method of selecting cells, analyzed by the instruments and determined interesting or unique, for comparison with traditional methods. This requirement has led to the development of cell sorting devices. The earliest of these (Fulwyler, 1965) sorted cells based on volume following the Coulter approach. However, cell volume appears to be most useful when combined with other parameters. Instruments, capable of cell sorting based on optical parameters as well, consequently have been developed (Hulett *et al.*, 1969; Steinkamp *et al.*, 1973a). Today increasing emphasis is placed on multiparameter analysis and cell sorting. In this regard, several commercially available instruments have been developed (Becton Dickinson, Biophysics Systems, Particle Technology, and Phywe).

B.   Advantages of Flow Systems

Flow systems offer several advantages when compared with the various static system which analyze cells on slides. In most flow instruments, each cell is exposed to the light beam for only a few microseconds; thus, problems with fluorescence decay are minimized. Since a large number of cells can be analyzed in a short time (1000–1500 cells per second), measurements of high statistical significance are possible. An important feature of the flow-systems approach is that quantitative measurements are made on each and every cell in a population. Rapid cell-cycle analysis and quantitative immunofluorescence studies are now a reality. Interesting cells which occur with a low frequency in a population can now be detected (Horan *et al.*, 1974). High resolution has been brought to biological measurements largely through a careful evaluation of those factors which control the instrument response (Holm and Cram, 1973).

In addition to detection and multiparameter analysis, cells can be physically sorted from a population for further study of their immunological, biochemical, or morphological features. Cell sorting also provides the opportunity for verification of machine identification—an important point in clinical diagnosis.

II.   Operational Features of the Cell Analysis and Sorting System

The multiparameter analysis and cell sorting system developed at the Los Alamos Scientific Laboratory (LASL) incorporates several preexisting analytical techniques used previously in flow systems and allows all the various measurements to be performed simultaneously *on the same cell*. Electronic processing of total or two-color fluorescence, cell volume, and light-scattering signals from each cell provides a direct method for establishing relationships between the various cellular properties. Cell-sorting techniques permit physical separation of particular cells of interest from a heterogeneous population for further examination. The flow-system methodology is described briefly here and more comprehensively elsewhere (Holm and Cram, 1973; Mullaney and West, 1973; Steinkamp *et al*., 1973a, 1974).

Cells to be measured are suspended in normal saline and introduced into a flow chamber (Fig. 1) from a pressurized sample reservoir. The cell suspension enters the chamber via a sample inlet tube which also serves as the cell

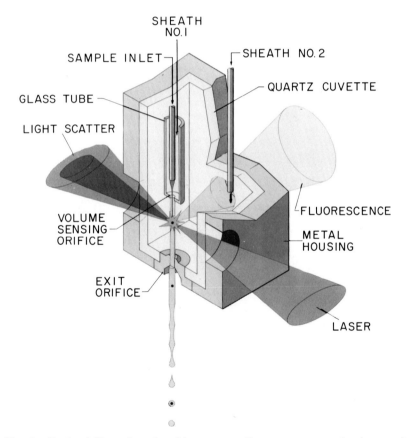

FIG. 1. Sectional illustration of multiparameter cell separator system showing sample and sheath fluid inlet tubes, cell stream, cell volume sensing orifice, laser illumination, and liquid jet from exit orifice.

volume signal electrode. Flowing coaxially around the inlet tube is a particle-free sheath fluid (sheath No. 1) of normal saline. Sheath No. 1 serves to center the cell stream as both streams flow through a 75-μm diameter volume sensing orifice (Coulter principle). This laminar flow arrangement improves the volume resolution (Merrill *et al.*, 1971). Upon leaving the volume sensing orifice, the cells enter a fluid-filled viewing region provided by a second saline sheath (sheath No. 2) where the cells intersect an argon-ion laser beam (488 nm wavelength). At the cell stream/light beam intersection, red and green fluorescence and light scattering measurements are made. The two fluid inlet tubes and the sample orifice comprise the volume sensing circuit. The relationship of the electrical signal produced and particle shape size and flow orientation has been discussed elsewhere (Coulter, 1965; Grover *et al.*, 1969a,b).

Optical signal detection is accomplished with a dual photomultiplier tube arrangement for fluorescence (Steinkamp *et al.*, 1973a) and a photodiode for light scattering (Mullaney *et al.*, 1969). All signals are processed using electronic techniques similar to those used in nuclear physics.

After optical measurement, the liquid stream carrying the suspended cells emerges into air as a jet from the exit orifice (Fig. 1). A piezoelectric trans-

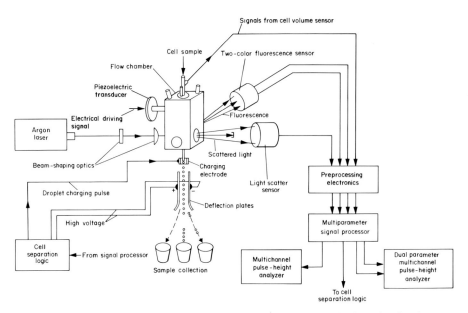

FIG. 2.   Multiparameter cell separator system diagram illustrating the flow chamber, laser illumination, cell sensors, signal processing and cell separation electronics, and droplet generation, charging, and deflection scheme.

ducer coupled to the flow chamber (Fig. 2) produces uniform liquid droplets. By adjusting the concentration appropriately, only 1% of the droplets contain a cell. Downstream from the chamber, droplets can be selectively charged. Charged drops are separated electrically from the main stream by a pair of deflection plates; thus, cell sorting is accomplished.

With the laser beam shaped as an ellipse with a minor axis less than the cell diameter, information can be obtained about cell dimensions (e.g., nuclear-to-cytoplasmic ratios) and the presence or absence of doublets (two cells stuck together) using pulse-shape analysis. For example, if two different-color fluorescent stains are used for specific staining of the nucleus (DNA content) and cytoplasm (total protein) (Crissman and Steinkamp, 1973), it is feasible to measure nuclear-to-cytoplasmic ratios (Wheeless and Patten, 1973a,b; Steinkamp and Crissman, 1974). By the use of signal threshold level discriminators, the *time duration* of the two simultaneous fluorescence signals is detected and converted to voltage signals. These signals, which are proportional to nuclear (N) and cytoplasmic (C) dimensions, are electronically processed to provide a N/C ratio signal (Steinkamp and Crissman, 1974). Single-color nuclear stains, coupled with light scatter for cytoplasmic diameter measurement, may also accomplish the same result.

The available signals are processed in several ways: as a single parameter, ratios of two signals, gated single parameters, and a complete two parameter analysis (Fig. 3). Processed signals are stored and displayed as pulse-amplitude frequency distribution histograms using a multichannel pulse-height analyzer. These signals are also used to trigger cell sorting. A two

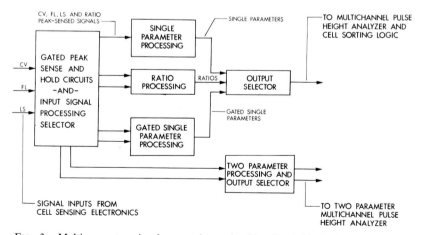

FIG. 3. Multiparameter signal processing unit abbreviated block diagram: CV (cell volume), FL (fluorescence), and LS (light scatter).

parameter pulse-height analyzer is used to produce pulse-amplitude frequency distribution histograms (isometric plot), or two-dimensional contour displays.

With single or multiple signal inputs to the processing logics, the maximum signal amplitudes are processed on a cell-by-cell basis. Ratios of parameters (e.g., DNA-to-protein) are computed similarly by selecting two signal inputs for the numerator and denominator of an electronic pulse divider. The ratio is then electronically analyzed as a single parameter.

Gated single-parameter analysis permits the examination of particular subclasses of cells within selectable ranges of the various cellular property values. This is accomplished by selecting two parameter combinations of the input signals. One signal is selected as the input to a single-channel, pulse-height analyzer (SCA) incorporated in the gated analysis processing block (Fig. 3). If its amplitude falls within a selected range, it triggers a gate open for the second signal to be routed to the pulse-height analyzer. For example, the protein content distribution for G_1 cells can be obtained by analyzing only those protein signals from cells showing G_1 DNA content (Crissman and Steinkamp, 1973). Gated single-parameter techniques also permit distributions to be obtained from weakly fluorescing cells by requiring coincidence with light scatter or cell volume signals (Mullaney and West, 1973; Steinkamp and Kraemer, 1974).

Processed signals can also activate cell sorting if the amplitudes fall within preselected ranges of single-channel, pulse-height analyzers located within the cell separation logic (Fig. 4). If cell sorting is initiated, an electronic time

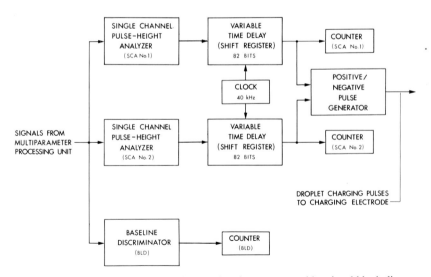

Fig. 4.   Cell separation logic and droplet charging generator abbreviated block diagram.

delay is activated which, after the appropriate time delay period between cell sensing and droplet formation, triggers an electrical pulse which is applied to the charging electrode (Fig. 2). Typical delay times between sensing and droplet formation point are 1000 μsec. The resulting electric field at the droplet formation point causes a group of about nine droplets, one of which contains the cell of interest, to be charged and then deflected by a static electric field into a collection vessel. An additional feature in the logics prevents sorting of cells which are too close in time by simply rejecting both signals. Cells are typically sorted at rates of a few hundred per second. Those cells not meeting the preset criteria can also be collected and reanalyzed. Sorted cells are deposited onto microscope slides using centrifugal cytology, counterstained, and examined microscopically.

III. Light-Scattering Properties of Cells and Cell Models

Investigations have shown that some information on cell sizing can be obtained from the light-scattering signal detected in the forward direction (0.5 → 4.0°) for reasonably spherical cells. At larger angles (4.0 → ~ 25.0°), the cellular internal structure does affect the scattering pattern. Thus, some combination of scattering signals measured at more than one angle can, in principle, provide a method for cell identification (Brunsting and Mullaney, 1972a; Salzman and Mullaney, 1974). However, much work must be done in this area before light scattering can be used reliably as a measure of cell size. The physics of the light scattering situation is not clearly defined as yet for mammalian cells. What is required is an exhaustive analysis somewhat akin to that conducted by Wyatt (1968) for bacteria. In principle, the scattering pattern contains information on the structural characteristics of the scattering object. However, the inference of these characteristics from the scattering data will require a solution to several of the difficult problems outlined below.

When a particle such as a mammalian cell is illuminated, light can be removed from the incident beam by two mechanisms: (1) absorption and (2) scattering. In the first case, the absorbed light may be reemitted as fluorescence if the cells have been stained with an appropriate fluorescent dye and there is a good match between the illuminating wavelength and the absorption spectra of the bound dye. In the second case, the presence of the particle (cell) in the light beam perturbs the beam and the incident light is spatially redistributed. This process is referred to as "light scattering" and can be due to several mechanisms. For particles $\gtrsim 2\,\mu$m illuminated with visible light, these mechanisms include (1) diffraction, (2) refraction, and (3) reflection

(Hodkinson and Greenleaves, 1963). The physical analysis of the light scattered by a cell in the size range suggested above is quite complex and can be properly approached only by exact electromagnetic theory considerations (Mullaney and Dean, 1970). Although not precisely accurate, we can dissect the scattering into these three mechanisms, which permits a physical evaluation of the scattering. This approach has been used for mammalian cells (Mullaney et al., 1969); however, it has certain limitations for the reasons given below.

In Fig. 5, a mammalian cell is modeled, to a first approximation, as a homogeneous sphere of refractive index m_2 immersed in a medium of index m_1 such that the refractive index relative to the medium is m. A ray from the incident beam undergoes transmission with refraction as well as reflection, as shown. In addition, light is diffracted about the particle, and this light is propagated within the geometrical shadow of the cell. Diffraction effects are independent of refractive index and are only a function of the silhouette presented to the incident beam. The diffraction intensity distribution (intensity as a function of scattering angle) consists of a series of lobes centered on the optical axis. Most of the diffracted energy is contained in the forward lobe, with a decreasing amount in the successive lobes. Figure 6 shows the

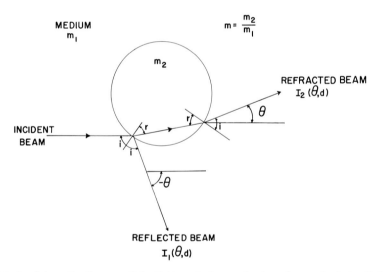

FIG. 5. Schematic diagram of the light scattering mechanisms for a spherical cell. The cell is assumed to be homogeneous with refractive index m_2 immersed in a medium of refractive index m_1 (relative refractive index $m = m_2/m_1$). Each incident ray undergoes reflection and refraction at the sphere, giving rise to a reflected beam I_1 (θ, d) and a refracted beam I_2 (θ, d). The scattering angle θ represents the angular deviation of the scattered light from the incident direction. Diffraction, which is responsible for most of the forward scattered light, is not depicted here.

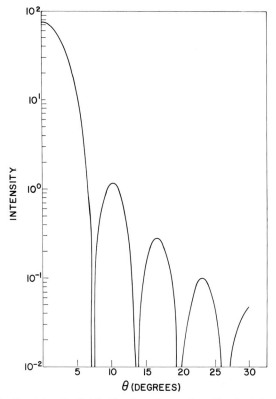

FIG. 6.   Diffraction intensity distribution for a 6-μm sphere illuminated with 6328 Å light (helium-neon laser). The vertical axis is the log of the intensity (in arbitrary units); and the horizontal axis, the scattering angle θ; $d = 6\mu$m. Sin $\theta_m = 0.7714/d$; $\theta_m = 7.4°$.

intensity distribution for a 6-μm sphere illuminated with 6328 Å light. The main forward lobe, which terminates at 7.4° for this situation, contains more than 80% of the diffracted light (Born and Wolfe, 1965). The position of the successive minima is a strong function of size, as shown in Table I. The position of the first minimum is inversely related to the diameter (i.e., the larger the particle, the less is the angular extent of the diffracted light).

Unlike diffraction, the reflection and refraction contributes are dependent upon the relative refractive index of the cell. Most viable cells have a

TABLE I

ANGLE OF FIRST DIFFRACTION MINIMUM, θ_m ($\lambda = 6328$ Å)

Particle diameter (μm)	2	4	6	8	10	12	14	16	20
θ_m (degrees)	22.8	11.2	7.4	5.6	4.4	3.6	3.0	2.8	2.2

TABLE II

INDEX OF REFRACTION FOR BIOLOGICAL SAMPLES

Sample	m_{air}	m_{water}	References
Human red blood cells	1.39	1.046	Barer (1952)
	1.40	1.05	MacRae *et al.* (1961)
Locust spermatocytes	1.40	1.05	
cytoplasm	1.37	1.03	Ross (1954)
Sea urchin eggs (*Psam-*	1.38	1.04	Mitchison and Swann (1953)
mechinus milaris)			
Snail spermatocytes, snail	1.37	1.03	Barer *et al.* (1953)
amoebocytes, *Euglena*,			
human oral epithelial cells			
Extremes observed for	1.42	1.07	
various cells	1.36	1.03	Barer (1956)
Various fixed biological	1.50	1.13	Ross (1967)
samples			
Plastics CR-39 (Columbia	1.5001	1.128	Gray (1957)
resin)			
Methyl methacrylate	1.4890	1.120	

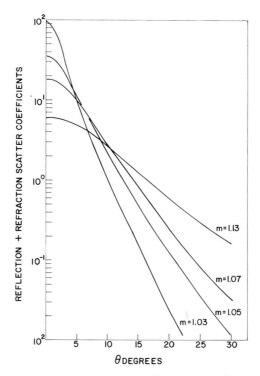

FIG. 7.   Reflection and refraction scattering coefficients (scattering intensity due to reflection and refraction divided by particle area) as a function of angle for several values of the relative refractive index. The values $1.03 < M < 1.07$ correspond to live cells; $m = 1.13$ corresponds to fixed cells.

refractive index close to that of water so that, when immersed in a buffer or other waterlike medium, the relative refractive index approaches the value of 1.00. However, if the material is fixed, the refractive index is raised somewhat; fixed cells in water have relative refractive indices similar to those of plastic spheres in water. Typical values for several cell types are given in Table II.

To compare the effects of various relative refractive index values on the reflection and refraction contributions, reflection and refraction scatter coefficients (intensity per unit area) were calculated (McK. Ellison and Peetz, 1959). These quantities are independent of cell size and serve to illustrate the refractive index effects only. Figure 7 is a plot of these quantities as a function of scattering angle for the following values of m: 1.03, 1.05, 1.07, and 1.13. The lower values are similar to those of viable cells. These functions are decreasing smoothly with increasing angle, and as m

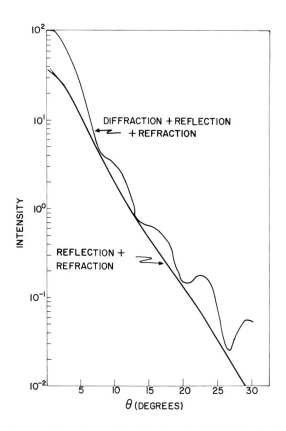

FIG. 8. Total scattered intensity as a function of angle calculated as the simple sum of the contributions shown in Figs. 6 and 7 for a 6-μm sphere (d). $\lambda = 6328$ A; $m = 1.05$.

approaches 1.00, the functions become more peaked in the forward direction. The refracted light at small angles becomes sizable when compared with the diffracted light. Most of the effect shown in Fig. 7 is due to refraction; reflection is not an important scattering mechanism for cells for scattering angles less than 30° (Mullaney and Dean, 1970).

If the three contributions are simply added together, neglecting phase relationships between them, the behavior shown in Fig. 8 is obtained. These results were calculated for a 6-μm cell-like object. The total scattering now has many of the maxima and minima significantly diminished owing to the effect of refraction and reflection. Also notice that the refraction and reflection contributions are competitive in amplitude with that due to diffraction. This becomes particularly true as m approaches 1.0, and this simple model fails here because constructive and destructive interference effects can no longer be neglected (Mullaney, 1970). However, the main point of this analysis is that the main forward lobe tends to be that part of the scattering pattern least sensitive to the refractive index of the cell and hence to the internal details. The diffraction-dominated forward lobe is mainly a function of particle size, and if pure size information is desired, measurements should be made in this region.

Following this approach, calculations on the relationship between particle size and the light scattered in the forward direction were performed using exact electromagnetic considerations (Mullaney and Dean, 1970). The results of this type of calculation, which represents an attempt to relate particle size with the scattering signal, are shown in Fig. 9. The solid curve is the result calculated by diffraction for a scattering angle of 0.5°. The solid dots represent the approximate model outlined above (diffraction plus refraction plus reflection) and the open symbols for two different exact electromagnetic theory methods. The vertical axis is the intensity and the horizontal axis the particle size parameter $\alpha = \pi d/\lambda$, where $\lambda =$ wavelength in microns and $d =$ the diameter in microns. Details of the method of computation and physical import of these theoretical studies are given elsewhere (Mullaney and Dean, 1970).

From this discussion, several important points can be derived. If a cell has properties that make it a diffracting object (relative refractive index $m = 1.10$, or absorbing), the scatter signal at this angle is related to the cell diameter by approximately $S = bd^n$ where $n \approx 3.0$; the light scattering signal is nearly proportional to cell volume. For low relative refractive indices as encountered with live cells, the scattering intensity oscillates somewhat with particle size and, in certain cases, relative size information is possible. In the case shown, $m = 1.05$, this is approximately true for $2 \times 10^3 < \alpha^3 < 2 \times 10^5$ or, at a wavelength of 6328 Å, 3 μm $< d < 11$ μm. A diffracting particle (plastic microsphere, for example) cannot always be used reliably

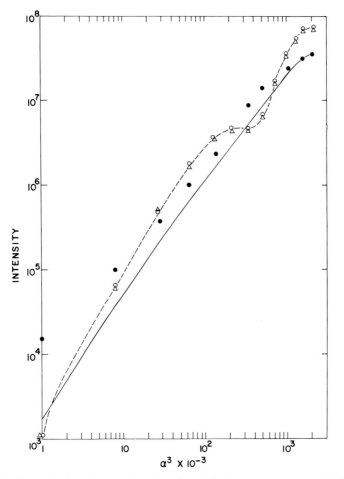

FIG. 9. Scattering intensity as a function of particle-size parameter α as calculated by several methods. Vertical axis is log intensity at scattering angle 0.5°; horizontal axis is the particle size parameter $\alpha = \pi d/\lambda$, where d = particle diameter and λ = wavelength. For $\lambda = 6328$ Å, the d range covered is 2 μm $< d <$ 20 μm in 2-μm steps. △, DAMIE; ○, DEMIE; ●, Hodkinson; $m = 1.05$. ——, Calculated by diffraction.

for calibration of an instrument for measurements on viable cells under the conditions illustrated. Published experimental data on various particles confirm these predictions (Mullaney *et al.*, 1969). The important point is that one should exercise caution when attempting to extract size information from the small-angle light scattering signal. The equation of the diffraction curve predicted by calculation in Fig. 9 has been essentially verified on the

LASL multiparameter cell analysis and sorting instrument (Steinkamp *et al.*, 1973a). Here the volume and light-scattering signal can be analyzed simultaneously on each particle in a population. The measurements were made under several optical situations, and, among other factors, the results were found to vary with shape of the incident beam and collection geometry for the small-angle detector. It has been observed previously (Hodkinson and Faulkner, 1965) that the response of aerosol counters also varies with the optical arrangement used for illumination and detection. Care should be exercised in using any light-scattering photometer for particle (cell) sizing in this respect. In the case of smaller (4.9 μm) yeast cells, it has been found that the forward scattering ($\theta \leq 13°$) obeyed quite well Mie theory predictions for transparent spheres (Latimer and Tully, 1968).

Outside the forward lobe, refractive effects become increasingly important. From Fig. 5 it is apparent that refracted rays will traverse the interior of the cell; hence, refracted light should manifest, to some degree, the internal structure of the cell. To explore this notion in a quantitative sense, the cell model outlined above was expanded to include a cell nucleus (Brunsting and Mullaney, 1972b). This new approach envisions a cell as a coated sphere, the core being the nucleus and the coating the cytoplasm. In this model, the relative refractive indices and size of core and coating can be varied. This model predicts that the coated sphere behaves as a homogeneous sphere in the forward lobe, again demonstrating that diffraction and thus gross size effects are most prominent in this angular regime. At larger angles, however, the structure in the scattering pattern becomes more complicated, and the character of the light-scattering curves will vary with changes (for example, in the ratio of nuclear diameter to whole-cell diameter). Certain cell types are suitable for comparison with this model, and experimental data have been obtained for Chinese hamster cells (line CHO) in G_1 phase and red blood cell (RBC) ghosts.

CHO cells growing in suspension are nearly spherical while in G_1 phase. The ratio of nuclear diameter to whole cell diameter is nearly constant for these cells and has a value = 0.8. In Fig. 10, the results of experimental measurements (Brunsting, 1972; Brunsting and Mullaney, 1974) on a suspension of these cells (CHO) are shown with the pattern calculated on the basis of an equivalent homogeneous sphere as well as a coated sphere. There is a very good agreement among all three curves in the forward direction. At larger angles, the homogeneous sphere model does not account for experimental results, whereas the coated sphere does, suggesting the influence of nuclear structure on the scattering. A similar study was conducted recently for RBC ghosts (Fiel and Mullaney, 1974). In this case,

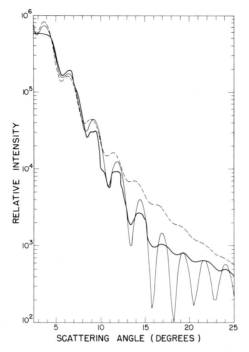

FIG. 10. Comparison of experimentally obtained light-scattering distributions for Chinese hamster cells (line CHO) in G_1 (——) with theory. The light solid curve is the result of the coated sphere model, and the dashed curve (- - -) is the result for an equivalent homogeneous sphere.

the RBC ghosts (which were essentially hollow) are best modeled as thin shells of thickness 150 Å and diameter 7.65 μm.

The evidence to date strongly suggests that gross internal details have an effect on the wide-angle scattering. This method is now being pursued as a technique for obtaining cell "signatures" with the flow system (Salzman and Mullaney, 1974).

There is much work still necessary before light-scattering properties can be used reliably for cell identification. To date, it has proved useful in studies where a clean light-scattering spectrum can be obtained, whereas the fluorescence spectrum contains a considerable amount of debris signals sufficient to obscure the true spectrum (Mullaney and West, 1973). Here one can gate on the fluorescence analysis only when a light-scattering pulse characteristic of a cell is present. This sort of gated analysis has proved useful in several applications, particularly when weakly fluorescing cells are being analyzed.

IV.   Cell Preparative Procedures for Flow-System Analysis

The successful application of flow systems for rapid, single-cell analysis is critically dependent upon preparative techniques which maintain the cells in a monodispersed state during fixation, staining, and measurement. In instances where fluorescent staining techniques are employed, the quality and specificity of cellular staining must also be evaluated to ascertain the reliability of analytical results. Automated analytical systems which are designed to perform rapid and precise measurement of individual cells cannot be totally relied upon to distinguish fluorescent cellular debris or cell clumps from properly stained single cells. Therefore, sample preparation involving both the disaggregation of tissue into single-cell entities and cell staining play an extremely important role in flow-system methodology. *Experience has shown that failure to obtain reliable and reproducible data has more often resulted from poorly prepared samples rather than faulty or defective instrumentation.* A flow system operating at peak performance can produce no better data than the input cell sample can provide.

During the past few years, considerable effort has been invested to devise adequate cell dispersal protocols and quantitative, reproducible cell staining methods which provide quality single-cell suspensions for subsequent flow instrument analysis (Kraemer *et al.*, 1972; Crissman, 1974a). Most of the material analyzed has consisted of cells grown in tissue culture; however, more recently efforts have been directed toward the adaptation of various techniques to develop protocols useful for dispersing and staining exfoliative cervical material as well as human and animal tumors.

A.   Methods for Preparing Single-Cell Suspensions

A large number of protocols of varying usefulness are available for disaggregation or dispersal of cells from portions of tissue or from monolayer cultures. For practical purposes, these can be divided into essentially three main categories: (1) enzymic, (2) chemical, and (3) mechanical. Certain protocols obviously involve combinations of these methods.

1.   ENZYMIC

Procedures involving the use of trypsin for dispersal of cells are probably the best known and most commonly employed. The enzyme is very effective in removal of adhesive factors on membranes of most cell types but has no deleterious effects on cell viability following commonly employed brief exposure. Although trypsin does not penetrate the membrane of living cells (DeLuca, 1965), it rapidly hydrolyzes the membranes of dead or degenerating cells as well as most cellular debris. Being a proteolytic enzyme, trypsin obviously has no effect on the viscous and "sticky" DNA released from dead cells which can cause cells to adhere to each other during subsequent centrifugation steps. Kraemer et al. (1971, 1972) have devised an effective method for dispersing cells from monolayer or suspension cultures (see Appendix A). It is extremely important that DNase be included in the protocol for the reasons mentioned above. Serum or medium containing serum should not be substituted for neutralization solutions.

Adapting the protocol to solid tissue involves mincing the tissue and rinsing it in Hank's balanced salt solution to remove blood cells. The tissue is then placed in a beaker containing the dispersal solution and stirred gently on a magnetic mixer at 37°C until the solution becomes opaque, the result of cells being released from the tissue. When a large quantity of single cells appears in the opaque suspension, as judged by periodic microscopic examination, the suspension is carefully decanted and poured through gauze into a chilled tube so that the larger portions of tissue left behind in the beaker can undergo repeated disaggregation following addition of fresh dispersal solution. A volume of warm neutralization solution equal to the original suspension volume is added to the tube containing the filtered cell suspension. The suspension is then pipetted about 10 times, allowed to stand at room temperature for about 5 minutes, and then placed on ice. The tubes can be accumulated and the cells centrifuged and combined at various intervals.

The above procedure obviously is an empirical approach to the problem of cell disaggregation. Time requirements for each step can be determined only by careful microscopic observation of the quantity and quality of the resulting cell suspension. Various tissues will react differently to this method, and modifications such as changes in trypsin concentration, replacement of trypsin by a different enzyme, as well as changes in pH may be necessary to obtain satisfactory results. However, repeated efforts and perseverance will lead eventually to a protocol which will become routine.

The use of collagenase in tissue dispersal techniques is presently showing considerable promise and, in many instances, has proved to be much more favorable than trypsin. Commercial preparations of crude collagenase contain clostripain, a trypsinlike activity, and other lytic enzymes. Chromatographically purified collagenase supposedly contains no clostripain or tryptic activity. Four types of crude collagenase have recently become available (Worthington), each with varying amounts of the contaminating enzymes mentioned above, and suggestions are presented for the use of each type, depending upon the type of tissue employed (Worthington catalog, 1974).

Collagenase has been reported effective for preparing isolated pancreatic islets (Lacy and Kostianovsky, 1967; Kemp et al., 1973), rat liver cells

(Seglen, 1973), beating heart cells (Vahouny *et al.*, 1970), and fat cells (Rodbell, 1961). The enzyme has also been shown effective in dispersing cells contained in vaginal and cervical samples (Romero and Horan, 1974). Although these samples contain large populations of degenerating cells, such cells must be preserved intact for cancer diagnosis, and the use of lytic enzymes such as trypsin must be avoided. The use of the various commercial forms of collagenase may prove to be quite effective, especially in combination with other enzymes, such as DNase.

Several other enzymes which have been used for cell dispersal or removal of cell aggregating factors include elastase (Phillips, 1972) and hyaluronidase alone (Pessac and Defendi, 1972) or in combination with collagenase (Müller *et al.*, 1972). A recent study on the effects of neuraminidase, trypsin, calcium, and magnesium on the aggregation of HeLa cells from suspension cultures (Deman *et al.*, 1974) showed that trypsin decreased but neuraminidase increased intercellular adhesion. Calcium ions promoted aggregation, but magnesium ions had no such effect under specified conditions.

There are many enzymes which may be better suited for a particular cell type than those suggested above; however, the adaptation of any specific enzymic protocol requires that preliminary studies be made to determine the enzyme concentration and time period of incubation which produce optimal cell dispersal and minimal cell loss.

2. CHEMICAL

There are certain instances where the use of enzymes for cell dispersal is prohibitive. Such is the case in studies where a particular biochemical component on the cell membrane may be removed by enzymic treatment. Kraemer *et al.* (1973a) have used 0.5 mM EDTA in saline GM to remove cells from tissue culture flasks for cell surface concanavalin A binding studies (see Appendix B). McCarty (1972) used the commercial nonenzymic product Mucolexx to disperse mucoid specimens. The use of tetraphenylboron (TPB) for the dissociation of liver, brain, kidney, and connective tissue has been suggested by Rapport and Howze (1966a,b,c). This compound specifically complexes potassium ions but not divalent cations. Several studies indicate that TPB inhibits cellular metabolism (Harris and Leone, 1966; Friedmann and Epstein, 1967); however, it was found also that culturing of liver cells 18 hours post-TPB treatment restores normal metabolic functions (Castanello and Gerschenson, 1970). Restoration of cellular metabolism is not generally an important consideration, since cells are usually fixed immediately following dispersal; however, the study of certain intracellular enzyme systems may indeed require concern for the dispersal protocol of choice.

3. MECHANICAL

Some form of mechanical dispersal is employed in each of the general methods discussed above. Shearing tissue with a pipette or a syringe containing a wire mesh are procedures commonly employed. Unit gravity sedimentation following preliminary dispersal is also a method that can be useful for separating single cells from debris and cell clumps. Use of homogenization or sonication procedures should be avoided if at all possible, or at least be used with extreme caution. Such techniques generally have deleterious effects on cell integrity and often result in significant cell fragmentation and cell loss.

No single dispersal technique obviously will be adequate for every cell sample, but whatever the choice of technique it must satisfy the criteria which demand minimizing clumping, cell doublets, cellular debris, cell loss, and especially any preferential loss of a given cell type. Microscopic examination of the cell sample during all phases of dispersal is a stringent requirement for assuring that these criteria are satisfied.

B. Fixation of Cells in Suspension

The "ideal" fixative is one which prevents cellular autolysis and bacterial growth, preserves all the biochemical and cellular components, but does not interfere with the specificity of subsequent staining reactions nor induce gross changes in cell morphology. Most fixatives presently in use fulfill but a few of these criteria. For instance, formalin is regarded as an all-purpose fixative for nucleic acids, proteins, and lipids; however, many carbohydrates including glycogen and some monosaccharides are not retained and leach from the cell. Alternatively, methyl and ethyl alcohol are excellent fixatives for glycogen but solubilize cellular lipids. The "ideal" fixative obviously has yet to be devised, but there are a number of excellent histochemical reviews available (Humason, 1967; Pearse, 1968; Hopwood, 1969) which suggest and prescribe the best available fixatives and protocols for specific biochemical moieties in cells. However, since most of the fixation procedures provided in these reviews are designed for pieces of tissue or for smears of cells on slides but not for cells in liquid suspension, the protocol for fixation of cells in suspension with various agents will be addressed as based on recent experiences gained with regard to compatibility with flow-system techniques.

1. PROCEDURES FOR CELL FIXATION

a. Formalin. Of all the fixatives available, formaldehyde is by far the most widely used. Formaldehyde is a gas and is commercially available as an aqueous solution (approximately 37–40%), referred to as formalin. Although it is generally overlooked, commercial formalin also contains 10–15% methanol. As a fixative, commercial formalin is generally diluted one part to nine, and thus contains about 4% formaldehyde and 1% methanol. In much of the older literature, this same concentration is also referred to as 10% formalin; however, referral to the final formaldehyde concentration leads to less confusion. Upon standing for long periods of time, commercial formalin may either polymerize to form paraformaldehyde or become oxidized into formic acid (Humason, 1967). A white precipitate indicates that polymerization has occurred. For this reason, it is recommended that small, 1-pound bottles of formalin be purchased periodically rather than large, 1-gallon bottles so that the stock remains fresh. Small amounts of the white precipitate are tolerable and can be filtered from the formalin.

A 4% formaldehyde solution in saline G (see Appendix B) has been used successfully for fixing cells in liquid suspension for subsequent staining by the Feulgen method (Kraemer et al., 1972). Cells fixed by this protocol after dispersal show little or no aggregation throughout staining and analysis. It is important that the fixative be used at neutral or slightly alkaline PH.

Use of an indicator such as phenol red and adjustment of pH with NaOH stock solution are suggested. The fixative reagents should be cold and fixation proceed at low temperatures, either on ice or in a refrigerator. Cells fixed for 12–18 hours at 4°C can stand unrefrigerated for 1–2 days before any noticeable reaggregation occurs so that transport of unrefrigerated fixed cell samples is feasible. Samples should never be frozen under any conditions.

A note of caution should be added concerning prolonged fixation of cells in formalin. Experience has shown that cells fixed in this medium for 2 weeks or more, depending on cell type, often tend to exhibit considerable nonspecific cytoplasmic staining following the Feulgen reaction. This could be due to the reaction of formaldehyde with unsaturated fatty acids (Jones, 1972). On the basis of cell analysis measurements, best results are obtained when cells are fixed no longer than 1 week in cold formalin. If cells must be stored for longer periods, it is advised that the samples be rinsed once in distilled water and transferred to 70% ethanol.

The essential feature of formaldehyde fixation is the formation of methylene bridges which form cross links between protein end groups (Pearse, 1968). Formalin fixation appears to interfere with the binding of intercalating fluorescent dyes, such as ethidium bromide (EB) and propidium iodide (PI) when used to stain cellular DNA (Crissman, 1974b). Cells fixed in a 4% formaldehyde solution for periods of 30 minutes, 1.5, 5, and 17 hours and rinsed extensively in 70% ethanol prior to staining with EB show a decrease in relative fluorescence of 52%, 54%, 59%, and 62%, respectively, as compared to cells fixed in 70% ethanol for the same time periods. It is suspected that the formaldehyde bridges in the chromatin interfere with intercalation of EB; therefore, in these instances, the use of 70% ethanol fixation is suggested.

b. Ethanol and Methanol. Cells fixed in absolute ethanol or methanol appear to have a greater tendency to reaggregate than cells fixed in formalin. However, fixation at concentrations of 50–70% in aqueous solutions alleviates problems of clumping during subsequent staining. Also, if the cell pellet is resuspended first in cold saline and then 95% cold ethanol ($-20°C$) added in small amounts with pipetting to final 70% ethanol, the initial clumping of cells is minimized (see Appendix C). A fixation time of just 30 minutes is adequate; however, cells can be stored for 1–2 weeks at $-20°C$. Ethanol can also be used to fix cells prior to staining total cellular protein or various other cellular components which are not alcohol-soluble, such as lipid. However, 70% ethanol is a poor fixative for cells to be stained by the Feulgen procedure, as cell loss and cell clumping following acid hydrolysis are excessive.

c. Acetone. Acetone seems to be the fixative favored for immunofluorescence studies of viral antigens. Cram *et al.* (1974) have used 85% acetone routinely in antigen–antibody studies of pig kidney cells (PK-15) infected with hog cholera virus using flow-system analysis techniques. Cells are suspended in cold saline, and cold acetone is added gradually with mixing to a final concentration of 85%. Considerable cell loss ensues if cells are fixed in absolute acetone. It is suspected that absolute acetone and also absolute alcohol solubilize some of the lipid components in the cell membrane, causing cells to become extremely fragile and susceptible to breakage during subsequent centrifugation.

d. Fixative Reagents To Be Avoided. Other fixatives or reagents to be avoided include acetic–alcohol (glacial acetic acid–ethanol, 1:3) which also causes excessive cell loss; picric acid containing fixatives, since this compound fluoresces and is extremely difficult to remove from the cells; and mercuric compounds, such as mercuric chloride, which form crystals that can be removed only by prolonged treatment of the cells with iodine solution.

2. CRITERIA FOR FIXATION PROTOCOLS

There is a real need for the development of fixation protocols for cells in suspension which minimize cell clumping and cell loss. Existing methods generally designed for cell preparation on slides often require certain modifications to be of value for flow-system analysis. New fixative procedures should always be tested and evaluated in cell-suspension systems in conjunction with specific staining methods, since the two protocols are most often interdependent. Finally, cells should also be examined with a fluorescence microscope prior to and following any new fixation protocol to determine whether any undesirable fluorescence has been imparted to the cells during the procedure.

V. Methods for Staining Cells in Liquid Suspension

In contrast to conventional methods for staining cells on slides, flow systems require that the cell samples be stained in liquid suspension. This involves centrifugation of the cells after each step of the staining protocol, removal of the supernatant, and resuspension of the cell pellet. Mild vortexing of the sample during addition of each subsequent solution aids in avoiding cell clumping. The choice of a particular protocol should be made with special regard to the number of steps and types of reagents employed, since numerous centrifugations in deleterious reagents can lead to excessive cell loss.

A. Choice of Fluorescent Dyes

The dyes used in the staining procedure must also meet certain criteria. Since lasers used in existing flow systems have only specific wavelength lines available, the dye of choice must be excitable near optimum at one of the given wavelengths. When excited, quantitative dyes should emit strong fluorescence emission intensities that are proportional to the amount of bound biochemical component. Instruments such as the Aminco-Bowman spectrophotofluorometer (American Instrument Co., Silver Spring, Maryland) are extremely useful for determining excitation and emission spectra of fluorescent dyes and for evaluating the potentiality of a dye in a given flow system. Other criteria dictate that the dye should bind tenaciously to its specific cellular component and should not leach out into the suspension fluid. Many biological dyes contain various contaminants, and detection, removal, and occurrence of such impurities have been covered extensively in a recent review by Horobin (1969).

B. Quantitative Staining Procedures for DNA

1. FEULGEN PROCEDURE

The Feulgen (nucleal) reaction is probably the most widely used protocol for quantitative staining of cellular DNA. The procedure involves removal of the purines from the DNA by acid hydrolysis, exposing the aldehydes of deoxyribose sugars which, in turn, react with an aldehyde-specific agent (Deitch, 1966). Proof that an aldehyde function is responsible for the reaction comes from the observation that aldehyde blocking reagents negate the Feulgen reaction when applied after acid hydrolysis (Lessler, 1951; Longley, 1952). RNA does not contribute to cell staining, since it is removed rapidly from the cells during acid hydrolysis (Ely and Ross, 1949). Deitch (1966) and Pearse (1968) have provided excellent reviews of the Feulgen procedure.

Feulgen staining of cells in liquid suspension using the fluorescent dye auramine 0 has been described in detail by Trujillo and Van Dilla (1972). This procedure has been modified more recently (see Appendix B), and the dye acriflavine has been substituted for auramine 0 (Kraemer et al., 1972; Crissman, 1974b). Acriflavine was first utilized by Ornstein et al. (1957) to demonstrate substances with vicinal glycol grouping (e.g., glycoprotein, glycogen) following oxidation of tissue sections in periodic acid. Kasten (1959) recommended the use of acriflavine as a suitable fluorescent dye for use in the Feulgen reaction; however, there is some doubt as to whether the mechanism of staining is the same as in the case of fuchsin-sulfurous acid. Pearse (1968) recommended that these fluorescent dyes in the presence of SO_2 and HCl be referred to as pseudo-Schiff reagents. Regardless of the precise mechanism of staining, Culling and Vassar (1961) demonstrated the specificity of acriflavine for DNA when used in the conventional Feulgen reaction. Sprenger et al. (1971) have also used acriflavine to stain cells in suspension. Conventional hydrolysis of cells in 1 N HCl at 60° C has proved to be undesirable for cells in suspension, and more reproducible results have been obtained when higher normalities of HCl were used at room temperature. Deitch et al. (1968) and Fand (1970) have previously shown similar results for hydrolysis of cells on slides. It was also shown (Deitch et al., 1968) that optimal hydrolysis time varied with the fixative employed. Maximal fluorescence intensity is achieved when formalin-fixed cells in suspension are hydrolyzed for 20 minutes in 4 N HCl at room temperature (Trujillo and Van Dilla, 1972). No noticeable effect occurs with further hydrolysis up to at least 1 hour.

Acriflavine-stained cells generally have been about 10 times brighter than those stained with auramine 0, improving the coefficients of variation by a factor of about two. Cell samples routinely examined by fluorescence microscopy exhibit golden yellow nuclei, together with very faint nonspecific cytoplasmic staining which fluoresces pale green. Chromosomes in mitotic figures also fluoresce golden yellow, with almost a complete absence of greenish fluorescence. Cells treated with DNase as well as non-hydrolyzed control cells exhibit only a dim fluorescence. Microscopic examination also provides a visual check for monodispersity.

Other fluorescent dyes which have been tested for use as possible Feulgen-DNA stains have been flavophosphine N (Benzoflavine, Chroma), coriphosphine, acridine yellow, and phenosafranine. Of these, only flavophosphine N proved as satisfactory as acriflavine when used in the LASL flow system.

A typical Feulgen-DNA distribution of WI38 cells stained with acriflavine and analyzed using the flow system with the laser tuned at 488 nm is shown in Fig. 11. The G_1 and G_2 + M peaks are designated and represent cells in these phases of the cell cycle. Lying between the two peaks are cells in the DNA synthetic phase of the cell cycle. The expanded portion of the spectrum shows the tetraploid cells which exist as a small percentage (about 1%) in most early-passage WI38 cell populations. Note that the relative positions of G_1, G_2 + M, and tetraploid peaks are in the expected 1:2:4 ratio, respectively, indicating the good linearity of the system.

It can also be observed that there are no extraneous peaks in the spectra, indicating the absence of excessive cell clumps. There is no indication either of fluorescent cellular debris as noted by the absence of structure in the spectrum to the left of the G_1 peak. The broadening of the G_1 and G_2 + M portions of the curve is due largely to variations in staining and instrumentation (Holm and Cram, 1973).

2. INTERCALATING DYES

Ethidium bromide (EB) is an orange-red fluorescent compound which selectively and quantitatively intercalates into the double-stranded regions of nucleic acids (LePecq and Paoletti, 1967). LePecq (1971) has stated that there are two binding complexes formed between EB and nucleic acids. Complex I, which occurs at high salt concentration, increases the quantum efficiency of the dye by about 25% and likewise increases the fluorescence intensity by approximately 50–100 times over that of unbound EB. The EB/polynucleotide interaction is independent of base composition or molecular weight and occurs in a proportionality of one fluorescent binding site to five nucleotides. A second EB/nucleic acid interaction (complex II) occurs only at low ionic strength and results in no increase in EB quantum efficiency. This nucleic acid/EB complex forms in a pH range of 4–10 and in NaCl concentrations not less than 10^{-2} M. Heavy metals such as mercury, copper, and silver inhibit the reaction, and anions such as ClO_4^- precipitate EB (see review by LePecq, 1971). Ethidium bromide also interacts more strongly with deoxynucleoprotein (DNP) lacking histone fraction f_1 than with whole DNP (Angerer and Moudrianakis, 1972).

Ethidium bromide has been used as a quantitative stain for cellular DNA after treatment of fixed cells with RNase (Dittrich and Göhde, 1969). This technique is simple and rapid, requiring only about 2 hours (see Appendix

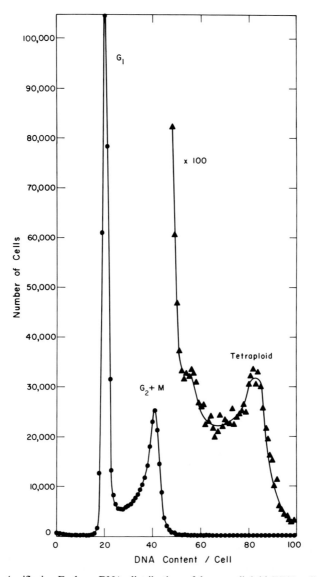

FIG. 11. Acriflavine-Feulgen DNA distribution of human diploid WI38 cells (passage 25). A portion of the distribution is expanded (100 ×) to demonstrate a small proportion (< 1%) of tetraploid cells in the population.

C). Since no HCl hydrolysis is employed as in the Feulgen procedure, cell loss is also minimized.

Cells stained with EB and analyzed in the flow system using the 488-nm laser line for excitation yield DNA distributions comparable to acriflavine-stained cells. However, EB-stained cells are at least five times brighter than cells stained by the acriflavine-Feulgen procedure. Ethidium bromide-

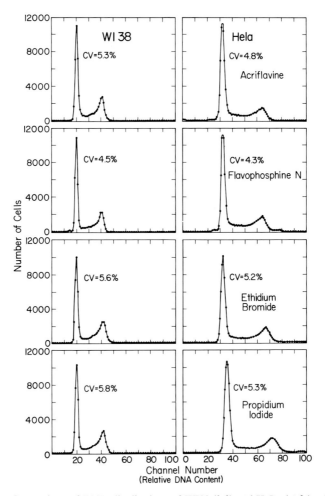

Fig. 12.   Comparison of DNA distributions of WI38 (*left*) and HeLa (*right*) cells stained by the Feulgen procedure (acriflavine, flavophosphine N) or with intercalating dyes (ethidium bromide, propidium iodide). The coefficient of variations (CV = 100 × standard deviation divided by the mean) for the G_1 populations are indicated.

stained cells stored in distilled water at refrigerated temperatures for at least one week yield acceptable fluorescent DNA distributions upon analysis. The main disadvantage of the EB protocol, as mentioned previously, is the occasional and inexplicable induction of clumps which occurs during the staining procedure after 70% ethanol fixation.

Propidium iodide (PI) is an analog of ethidium bromide and apparently possesses similar binding properties for reacting with double-stranded nucleic acids (Hudson *et al.*, 1969). The staining protocol for DNA is similar to that for EB. Propidium iodide bound to cellular DNA has an emission peak 10 nm longer than EB and, for this reason, provides better color separation when used in two-color staining involving green-yellow fluorescing compounds, such as fluorescein isothiocyanate (FITC) (Crissman and Steinkamp, 1973).

The fluorescent DNA distributions of WI38 and HeLa cells stained for DNA by either the Feulgen procedure (acriflavine, flavophosphine N) or with intercalating dyes (ethidium bromide, propidium iodide) are shown in Fig. 12. It is apparent that the distributions are all strikingly similar irrespective of the dye or staining protocol employed. The results obtained with the intercalating dyes not only demonstrate alternative methods for staining DNA but also substantiate by an independent method the accuracy of the Feulgen procedure for assessing the relative DNA content of cells.

3. MITHRAMYCIN TECHNIQUE

An extremely rapid method requiring only 20 minutes has been devised recently for staining DNA using the compound mithramycin (Crissman and Tobey, 1974). Mithramycin is highly selective for DNA and does not interact significantly with RNA (Ward *et al.*, 1965). With the exception of the requirement of magnesium, the mechanism of mithramycin-DNA binding is believed similar to that for actinomycin. No hydrolysis or RNase treatment is required in the staining protocol. Cells unfixed prior to staining or ethanol-fixed cells treated for 20 minutes with the staining solution (100 μg/ml mithramycin in 25% ethanol and 15 mM MgCl$_2$) yield DNA distributions comparable to those obtained with Feulgen-stained cells (see Fig. 13). Prior DNase treatment negates straining, while RNase has no effect on cellular staining.

Mithramycin has a maximum excitation peak at about 395 nm and an emission peak at 530 nm. Since excitation of mithramycin-stained cells in Fig. 13 was made suboptimally at 457 nm, the lowest efficient wavelength available in the argon-ion laser, the coefficient of variation (CV = 100 × standard deviation divided by the mean) for the cell population was somewhat higher than that of the acriflavine-stained cells. However, the fraction of cells in G$_1$, S, and G$_2$ + M obtained by computer-fit analysis (Dean and

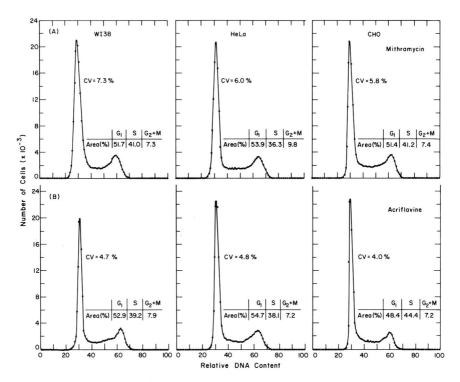

FIG. 13. Comparison of DNA distributions of WI38, HeLa, and CHO cells stained with mithramycin (A) or by the acriflavine-Feulgen procedure (B). The percentage of cells in the various phases of the cell cycle as well as the coefficients of variation (CV) for the G_1 peaks were obtained by computer-fit analysis.

Jett, 1974) compared well. The mithramycin technique allows cell-cycle analysis to be performed in just 20 minutes and permits the constant monitoring of cell populations throughout the course of an experiment.

4. IMPLICATION FOR ACRIDINE ORANGE STAINING

Use of the metachromatic dye acridine orange (AO) as a quantitative vital stain for DNA and RNA in cells has been proposed by a number of investigators (Rigler, 1969; West, 1969). However, Kasten (1967) warned against such premature assumptions without proper controls for assuring DNA and RNA specificity of staining. Attempts to use AO supravitally at concentrations of 10^{-6} g/ml or less, at which cells exhibit only green fluorescence, have failed to yield DNA distributions similar to those obtained for cells stained by the Feulgen procedures or by the intercalating dyes mentioned above (Tobey, 1973). Cellular AO distributions are more often strikingly similar to cell volume distributions. In Fig. 14 are AO green

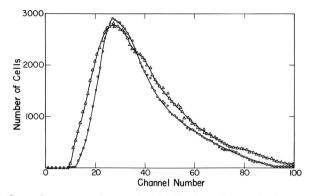

FIG. 14. Green fluorescence (O——O) and cell volume (△——△) distributions of HeLa cells supravitally stained with acridine orange (10^{-6} gm/ml in phosphate-buffered saline). Analysis of only the green fluorescence was achieved through the use of appropriate color filters.

fluorescence and cell volume frequency distributions for HeLa cells. This figure demonstrates a good, but not necessarily direct, correlation between green fluorescence of acridine orange and cell volume. Microscopic examination of AO-stained cells reveals significant green fluorescence over the entire cell, indicating the absence of nuclear specificity. Mullaney and West (1973) have stained CHO cells with AO at concentrations as low as 10^{-9} M and analyzed light scattering and green fluorescence from cells to obtain a higher signal-to-noise ratio. Small-angle light scattering which can be related to particle size also shows a fluorescence-to-cell size correlation.

It appears that, at least for the present, AO cannot be relied upon to serve as a quantitative stain for DNA or RNA in cells even though the stain seems to react in a quantitative manner with nucleic acids in solution (Rigler, 1966; Ichimura et al., 1971). However, use of AO as a vital stain does appear to have application in studies involving differential analysis of human leukocytes. Adams and Kamentsky (1971) demonstrated that various human leukocyte types in whole blood could be resolved into three distinct populations on the basis of red fluorescence after staining in AO (10^{-6} gm/ml). Red blood cells do not appear to stain significantly at this concentration. More recently Steinkamp et al. (1973b) confirmed these findings by electronically sorting cells stained with AO from each of the three populations.

Another interesting application employing AO staining and electron microscopy (EM) techniques has been demonstrated recently by Frenster (1971, 1972). Human lymphocytes supravitally stained with AO (10^{-3} M), fixed briefly in glutaraldehyde, and then treated with DNase (1 mg/ml), show an electron density reaction product as a result of the interaction of AO with DNA binding sites specifically within the extended euchromatin of the living cell nucleus (Frenster, 1971). Use of AO as a chromatin probe should have important implications for the study of gene activation and repression.

Other recent uses of AO have been in studies of chromatin thermal stability of normal and neoplastic cells (Alvarez, 1973) and activation of nuclear chromatin (Smets, 1973); however, neither of these studies demonstrated DNA or RNA specificity of AO staining. Furthermore, the study by Smets (1973) indicated that cells treated with both DNase and RNase still bound 20% of the dye as compared to RNase-treated cells.

5. STAINING ARTIFACTS

If the dyes used for cell staining are excited near optimal, then the coefficient of variation (CV) of the G_1 and $G_2 + M$ peaks, in most cases, should be relatively small. A fluorescence distribution with extremely large CV values should lead to suspicion of excessive nonspecific cytoplasmic staining. Routine examination of stained cell samples under the fluorescence microscope prior to flow analysis should be made to determine the extent of any staining artifacts as well as the presence of fluorescent debris and cell clumps. In many instances, such staining artifacts can be alleviated either by avoiding prolonged fixation or by rinsing cells extensively after staining until the supernatant is clear. However, in some cell types, Feulgen reagents do bind the various cytoplasmic aldehydes, referred to as "plasmals," even in the absence of acid hydrolysis. Thus, the reaction leading to this phenomenon is commonly referred to as the "plasmal reaction" and occurs particularly in cell types having acetal phosphatides commonly associated with ester phosphatides (see review by Pearse, 1968). Brain and fat tissues which have a high lipid content are quite susceptible to the plasmal reaction. Pearse (1968) suggested the use of various aldehyde blocking agents prior to acid hydrolysis to aid in minimizing the plasmal reaction. However, many of these compounds are also fluorescent, and care should be exercised in using these reagents.

Adams and Bayliss (1971a) and Elleder and Lujda (1971) have shown recently that the intensity of the plasmal reaction is substantially reduced when the final rinse is made with 3 N HCl. However, there was considerable controversy concerning the length of time the rinse or wash was applied and resulting reduction in intensity of the reaction. Adams and Bayliss (1971b) found that the plasmal reaction was more substantially reduced in rat brain tissue after a 10-minute wash compared to a 1-minute rinse.

6. CRITERIA FOR QUANTITATIVE DNA-SPECIFIC STAINING

A number of requirements must be satisfied before assuming staining specificity for cellular DNA. First, the fluorescent dye used in a given protocol should be highly specific for DNA, and the fluorescence intensity of the dye must be proportional to the amount of bound DNA. Microscopic examination of stained cells should show nuclear specificity with minimal cytoplasmic fluorescence, if any. Treatment of cells with DNase should negate nuclear staining, while treatment with RNase should not change significantly the fluorescence intensity of staining. It should be noted that enzymic removal of DNA or RNA from formalin-fixed cells is often extremely difficult; therefore, cells should be first fixed in 70% ethanol, enzymically treated, rinsed, and postfixed in formalin.

Other tests for DNA specificity include the near absence of Feulgen staining with omission of acid hydrolysis. The ratio of the mode of the $G_2 + M$ peak to the G_1 peak should be near a factor of two, and the fraction of cells in the S phase as determined by computer-fit analysis of the DNA distributions should be in good agreement with results obtained with thymidine-^{3}H pulse-labeling and conventional autoradiographic techniques. Likewise, the percentage of cells in G_1 and $G_2 + M$ should be comparable to those proportions as determined by biochemical analysis (Puck and Steffen, 1963). The degree of accuracy of experimental results will be proportional to the degree to which the above critieria are satisfied. Thus, studies such as those involving cell-cycle analysis must begin with the critical considerations outlined above. If the fluorescence distributions do not truly reflect the proportionality of cells in the various phases of the cell cycle, then the best designed experiments can yield only results that lead to false and erroneous conclusions.

C. Procedure for Polysaccharide Staining

The periodic acid–Schiff (PAS) reaction has long been used for staining polysaccharides, mucopolysaccharides, and mucoproteins (Humason, 1967; Pearse, 1968). The procedure is quite similar to the Feulgen procedure except that periodic acid is used in place of HCl in the protocol. Periodic acid is an oxidant that breaks the C–C bonds in vicinal glycol groups, converting them to dialdehydes. The Schiff reagent is then reacted with these aldehyde groups as in the conventional Feulgen procedure. The DNA is not affected by periodic acid oxidation and does not react in this procedure. Line CHO cells stained by the PAS reaction yield fluorescence distributions quite similar to the cell volume distributions, indicating a close correlation between the increase in cell size and accumulation of cellular polysaccharide materials (H. A. Crissman, unpublished observations). The presence and accumulation of acid mucopolysaccharides in cells are of diagnostic value in certain disease conditions (Cleland, 1970).

D.   Protein Staining Techniques

Many fluorescent dyes specific for staining proteins have been described (Nairn, 1969; Udenfriend, 1969). Among the most commonly used is the compound fluorescein isothiocyanate (FITC). When excited, the dye emits a greenish yellow fluorescence. Maximum excitation and emission peaks lie at about 490 and 530 nm, respectively. Dittrich *et al.* (1971) have demonstrated the usefulness of the dye for staining protein in cells in liquid suspension, and Crissman and Steinkamp (1973) have validated the technique by demonstrating a nearly direct correlation between cellular fluorescence and cell volume. A number of protein dyes have been investigated and evaluated recently as to their usefulness in flow-system analysis

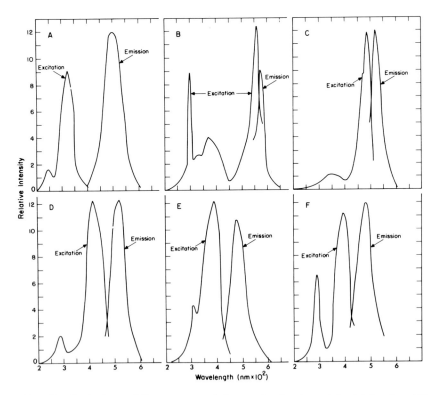

FIG. 15.   Excitation and fluorescence emission spectra for six protein staining dyes: (A) dansyl; (B) rhodamine B isothiocyanate (RITC); (C) fluorescein isothiocyanate (FITC); (D) brilliant sulfaflavine; (E) 8-anilino-1-naphthalene sulfonic acid (ANSA); and (F) fluorescamine. The spectra were obtained using the Aminco-Bowman spectrophotofluorometer (American Instrument Co., Silver Spring, Maryland).

(Freeman and Crissman, 1974). Dyes examined included FITC, rhodamine B isothiocyanate (RITC), ANSA, dansyl chloride, brilliant sulfaflavine, and fluorescamine. Excitation and emission spectra for these compounds are provided in Fig. 15, and the fluorescence distributions for WI38 cells stained with each of the above dyes, except brilliant sulfaflavine, are shown in Fig. 16 along with a cell volume distribution. Evaluation of the dyes and/or staining techniques was based on the following criteria: (1) agreement between fluorescence and cell volume distributions; (2) compatibility of staining protocol with cells in suspension; (3) minimal cell loss; (4) dye brightness when excited at existing laser wavelengths; and (5) tenacity of dye-protein binding. Based on these criteria, both FITC and RITC proved immediately useful for flow-systems analysis.

Fluorescamine (Udenfriend et al., 1972) is nonfluorescent until bound to primary amino groups. The unreacted portion is autohydrolyzed in several seconds in aqueous solution. The dye/protein interaction time is less than 1 second. The compound reacted rapidly with cellular protein; however, because the excitation maximum of fluorescamine is about 395 nm, the fluorescence signal was rather low when excited at 457 nm. The dye ANSA was also found to be unsatisfactory for a similar reason. The protocol employing dansyl chloride in 95% ethanol saturated with $NaHCO_3$ (Rossilet and Ruch, 1968) caused excessive cell loss, and brilliant sulfaflavine leached significantly from cells into the suspension medium with time.

VI.   DNA Content Measurements of Diploid and Heteroploid Cells

A.   Variations in DNA Content Resulting from Abnormal Mitoses

Chromosomal nondisjunction is a well-known biological phenomenon which introduces DNA content variability by yielding daughter cells with disproportionate chromosome numbers and thus unequal DNA contents. Feulgen-DNA distributions of CHO cell populations undergoing Colcemid-induced nondisjunction (Kato and Yoshida, 1970) are shown in Fig. 17. The increasing widening at the base of the G_1 peak indicates the presence of abnormal cells having more or less DNA than the modal value for the population, an observation that is in good agreement with expected results. This illustration demonstrates the sensitivity of flow analysis for determining DNA content dispersions in cell populations.

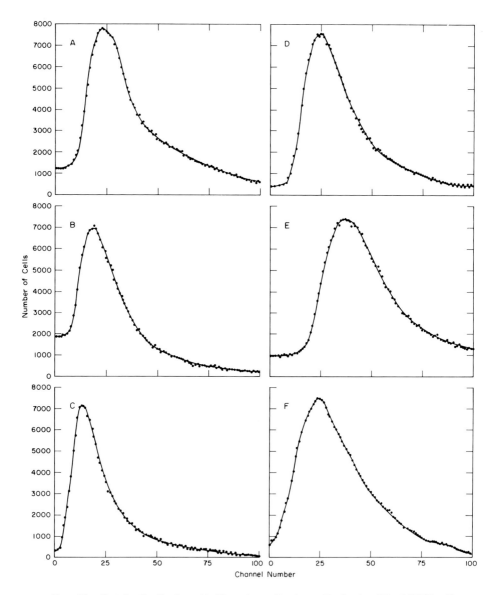

FIG. 16. Protein distributions (A–E) and a cell volume distribution (F) of WI38 cells grown on monolayer, fixed 30 minutes in 70% ethanol, and stained with various dyes: (A) rhodamine B isothiocyanate (RITC); (B) fluorescein isothiocyanate (FITC); (C) dansyl; (D) 8-anilino-1-naphthalene sulfonic acid (ANSA); (E) fluorescamine. The cell volume measurement (F) was made on unfixed cells. The concentrations used for RITC (A) and FITC (B) staining were 0.10 mg/ml and 0.03 mg/ml, respectively, in 0.5 M sodium bicarbonate. Analysis was performed using the 488-nm laser line for excitation.

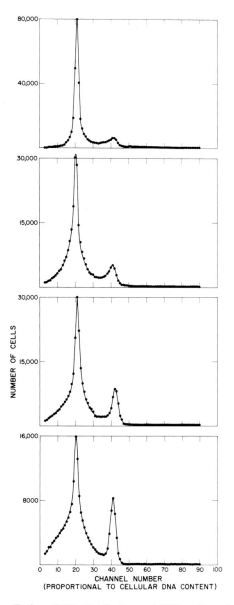

FIG. 17. Acriflavine-Feulgen DNA distributions of CHO cells undergoing progressive Colcemid-induced nondisjunction. Cell cultures were blocked with 0.06 μg/ml Colcemid; mitotic cells were selectively dislodged, reversed in fresh medium for 4 hours, collected, and stained. The top three distributions represent mitotic cells collected from the same monolayers at 2, 4, and 6 hours after addition of Colcemid. In the bottom panel, distribution is from a different monolayer blocked for 6 hours continuously, shaken, and then reversed as above.

B.   DNA Content and Chromosome Number Analysis

Comparisons of the DNA distributions and chromosome numbers in diploid and heteroploid cell lines have shown that karyotypic variability in heteroploid cell populations did not result in a proportionate variability of the DNA contents of these populations (Kraemer *et al.*, 1971, 1972,

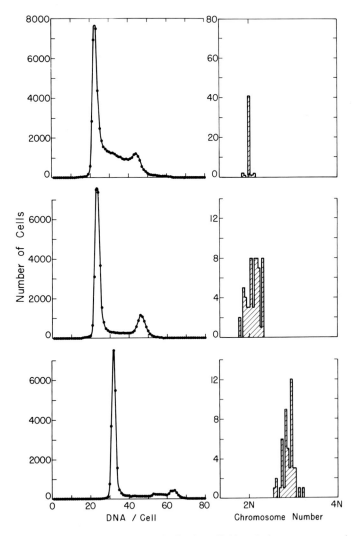

FIG. 18.   Acriflavine-Feulgen DNA distributions (left) and chromosome number histograms (right) of cultured human diploid and heteroploid cells: *top*, WI-38, diploid; *middle*, human melanoma, low heteroploid; *bottom*, HeLa, high heteroploid.

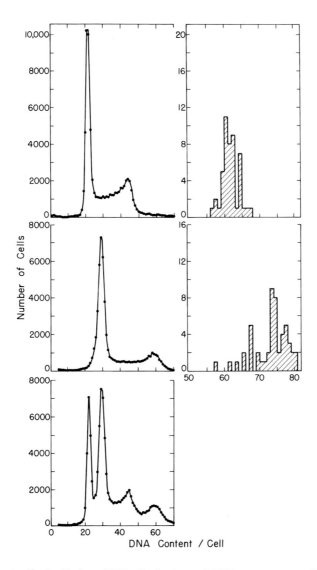

FIG. 19. Acriflavine-Feulgen DNA distributions of DNA content per cell (*left*) and chromosome number histograms (*right*) of two HeLa sublines: *top*, HeLa 4; *middle*, HeLa 229. The third DNA distribution (*bottom*) was obtained from mixed populations of HeLa 4 and HeLa 229 which were stained and analyzed together to eliminate any potential preparative differences.

1973b). Figure 18 shows a significant dispersion in chromosome number of the human heteroploid cell lines as compared to diploid cells; however, the DNA content dispersion as expressed in terms of the G_1 coefficient of variation (%) was only 4.9, 3.2, and 4.8 for the diploid WI38, heteroploid HeLa, and human melanoma, respectively. If the chromosome number dispersion were truly reflective of variation in DNA contents of heteroploid cells, then results similar to the nondisjunctive experiment above should have been observed. It may be noted that the relative DNA content of the HeLa cells is indeed higher than either WI38 or human melanoma populations which are quite similar. However, despite the higher DNA content of HeLa cells, the *variability* in DNA content from cell-to-cell is no greater than found in diploid human cells. Chromosome number is more variable, leading to the conclusion that *number* variability is a packaging defect rather than a reflection of variable DNA content. More details on the consequences of this phenomenon are given elsewhere (Kraemer *et al.*, 1971, 1972, 1973b).

Studies by Kraemer *et al.* (1973b) have shown that HeLa sublines, carried separately in different laboratories, displayed discernible differences in both chromosome number and relative DNA content. Figure 19 demonstrates this phenomenon for the two HeLa populations. It can be noted that both lines show significant perimodal dispersion of chromosome number and yet constancy of cellular DNA content. Differences in relative DNA contents are apparent.

VII.   Cell-Cycle Analysis

In the past, analysis of cell-cycle composition has relied largely on biochemical parameters such as the average DNA content of a given cell population as determined by ultraviolet absorption measurements or by the diphenylamine technique. Other methods used include the use of thymidine-^{3}H in autoradiographic methods for determining the fraction of the population synthesizing DNA. Application of high-speed cell analysis offers a rapid, simplified alternative for determining the cell-cycle distributions of all cells in a given population based upon population relative DNA contents. Flow-analysis techniques do not rely upon cell-cycle progression parameters and are, therefore, extremely useful for studying cell populations undergoing experimental perturbations such as those induced by drugs, radiation, and other causes.

A.   Life-Cycle Analysis by Computer-Fitting Techniques

Efforts to deduce the fraction of cells in G_1, S, and G_2 + M from cellular DNA distributions have led to the development of a computer-fit program (Dean and Jett, 1974) which readily deduces the fraction of cells in each phase of the cell cycle. The computer program was validated by comparison of results obtained by biochemical and flow-system techniques (Crissman, *et al.*, 1974). Figure 20 shows a computer fit to the data points provided by flow analysis and a comparison of the percentage of cells in S phase obtained by this method, as well as percentages derived from pulse-labeling with thymidine-^{3}H and autoradiographic techniques. It is readily apparent that the agreement between the two methods is excellent.

B.   Analysis of Synchronized Cell Populations

Other studies also support and validate flow-system methods for accurate and reliable analysis of the cell cycle. Figure 21 shows that cells synchro-

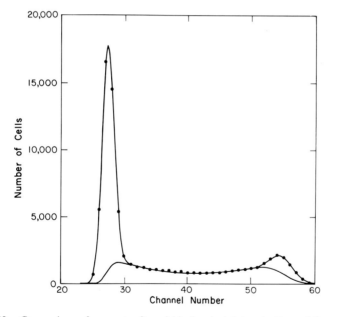

FIG. 20. Comparison of computer-fit and biochemical determinations of the percentage of cells in S phase. The fraction of cells in S phase was obtained either by computer fit of acriflavine-Feulgen DNA distributions (FMF) of CHO cells or by conventional autoradiographic analysis of 10 replicates pulse-labeled for 15 minutes with thymidine-^{3}H. Percentage of cells in S: pulse label, 36.7 ± 2.1; FMF, 37.2 ± 1.0.

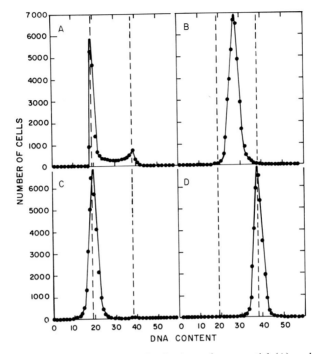

FIG. 21. Acriflavine-Feulgen DNA distributions of exponential (A) and synchronized populations of CHO cells. The dashed lines represent relative G_1 and $G_2 + M$ positions in the distributions with S phase lying essentially between the dashed lines. (B) Double thymidine blockade; (C) 30 hours in isoleucine-deficient medium; (D) mitotic selection.

nized in various phases of the cell cycle by biochemical techniques [e.g., yielding cells in isoleucine deprivation G_1 arrest (Tobey and Ley, 1971); double thymidine blockade, causing accumulation of cells in early S (Galavazi and Bootsma, 1966; Bostock et al., 1971; Tobey et al., 1972)] or mechanical methods [e.g., mitotic selection (Terasima and Tolmach, 1963; Tobey et al., 1967)] yield DNA distributions in good agreement with results predicted by independent biochemical methods.

A recent report (Haag, 1973) has suggested that dense chromatin such as that found in mitotic cells interferes with quantitative Feulgen staining and/or fluorescence analysis. That study showed that the acriflavine fluorescence of stained mitotic cells was decreased 15–26% from the expected G_2 DNA content values. In contrast to those findings, our data (Fig. 21) clearly show the mitotic cells at twice the value for G_1 cells, indicating no problems in either staining or analysis as described here. Only in instances where cells were Colcemid-arrested in mitosis for prolonged time periods have results been obtained (L. S. Cram and R. A. Tobey, unpublished observation) similar to those reported by Haag (1973). In support of the findings presented here, Fujita (1973), using conventional Feulgen staining (pararosaniline–Schiff) and cytofluorometry, found that stained small and large neurons of the human cerebellum had similar Feulgen-DNA values, irrespective of nuclear size or pattern of chromatin distribution.

Flow analysis of cell populations synchronized by the three protocols described in Fig. 21 revealed that, after synchronization, a significant fraction of each population was unable to resume cycle progression (Tobey et al., 1972). The term "traverse perturbation index" has been used to define the fraction of cells converted to a slowly progressing or nontraversing state due to experimental manipulation. Examination of G_1, early S phase, and mitotic synchrony induction protocol revealed traverse perturbation indices of 12, 17, and 5%, respectively. These findings, which have important implications for studies employing these synchrony protocols, could not have been obtained readily by conventional biochemical techniques.

C. Effects of X-Irradiation

Figure 22 illustrates flow-analysis confirmation of the G_2 accumulation of cells (e.g., division delay) after exposure of cells to X-irradiation. After exposure to 300 rad of X-irradiation, a progressive decrease in the fraction of cells in the G_1 and S phases occurred; cell division ceased for over 8 hours, as indicated both by cell count and DNA distribution. The resumption of division could be easily followed with flow analysis. It may be noted from the control distributions that the radiosensitive mutant of L5178Y cells used in these experiments has a relatively short G_1 period in comparison to the CHO cell.

D. Simultaneous Analysis of DNA, Protein, and Cell Volume

Double-staining of cells for both DNA and protein can be accomplished by using two fluorescent dyes having similar excitation but differing and separable emission spectra. Dittrich et al. (1971) have used ethidium bromide and fluorescein isothiocyanate for staining cellular DNA and protein, respectively. Ethidium bromide (EB) and propidium iodide (PI) fluoresce orange-red; and fluorescein isothiocyanate (FITC), a greenish yellow. Since PI bound to cells has an emission maximum near 590 nm compared to 580 nm for EB and 530 nm for FITC, better spectral separation is obtained using the PI/FITC combination (Crissman and Steinkamp, 1973). Although some overlapping of the two colors occurs, adequate resolution is obtained by the use of appropriate color filters. The protocol involves staining of cells first with PI and then with FITC by techniques mentioned earlier.

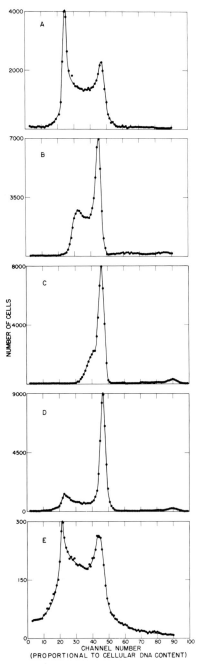

FIG. 22. Acriflavine-Feulgen DNA distributions of L5178Y (radiation-sensitive mutant) at various times after X-irradiation (300 rad): (A) control; (B) 4.5 hours; (C) 7.5 hours; (D) 17 hours; (E) 37.5 hours. The small but significant radiation-induced tetraploid population is vividly displayed.

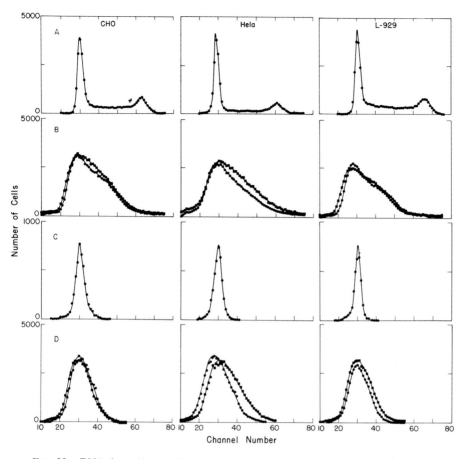

FIG. 23. DNA (row A), red fluorescence; protein (B, green-fluorescence, ■) and cell volume (B, ●) distributions (rows A and B) for CHO (*left*), HeLa (*middle*), and L-929 (*right*). cells. Cells were doubly stained for DNA and protein with propidium iodide and fluorescein isothiocyanate, respectively (Crissman and Steinkamp, 1973). Protein-to-cell colume ratios (C) were obtained by utilizing both the green fluorescence and cell volume signals from each cell. DNA-to-protein ratios (D) were obtained similarly by analysis of both the red and green fluorescence signals for each cell; ●, DNA/cell volume; ■, DNA/protein.

Figure 23 shows pulse-amplitude distributions obtained from DNA, protein, and cell volume measurements made on PI/FITC-stained Chinese hamster, HeLa, and L-929 cells. DNA distributions (Fig. 23A) were obtained by analyzing only red fluorescence. The relative protein content and cell volume distributions (Fig. 23B) were obtained by measuring either green fluorescence or volume signals separately. Ratios of protein content-

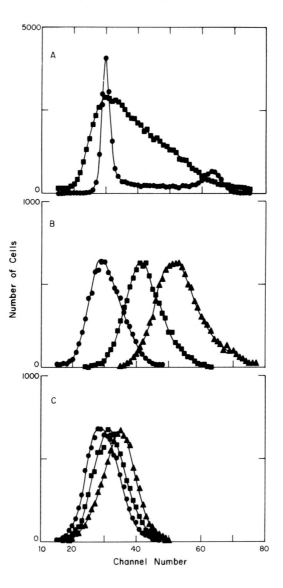

Fɪɢ. 24.   Gated single-parameter analysis of propidium iodide/fluorescein isothiocyanate-stained CHO cells. DNA (●) and protein (■) distributions (A) of stained CHO cells were obtained by measuring either the red or green fluorescence signal separately. Protein distributions (B) and DNA-to-protein ratios (C) for the G_1 (●), S (■), and G_2 + M (▲) phases of the cell cycle were obtained by electronic gated analysis.

to-cell volume (Fig. 23C) were obtained by processing simultaneously the green fluorescence and cell volume signal in a manner to form a ratio signal for each individual cell. The ratio histograms for protein-to-cell volume demonstrate vividly a high degree of correlation between these two cellular properties. In comparison, the ratios of both DNA-to-protein and DNA-to-cell volume (Fig. 23D) are not as well correlated as protein and cell volume.

Figure 24A shows frequency distributions of DNA and protein for CHO cells obtained as described in Fig. 23. Figures 24B and 24C represent the protein content distributions and DNA-to-protein ratios, respectively, for populations in G_1, S, and G_2 + M obtained by gated single-parameter analysis techniques. This method is extremely useful for selective and quantitative analysis of biochemical processes occurring at specific phases of the cell cycle.

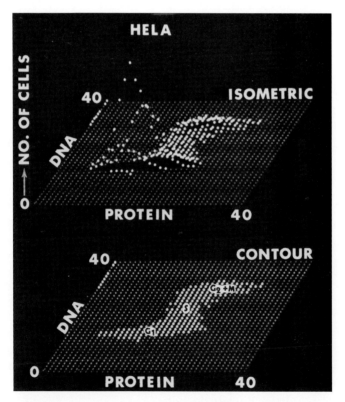

Fig. 25. Complete two-parameter analysis display of DNA and protein content of stained HeLa cells. The top figure is an isometric display which better shows cell density in particular regions. The lower figure, a contour display of the same distribution, is labeled to show various regions of the cell cycle.

Complete two-parameter cell analysis can be performed also by using a dual parameter, multichannel, pulse-height analyzer. Figure 25 shows an example of the additional fine structure which can be obtained from this analysis. In particular, the contour data display reveals wide variations in protein content of both G_1 and G_2 + M cells. However, in any given region in S, these variations are notably diminished. Numerous combinations of two-parameter data such as DNA-to-protein ratios vs either protein, DNA, or cell volume also can be generated and displayed. In addition, multiparameter analysis systems can be useful for investigating the cell-cycle dependence of antigen production using PI for DNA in combination with FITC conjugated to a given antibody. Other potential applications have been suggested elsewhere (Crissman and Steinkamp, 1973).

E.   DNA, Protein, and Nuclear-to-Cytoplasmic Ratio Analyses

Ratio analysis of the time duration of the red and green fluorescence of cells stained for DNA (red) and protein (green) provides a new and direct method for determining the nuclear-to-cytoplasmic size relationship in single cells (Steinkamp and Crissman, 1974). The rationale for this approach is based on the good correlation between total protein and cell volume, as demonstrated in the above study, and the fact that in many instances DNA content is likewise proportional to nuclear size (Baetcke *et al.*, 1967; Epstein, 1967). The technique termed "time-of-flight" derives the proportionality of the nuclear and cytoplasmic diameters from analysis of the time duration of the nuclear (red) and cytoplasmic (green) fluorescence.

The DNA and protein content distributions for a methylcholanthrene-induced (MCA-1) squamous cell carcinoma cultured *in vitro* and doubly stained with propidium iodide (DNA) and fluorescein isothiocyanate (protein) are shown in Fig. 26 along with the nuclear (N), cytoplasmic (C) diameter, and N/C ratio distribution. Peaks 1 and 2 (panel A) represent the relative DNA contents, and peaks 3 and 4 the relative diameter size, of G_1 and G_2 + M cells, respectively. The N/C ratio distribution (Fig. 26C) was obtained by simultaneous analysis of the duration of red (DNA) and green (protein) fluorescence as previously described.

Verification of the "time-of-flight" duration method for measuring nuclear and cytoplasmic size relationships was described previously (Steinkamp and Crissman, 1974). This new methodology offers a technique for establishing relationships among DNA, protein, and nuclear and cytoplasmic size in both normal and abnormal cell systems and has potential application for use in automated cytopathology. In particular, abnormal and malignant cervical cells often exhibit elevations in both DNA content and N/C ratios. Simultaneous analysis of both these relationships using

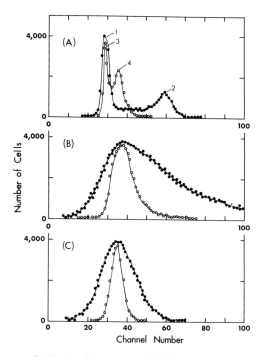

FIG. 26.   Frequency distribution histograms of MCA-1 tumor cells cultured *in vitro* and stained with propidium iodide (DNA content) and fluorescein isothiocyanate (total protein): (A) DNA content (●) and nuclear diameter (○) distributions; (B) total protein (●) and cytoplasmic diameter (○) distributions; and (C) DNA-to-protein (●) and nuclear-to-cytoplasmic ratio (○) distributions.

two-parameter analysis, as described in Fig. 25, should be extremely valuable for clinical diagnosis.

F.   Effects of Chemotherapeutic Agents

Determination of the effects of chemotherapeutic agents on cell-cycle progression is a difficult and tedious task. Cell enumeration and autoradiographic data can reveal cell division and DNA synthetic activity but do not indicate readily the presence of nonprogressing or arrested cells. Autoradiography may not reflect the percentage of the cell population in the S phase if cellular thymidine pools or cell membranes are seriously affected by the drug. However, use of flow-system analysis, which is independent of progression capacity, in conjunction with autoradiography and cell enumeration can be extremely valuable for providing the clinician with information that could be useful for design of regimens for drug administration. For example,

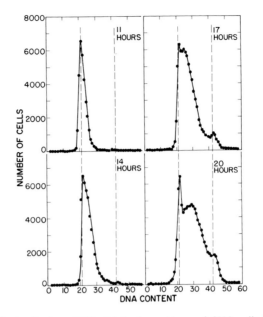

FIG. 27. Acriflavine-Feulgen DNA distribution patterns of CHO cells initially arrested in G_1 and then treated with sodium camptothecin (1 μg/ml) for various time periods following synchrony release. DNA distributions of G_1-arrested cells at $t = 0$ hour were similar to those shown in Fig. 21. Under conditions of this experiment, the CHO control cells had about a 17-hour generation time and exhibited considerable synchrony decay by 20 hours.

the DNA distributions of cells at various times after release from G_1 arrest (Tobey and Ley, 1971) while in the presence of sodium camptothecin (1 μg/ml) are shown in Fig. 27. Control G_1-arrested and exponential DNA distributions were similar to those in Fig. 21. This study indicated that the drug grossly reduces the rate of progression through all phases of the cell cycle (Tobey and Crissman, 1972a). However, since there is a large percentage of cells in S phase after 17 hours of continuous drug treatment, it might be predicted that addition near that time period of a second drug, such as hydroxyurea, which preferentially kills cells in S phase (Tobey and Crissman, 1972b), would be of therapeutic value.

Simultaneous analysis of the DNA content and volume of cells provides another powerful method for studying the effects of drugs. DNA distributions as indicated above reveal capability for cell-cycle traverse, while cell volume analysis yields information relating to the gross biosynthetic capacity of cell populations. Figure 28 shows both separate and simultaneous DNA and cell volume analyses for CHO cells treated continuously with 1-(2-chloroethyl)-3-(4-methylcyclohexyl)-1-nitrosourea (MeCCNU) for 24 hours

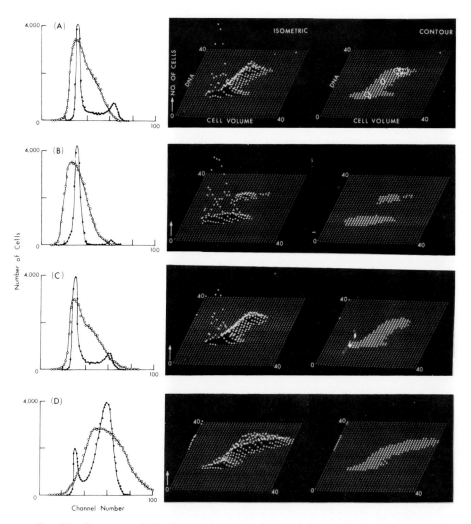

FIG. 28. Separate and simultaneous analyses of DNA and cell volume of control and 1-(2-chloroethyl)-3-(4-methylcyclohexyl)-1-nitrosourea (MeCCNU)-treated CHO cells. Untreated asynchronous (A), G_1-arrested ($t = 0$ hour) (B), and control cell population (C) at 24 hours after release from G_1 were fixed and stained by the acriflavine-Feulgen procedure. Cells, synchronized in G_1 and released in the presence of MeCCNU (10 μ g/ml), were collected after 24 hours (D) and analyzed similarly. ●, DNA content; ○, cell volume.

immediately after release from G_1-arrest (Tobey and Crissman, 1974). Similar DNA and cell volume data are also shown for cultures of asynchronous, G_1-arrested, and control CHO cell populations at 24 hours after synchrony release. At 24 hours, control cells had undergone synchrony decay and were quite similar in both DNA content and cell volume to the asynchronous population. However, MeCCNU-treated cells had accumulated in late interphase and likewise showed significant increase in cell volume. Simultaneous analysis indicates vividly that it is only those cells in late S and G_2 that continue to increase in size, while the volumes for cells in G_1 and early S remain near control values. It is clear from simultaneous analyses that drug-treated cells arrested in late interphase had lost cell division capacity but were not grossly inhibited for synthesis of other cellular macromolecules. However, these unbalanced growth conditions (e.g., excessively high cell volume/DNA ratio) will result eventually in cell death.

VIII.   Analyses Employing Immunofluorescence Techniques

Julius et al. (1972) and Cram et al. (1974) have demonstrated recently the application of flow systems for investigations involving immunofluorescence techniques. These techniques provide a rapid, quantitative means for determining the relative amount of antigen or antibody per cell on a cell-to-cell basis. Cell sorting provides a method for further studying the antigen binding properties of a specific cell type (Julius et al., 1972). Potential application of these and similar techniques will be extremely important in the area of animal and human disease diagnosis.

Analysis of fluorescence and fluorescence-to-cell volume ratios of PK-15 cells maximally infected with hog cholera virus (HCV) and reacted with fluorescein-labeled hog cholera antibodies is shown in Fig. 29 along with noninfected control cells (Cram et al., 1974). Comparison of the modal values of the fluorescence distributions shows the major portion of infected cells to be about three times brighter than controls, although some overlap of the distributions is apparent. The broadness of the fluorescence distribution of infected cells may reflect a differential sensitivity of PK-15 cells to HCV infection. Fluorescence-to-cell volume ratio analysis, which tends to eliminate variances in fluorescence intensity due to differences in cell size, further improves the resolution of the technique. Cell volume analysis of control and infected PK-15 cells showed no significant size differences between these cell populations (Cram and Brunsting, 1973).

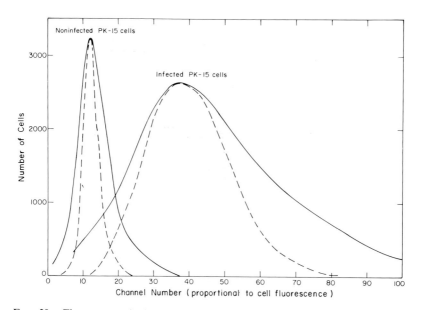

FIG. 29. Fluorescence (——) and fluorescence-to-cell volume ratio (- - -) analysis of control and hog cholera virus (HCV)-infected PK-15 cells reacted with the fluorescein-conjugated HCV antibody.

IX. Determination of the Cell Surface Density Binding of Concanavalin A

The extent and variation of lectin binding are of particular concern in studies relating to malignant transformation (Meezan *et al.*, 1969; Buck *et al.*, 1970; Nicholson, 1971). Studies on the binding of the lectin concanavalin A (Con A) to the cell surface have involved many different techniques. One unique technique is the flow analysis approach which derives the density of Con A binding for large populations of cells. Kraemer *et al.* (1973a) have demonstrated previously the use of fluorescein-conjugated Con A (i.e., Con A-F) in flow systems with saturation binding of Con A-F to CHO cells and analysis of the fluorescence distributions. A more recent study (Steinkamp and Kraemer, 1974) has shown that a significant improvement in the sensitivity of analysis could be achieved by employing both gating and fluorescence-to-cell surface area ratio techniques.

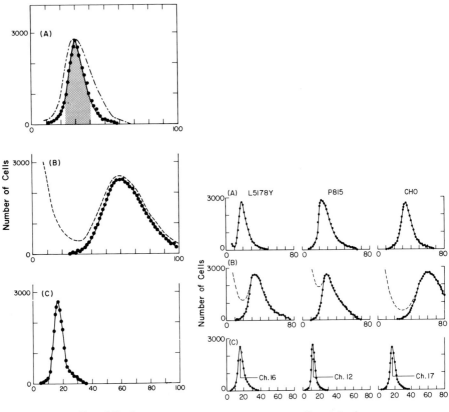

FIG. 30. Multiparameter analysis of saturation binding of Con A-F to the surface of asynchronous CHO cell populations grown in suspension culture. Frequency distribution histograms: (A) cell volume (-●-) and surface area (-··-) distributions; (B) ungated (---) and gated (-●-) cell surface fluorescence distributions; and (C) gated cell surface fluorescence-to-surface area ratio distribution.

FIG. 31. Multiparameter analysis of saturation binding of Con A-F to the surface of L5178Y (*left*), P815 (*center*), and CHO (*right*) cell lines grown exponentially in spinner culture. Frequency distribution histograms: (A) cell surface area distributions; (B) ungated (---) and gated (-●-) cell surface fluorescence distributions; and (C) gated cell surface fluorescence-to-surface area ratio distributions. The two-thirds volume, fluorescence, and ratio amplifier gains were fixed during these measurements.

Since enzymic cell dispersal methods are prohibitive in these cell surface studies, the cell samples often contain considerable amounts of cellular debris and cell clumps. Improvements in instrument design have tended to alleviate many of these problems. For instance, by electronically gating on the shaded area of the cell surface area distribution shown in Fig. 30A, the more accurate cell surface fluorescence distribution was obtained from the ungated distribution in Fig. 30B. The cell surface fluorescence-to-area ratio analysis (Fig. 30C) could then be achieved without distortion due to undesirable cellular material.

Comparisons between different cell lines are also feasible using this type of analysis. In such comparisons, differences in binding due to characteristic cell size differences can be eliminated using ratio analysis techniques. For example, Fig. 31 illustrates data derived from three animal cell lines quite different in both modal cell surface area and surface fluorescence under conditions of saturation binding. The ratio distributions (Fig. 31C) show that some of the binding differences are accountable on the basis of cell size but that differences in binding site density are also prominent. The methodology presented here has broad potential application for study of cell/lectin interactions in such systems as viral transformation, *in vitro* differentiating models, and cell-cycle mechanisms.

X. Applications of Cell-Sorting Techniques

A. Sorting and Autoradiographic Analysis of Thymidine-^{3}H-Labeled Cells

The DNA content distribution of CHO cells pulse-labeled with thymidine-^{3}H is shown in Fig. 32. Equal numbers of cells were sorted on the basis of G_1, S, G_2 + M content and were examined using conventional autoradiographic techniques. The percentage of labeled cells obtained from each sorted region is shown in Fig. 32. The small percentage of labeled cells in the sorted G_1 fraction is due to early S cells, while the labeled cells in the G_2 + M sorted fraction are predominantly late S cells and cells which had moved from S into G_2 during the pulse-labeling period. The cell population had a total labeling index of 33%. Using similar techniques, the assessment of nontraversing cells can be obtained by sorting cells that have been continuously labeled with thymidine-^{3}H for appropriate time periods. The occurrence and number of unlabeled G_1, S, and G_2 + M cells can reflect the extent of any phase-specific arrest of cells in the population. Similar methods have been used in the study of the nature of cell-cycle kinetics for populations of human diploid cells maintained in low serum containing medium (Dell'Orco *et al.*, 1974).

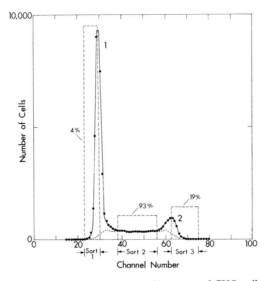

FIG. 32.   DNA content frequency distribution histogram of CHO cells pulse-labeled 15 minutes with 5 μCi/ml of thymidine-^{3}H, stained with propidium iodide, and sorted on the basis of DNA content. The dotted line is a second-degree polynomial computer curve fit representing cells in S phase. The percentage of thymidine-^{3}H-labeled cells from each sorted region is indicated.

B.   DNA Analysis and Sorting of Diploid and Tumor Cell Populations

A recent study has demonstrated the feasibility of detecting and sorting tumor cells based on DNA content (Horan *et al.*, 1974). Cell separations

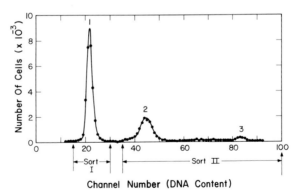

FIG. 33.   DNA content frequency distribution histogram showing the selected regions used for sorting methylcholanthrene-induced mouse (MCA-1) tumor cells stained by the acriflavine–Feulgen technique. The MCA-1 tumor was grown in C3H/Hej mice.

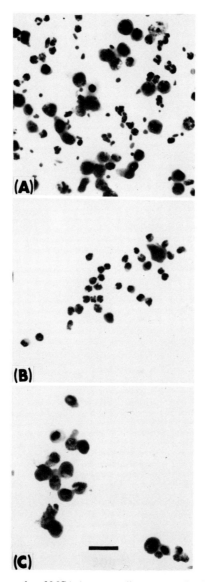

Fig. 34.   Photomicrographs of MCA-1 tumor cells counterstained using the Papanicolaou method: (A) cells prior to sorting; (B) sorted mouse leukocytes from sort region 1; and (C) sorted tumor cells from sort region 2 in Fig. 33. Bar represents 20 μm.

were achieved with sufficient preservation of cytological detail to permit visual identification. Figure 33 shows the DNA content distribution for methylcholanthrene-induced mouse tumor (MCA-1) cells. The DNA distribution shows three peaks, the first peak being cells having a normal 2C diploid DNA content and the second and third peaks representing cells having 4C and 8C DNA content, respectively. The MCA-1 tumor cell population was sorted on the basis of DNA content into two groups, designated by sort regions 1 and 2. Figure 34A is a photomicrograph of dispersed and counterstained tumor cells prior to sorting. Cells having normal DNA content from sort 1 region are identified as leukocytes (Fig. 34B), whereas those cells containing an elevated or abnormal DNA content from the sort 2 region are tumor cells (Fig. 34C). The two sorted subpopulations are quite morphologically distinct. The second and third peaks of Fig. 33 have been shown to be G_1 and $G_2 + M$ phase tumor cells, with S phase cells contained between the peaks (Horan *et al.*, 1974). Other tumor cells with nonelevated DNA content will most certainly not be identifiable on the basis of DNA alone. Other parameters which characterize cell abnormalcy such as nuclear-to-cytoplasmic size ratios and antigenic properties are under investigation.

C.   Differential Analysis and Sorting of Human Leukocytes

Melamed *et al.* (1973) have used acridine orange metachromatic staining of human leukocytes and flow-system analysis for differential leukocyte counting of leukemic patients undergoing chemotherapy. This method has potential application for many other clinical studies involving differential leukocyte enumeration. The use of two-parameter analysis as presented here further enhances the resolution of the methodology.

The green and red fluorescence distributions of normal human leukocytes vitally stained with acridine orange according to the method of Adams and Kamentsky (1971) are shown in Fig. 35. The green fluorescence distribution is unimodal, illustrating similarities in leukocyte nuclear staining, while the red fluorescence distribution shows distinct peaks which characterize human leukocytes into three groups on the basis of red fluorescence of cytoplasmic granules. The green-to-red fluorescence ratio distribution (Fig. 35C), which contains four leukocyte groups (peaks 1–4), demonstrates additional leukocyte characterization from analysis of nuclear-to-cytoplasmic staining properties. Leukocytes corresponding to peaks 1, 2, 3, and 4 have been separated and identified as granulocytes, monocytes, large lymphocytes, and small lymphocytes, respectively (Steinkamp and Romero, 1974). The small shoulder located on the right side of peak 1 (lymphocytes) of the red fluorescence distribution (Fig. 35B) is responsible for peak 3 of the green/red

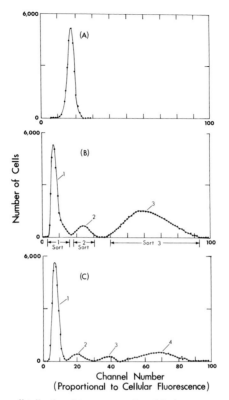

FIG. 35. Frequency distribution histograms of acridine orange-stained normal human leukocytes: (A) green fluorescence distribution; (B) red fluorescence distribution showing sort regions 1, 2, and 3; and (C) green-to-red fluorescence ratio distribution.

fluorescence distribution as verified by two parameter analysis. Two parameter analysis (Fig. 36) better illustrates the relationship between green and red fluorescence staining properties of normal human leukocytes. Peaks 1, 2, and 3 had been predicted earlier by Adams and Kamentsky (1971) to consist primarily of lymphocytes, monocytes, and granulocytes, respectively. Sorting of human leukocytes on the basis of red fluorescence has confirmed these results (Steinkamp et al., 1973b). Figure 37 shows photomicrographs of an enriched leukocyte population obtained by sorting leukocytes based on green fluorescence only as well as sorted and counterstained lymphocyte, monocyte, and granulocyte fractions obtained from sort regions 1, 2, and 3, respectively, of the red fluorescence distribution (Fig. 35). Erythrocytes, which are in excess to leukocytes by approximately 1000:1, were contained in all droplets and thus were sorted along with selected leukocytes.

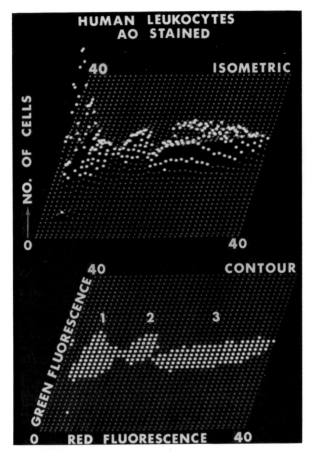

FIG. 36. Two-parameter green/red fluorescence plots of acridine orange-stained normal human leukocytes. (*Top*): Three-dimensional frequency distribution histogram, isometric plot; *bottom*: two-dimensional contour view. Regions 1–3 consist of lymphocytes, monocytes, and granulocytes, respectively.

XI.   Potential Applications for Fluorescent Enzyme Assay

Rotman and Papermaster (1966) have described the use of the fluorogenic substrate fluorescein diacetate (FDA) for analyzing general esterase activity in viable cells. These authors proposed that the nonfluorescent FDA, being a nonpolar compound, penetrates readily into the cell and is hydrolyzed by esterases producing the fluorescent product fluorescein. Since fluorescein is a polar compound, it is not released from the cell as rapidly as FDA enters, resulting in an accumulation of intercellular fluorescein. Applica-

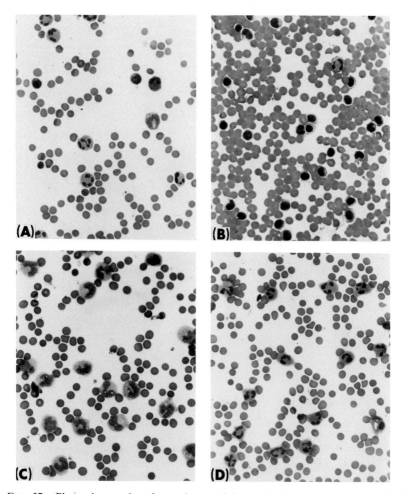

FIG. 37. Photomicrographs of sorted normal human leukocytes counterstained with Wright's Giemsa: (A) enriched leukocyte sort (channels 5–30 of Fig. 35A); (B) sorted lymphocytes (sort 1 region of Fig. 35B); (C) sorted monocytes (sort 2 region of Fig. 35B); and (D) sorted granulocytes (sort 3 region of Fig. 35B). Erythrocytes contained in the photomicrographs were separated along with selected leukocyte fractions.

tions of this technique in flow systems would include cell viability studies as well as studies which may be of diagnostic value. For instance, Willighagen *et al.* (1963), using cytochemical techniques, reported differential elevations in esterase activity of squamous cell carcinoma of the lung associated with survival.

Enzyme studies for the fluorescent assay of cellular acid or alkaline phosphatase can be made using substrates such as 3-methyl-O-fluorescein phosphate (Tierney, 1964). Since Kaplow (1968) has shown that alkaline phosphatase in leukocytes is reduced in myeloid leukemia patients, this fluorescence technique coupled with flow analysis should have immediate application in the field of medicine. In addition, Rotman *et al.* (1963) made a fluorometric assay of β-D-galactosidase using the substrate fluorescein-di-β-D-galactopyranoside. Guilbaunt (1973) has compiled numerous other enzymic studies which could also have potential applications.

Thus, flow systems can play an important role in reestablishing usage of cytochemical enzyme assay methods by providing sensitivity, quantitation, and speed of measurement to the technique, the converse of which has deterred this approach for many years. Through the use of biochemical standards, absolute values of measurement may be obtained. Rapid methods for enzyme kinetic studies are also possible.

XII.   Summary and Conclusions

At the present time, the application of flow systems for cell analysis is limited mainly by the cell preparative aspects of the methodology. Although automated analytical capabilities are now available for making a large number of biological studies, such investigations must await new or modified cell preparation methods which will permit the initiation of these studies. Techniques for improved cell dispersal and rapid cell staining are urgently needed to expand the usefulness of flow systems for various cytological investigations. In addition to the biochemical cellular components which are presently studied, fluorescent staining protocols should be devised for staining cellular RNA, histones, and lipids, since these moieties also appear to play an important role in cell metabolism. In other studies, various fluorogenic substrates could be used for quantitating specific enzyme systems in cells, providing optimal conditions for incubating cells in suspension are ascertained. The number of applications of flow systems certainly seem theoretically almost endless; however, new cytological studies will require the same diligent efforts which have produced those techniques used routinely today.

ACKNOWLEDGEMENTS

We thank the many individuals at LASL whose generous contributions made this chapter possible: in particular, L. S. Cram (immunofluorescence applications), R. A. Tobey (cell synchrony and drug studies), P. M. Kraemer (heteroploidy and cell surface studies), L. L.

Deaven and D. F. Petersen (heteroploidy and karyology analysis), P. K. Horan (tumor studies), D. Freeman (protein staining), P. N. Dean, J. H. Jett, and E. C. Anderson (computer-fit programming), M. R. Raju (radiobiology), T. T. Trujillo (suggestions for cell staining), R. A. Tobey, P. M. Kraemer, G. C. Salzman, M. S. Oka, C. R. Richmond, and L. S. Cram (helpful criticisms of the manuscript), and J. L. Horney (preparation of several illustrations). We are grateful to P. C. Sanders (autoradiography), J. Grilly (photography), A. Romero (tumor cell preparation), and E. Sullivan, E. Bain, H. Barrington, and T. Williams (manuscript preparation).

Appendix A: Cell Dispersal

Cells obtained from suspension culture or cells on monolayer are washed once in cold saline GM (gm/liter: glucose 1.1, NaCl 8.0, KCl 0.4, $Na_2HPO_4 \cdot 12H_2O$ 0.39, KH_2PO_4 0.15) and incubated at 37°C for 10 minutes in 10 ml of cell dispersal solution [0.5 mM EDTA, 0.1 mg/ml crystalline trypsin (Worthington) in saline GM]. After incubation, 10 ml of neutralization solution [0.2 mg/ml soybean trypsin inhibitor, 0.01 mg/ml DNase I (Worthington), 1 mg/ml bovine serum albumin in saline G (i.e., saline GM plus 1.54 gm/liter $MgSO_4 \cdot 7H_2O$, 0.16 gm/liter $CaCl_2 \cdot 2H_2O$), Puck *et al.* (1958)] is added to the incubation tube, and the suspension is worked moderately with pipetting. The tube is chilled on ice for about 5 minutes and centrifuged, and the supernatant is siphoned off prior to fixation.

Appendix B: Acriflavine-Feulgen Procedure

A. Fixation

Following the cell dispersal protocol as outlined above, cells are re-suspended in 10 ml of cold saline G by pipetting. An equal volume of saline GF (i.e., saline G containing 8% formaldehyde) is added, and the suspension is pipetted. Cells are fixed on ice for at least 12–18 hours prior to staining.

B. Staining

Formalin-fixed cells are centrifuged at low speed, and the supernatant fixative is siphoned off. Cells are washed once in distilled water, hydrolyzed in 4 N HCl at room temperature for 20 minutes, and then rinsed once in distilled water. The cells are then resuspended in filtered staining solution (0.2 mg/ml acriflavine, 5 mg/ml potassium metabisulfite dissolved in 100 ml of distilled water, after which 10 ml of 0.5 N HCl is added) and stained for 20 minutes at room temperature. Samples are rinsed at least three times in acid–alcohol (1 ml of concentrated HCl in 99 ml of 70% ethanol). The final supernatant rinse should be clear. Cells are resuspended in distilled water or, if cell volume measurements or cell sorting is to be performed, in saline for flow analysis.

Appendix C: Ethidium Bromide (EB) and Propidium Iodide (PI) Protocols

A.   Fixation

After cell dispersal, cells are resuspended in 5 ml of cold saline G, and then 15 ml of cold 95% ethanol ($-20°$ C) is gradually added to the suspension with vigorous pipetting. Fixation is achieved after 30'minutes on ice; however, the cells may be stored for at least 1 week at temperatures slightly above freezing in 70% ethanol.

B.   Staining

The technique for staining ethanol-fixed cells involves treating the samples initially with RNase (1 mg/ml) for 30 minutes at 37° C. After incubation, the cells are rinsed once in cold distilled water and then stained in EB or PI (5 mg/100 ml of 1.12% sodium citrate) for 20 minutes on ice. The cells are rinsed once and resuspended in distilled water or in saline for analysis.

REFERENCES

Adams, C. W. M., and Bayliss, O. B. (1971a). *Histochemie* **28**, 220.
Adams, C. W. M., and Bayliss, O. B. (1971b). *Histochemie* **28**, 229.
Adams, L. R., and Kamentsky, L. A. (1971). *Acta Cytol.* **15**, 229.
Alvarez, M. R. (1973). *Cancer Res.* **33**, 786.
Angerer, L. M., and Moudrianakis, E. M. (1972). *J. Mol. Biol.* **63**, 505.
Baetcke, K. P., Sparrow, A. H., Nauman, C. H., and Schwemmer, S. S. (1967). *Proc. Nat. Acad. Sci. U.S.* **58**, 533.
Barer, R. (1952). *Nature (London)* **169**, 367.
Barer, R. (1956). *Phys. Tech. Biol. Res.* **3**, 1–56.
Barer, R., Ross, K. F. A., and Tkaczyk, S. (1953). *Nature (London)* **171**, 720.
Born, M., and Wolfe, E. (1965). "Principles of Optics," 3rd rev. ed., p. 398. Pergamon, Oxford.
Bostock, C. J., Prescott, D. M., and Kirkpatrick, J. B. (1971). *Exp. Cell Res.* **68**, 163.
Brunsting, A. (1972). Ph.D. Thesis, Department of Physics and Astronomy, University of New Mexico, Albuquerque (Los Alamos Scientific Laboratory report LADC-72-974).

Brunsting, A., and Mullaney, P. F. (1972a). *J. Colloid Interface Sci.* **39**, 492.
Brunsting, A., and Mullaney, P. F. (1972b). *Appl. Opt.* **11**, 675.
Brunsting, A., and Mullaney, P. F. (1974). *Biophys. J.* **14**, 439.
Buck, C. A., Glick, M. C., and Warren, L. (1970). *Biochemistry* **8**, 4567.
Casperson, T. (1936). *Skand. Arch. Physiol.* **73**, 8.
Castanello, D. E., and Gerschenson, L. E. (1970). *Exp. Cell Res.* **59**, 283.
Cleland, R. L. (1970). *In* "Chemistry and Molecular Biology of the Intercellular Matrix" (E. A. Balaz, ed.), Vol. 2, pp. 1095–1150. Academic Press, New York.
Coulter, W. H. (1965). *Proc. Nat. Electron. Conf.* **12**, 1034.
Cram, L. S., and Brunsting, A. (1973). *Exp. Cell Res.* **78**, 209.
Cram, L. S., Forslund, J. C., Horan, P. K., and Steinkamp, J. A. (1974). *In* "Automation in Microbiology and Immunology" (C. G. Heden and T. Illeni, eds.). Wiley, New York (in press).
Crissman, H. A. (1974a). *In* "Proceedings of the First Annual Life Sciences Symposium on Mammalian Cells: Probes and Problems." AEC Symp. Ser., Technical Information Center, Oak Ridge, Tennessee, (in press).
Crissman, H. A. (1974b). In preparation.
Crissman, H. A., and Steinkamp, J. A. (1973). *J. Cell Biol.* **59**, 766.
Crissman, H. A., and Tobey, H. A. (1974). *Science* **184**, 1297.
Crissman, H. A., Jett, J. H., and Petersen, D. F. (1974). In preparation.
Culling, C., and Vassar, P. (1961). *AMA Arch. Pathol.* **71**, 88.
Dean, P. N., and Jett, J. H. (1974). *J. Cell Biol.* **60**, 523.
Deitch, A. D. (1966). *Introd. Quant. Cytochem.* **1**, 327–354.
Deitch, A. D., Wagner, D., and Richart, R. M. (1968). *J. Histochem. Cytochem.* **16**, 371.
Dell'Orco, R. T., Crissman, H. A., Steinkamp, J. A., and Kraemer, P. M. (1974). *Exp. Cell Res.* (in press).
DeLuca, C. (1965). *Exp. Cell Res.* **40**, 186.
Deman, J. J., Bruyneel, E. A., and Mareel, M. M. (1974). *J. Cell Biol.* **60**, 641.
Dittrich, W., and Göhde, W. (1969). *Z. Naturforsch.* **3**, 360.
Dittrich, W., Göhde, W., Severin, E., and Reiffenstuhl, G. (1971). *Lec. Int. Congr. Cytol.*, *4th.*
Elleder, M., and Lujda, Z. (1971). *Histochemie* **25**, 286.
Ely, J. O., and Ross, M. H. (1949). *Anat. Rec.* **104**, 103.
Epstein, C. J. (1967). *Proc. Nat. Acad. Sci. U.S.* **57**, 327.
Fand, S. (1970). *Introd. Quant. Cytochem.* **2**, 209–221.
Fiel, R., and Mullaney, P. F. (1974). *Amer. Chem. Soc. Abstr., Colloid Surface Chem.* No. 19.
Freeman, D., and Crissman, H. A. (1974). In preparation.
Frenster, J. H. (1971). *Cancer Res.* **31**, 1128.
Frenster, J. H. (1972). *Nature (London), New Biol.* **236**, 175.
Friedmann, T., and Epstein, C. J. (1967). *Biochim. Biophys. Acta* **138**, 622.
Fujita, S. (1973). *Histochemie* **36**, 193.
Fulwyler, M. J. (1965). *Science* **150**, 910.
Galavazi, G., and Bootsma, D. (1966). *Exp. Cell Res.* **41**, 438.
Gray, D. E., ed. (1957). "American Institute of Physics Handbook," 2nd ed. pp. 6–93. McGraw-Hill, New York.
Grover, N. B., Naaman, J., Ben-Sasson, S., and Doljanski, F. (1969a). *Biophys. J.* **9**, 1398.
Grover, N. B., Naaman, J., Ben-Sasson, S., Doljanski, F., and Nadav, E. (1969b). *Biophys. J.* **9**, 1415.
Guilbaunt, G. G. (1973). "Practical Fluorescence," 2nd ed. Dekker, New York.
Haag, D. (1973). *Histochemie* **36**, 283.
Harris, C. C., and Leone, C. A. (1966). *J. Cell Biol.* **28**, 405.

Hodkinson, J. R., and Faulkner, J. R. (1965). *Appl. Opt.* **4**, 1463.
Hodkinson, J. R., and Greenleaves, I. (1963). *J. Opt. Soc. Amer.* **53**, 577.
Holm, D. M., and Cram, L. S. (1973). *Exp. Cell Res.* **80**, 105.
Hopwood, D. (1969). *Histochem. J.* **1**, 323.
Horan, P. K., Romero, A., Steinkamp, J. A., and Petersen, D. F. (1974). *J. Nat. Cancer Inst.* **52**, 843.
Horobin, R. W. (1969). *Histochem. J.* **1**, 231.
Hudson. B., Upholt, W. B., Divinny, J., and Vinograd, J. (1969). *Proc. Nat. Acad. Sci. U.S.* **62**, 813.
Hulett, H. R., Bonner, W. A., Barrett, J., and Herzenberg, L. A. (1969). *Science* **166**, 747.
Humason, G. L. (1967). "Animal Tissue Techniques," 2nd ed., p. 569. Freeman, San Francisco, California.
Ichimura, S., Zama, M., and Fujita, H. (1971). *Biochim. Biophys. Acta* **240**, 485.
Jones, D. (1972). *Histochem. J.* **4**, 421.
Jovin, T., and Jovin, D. (1974). *J. Histochem. Cytochem.* **22**, 622.
Julius, M. H., Masuda, T., and Herzenberg, L. A. (1972). *Proc. Nat. Acad. Sci. U.S.* **69**, 1934.
Kamentsky, L. A., Melamed, M. R., and Derman, H. (1965). *Science* **150**, 630.
Kaplow, L. W. (1968). *Ann. N. Y. Acad. Sci.* **155**, 911–947.
Kasten, F. H. (1959). *Histochemie* **1**, 466.
Kasten, F. H. (1967). *Int. Rev. Cytol.* **21**, 141–202.
Kato, H., and Yosida, T. H. (1970). *Exp. Cell Res.* **60**, 459.
Kemp, C. B., Knight, M. J., Scharp, D. W., Lacy, P. E., and Ballinger, W. F. (1973). *Nature* (*London*) **244**, 447.
Kraemer, P. M., Petersen, D. F., and Van Dilla, M. A. (1971). *Science* **174**, 714.
Kraemer, P. M., Deaven, L., Crissman, H., and Van Dilla, M. A. (1972). *Advan. Cell Mol. Biol.* **2**, 47–107.
Kraemer, P. M., Tobey, R. A., and Van Dilla, M. A. (1973a). *J. Cell. Physiol.* **81**, 305.
Kraemer, P. M., Deaven, L. L., Crissman, H. A., Steinkamp, J. A., and Petersen, D. F. (1973b). *Cold Spring Harbor Symp. Quant. Biol.* **38**, 133–144.
Lacy, P. E., amd Kostianovsky, M. (1967). *Diabetes* **16**, 35.
Latimer, P., and Tully, B. (1968). *J. Colloid Interface Sci.* **27**, 475.
LePecq, J. F. (1971). *Methods Biochem. Anal.* **20**, 41–86.
LePecq, J. F., and Paoletti, C. (1967). *J. Mol. Biol.* **27**, 87.
Lessler, M. A. (1951). *Arch. Biochem. Biophys.* **32**, 42.
Longley, J. B. (1952). *Stain Technol.* **27**, 161.
McCarty, S. A. (1972). *Acta Cytol.* **16**, 221.
MacRae, R. A., McClure, J. A., and Latimer, P. (1961). *J. Opt. Soc. Amer.* **51**, 1366.
McK. Ellison, J., and Peetz, C. V. (1959). *Proc. Phys. Soc., London* **74**, 105.
Meezan, E., Wu, H. C., Black, P. H., and Robbins, P. W. (1969). *Biochemistry* **8**, 2518.
Melamed, M. R., Adams, L. A., Traganos, F., and Kamentsky, L. A. (1973). *Eur. J. Cancer* **9**, 181.
Mendelsohn, M., Mayall, B. H., Bogart, E., Moore, D. H., and Perry, B. H. (1974). *Science* **179**, 1126.
Mendelsohn, M., Rolman, W. A., and Bostrom, R. C. (1964). *Ann. N. Y. Acad. Sci.* **99**, 998–1009.
Merrill, J. T., Veizades, N., Hulett, H. R., Wolf, P. L., and Herzenberg, L. A. (1971). *Rev. Sci. Instrum.* **42**, 1157.
Mitchison, J. M., and Swann, M. M. (1953). *Quart. J. Microsc. Sci.* **94**, 381.
Mullaney, P. F. (1970). *J. Opt. Soc. Amer.* **60**, 573.
Mullaney, P. F., and Dean, P. N. (1970). *Biophys. J.* **10**, 764.
Mullaney, P. F., and West, W. T. (1973). *J. Phys. E* **6**, 1006.

Mullaney, P. F., Van Dilla, M. A., Coulter, J. R., and Dean, P. N. (1969). *Rev. Sci. Instrum.* **40**, 1029.

Müller, M., Schreiber, M., Kartenbeck, J., and Schreiber, G. (1972). *Cancer Res.* **32**, 2568.

Nairn, R. C. (1969). "Fluorescent Protein Tracing," 3rd ed., Williams & Wilkins, Baltimore, Maryland.

Nicholson, G. L. (1971). *Nature (London), New Biol.* **233**, 244.

Ornstein, L., Mautner, W., Davis, B. J., and Tamura, R. (1957). *J. Mt. Sinai Hosp., New York* **24**, 6.

Pearse, A. G. E. (1968). "Histochemistry, Theoretical and Applied," 3rd ed., Vol. 1. Little, Brown, Boston, Massachusetts.

Pessac, B., and Defendi, V. (1972). *Science* **175**, 898.

Phillips, H. J. (1972). *In Vitro* **8**, 101.

Puck, T. T., and Steffen, J. (1963). *Biophys. J.* **3**, 379.

Puck, T. T., Ceiciura, S. J., and Robinson, A. (1958). *J. Exp. Med.* **108**, 945.

Rapport, C., and Howze, G. B. (1966a). *Proc. Soc. Exp. Biol. Med.* **121**, 1010.

Rapport, C., and Howze, G. B. (1966b). *Proc. Soc. Exp. Biol. Med.* **121**, 1016.

Rapport, C., and Howze, G. B. (1966c). *Proc. Soc. Exp. Biol. Med.* **121**, 1022

Rigler, R. (1966). *Acta Physiol. Scand.* **67**, Suppl. 267, 122.

Rigler, R. (1969). *Ann. N. Y. Acad. Sci.* **157**, 211.

Rodbell, M. (1961). *J. Biol. Chem.* **229**, 375.

Romero, A., and Horan, P. K. (1974). In preparation.

Ross, K. F. A. (1954). *Quart. J. Microsc. Sci.* **95**, 425.

Ross, K. F. A. (1967). "Phase Contrast and Interference Microscopy for Cell Biologists." Arnold, London.

Rossilet, A., and Ruch, F. (1968). *J. Histochem. Cytochem.* **16**, 459.

Rotman, B., and Papermaster, B. W. (1966). *Proc. Nat. Acad. Sci. U.S.* **55**, 134.

Rotman, B., Zdevic, J., and Edelstein, M. (1963). *Proc. Nat. Acad. Sci. U.S.* **50**, 1.

Salzman, G. C., and Mullaney, P. F. (1974). *Amer. Chem. Soc. Abstr., Colloid Surface Chem.* No. 18.

Seglen, P. O. (1973). *Exp. Cell Res.* **82**, 391.

Smets, L. A. (1973). *Exp. Cell Res.* **79**, 239.

Sprenger, E., Böhm, N., and Sandritter, W. (1971). *Histochemie* **26**, 238.

Steinkamp, J. A., and Crissman, H. A. (1974). *J. Histochem. Cytochem.* **22**, 616.

Steinkamp, J. A., and Kraemer, P. M. (1974). *J. Cell Physiol.* **84**, 197.

Steinkamp, J. A., and Romero, A. (1974). *Proc. Soc. Exp. Biol. Med.* **146**, 1061.

Steinkamp, J. A., Fulwyler, M. J., Coulter, J. R., Hiebert, R. D., Horney, J. L., and Mullaney, P. F. (1973a). *Rev. Sci. Instrum.* **44**, 1301.

Steinkamp, J. A., Romero, A., and Van Dilla, M. A. (1973b). *Acta Cytol.* **17**, 113.

Steinkamp, J. A., Romero, A., Horan, P. K., and Crissman, H. A. (1974). *Exp. Cell. Res.* **84**, 15.

Terasima, T., and Tolmach, L. J. (1963). *Exp. Cell. Res.* **30**, 344.

Tierney, J. H. (1964). *Fed. Proc., Fed. Amer. Soc. Exp. Biol.* **23**, 2381.

Tobey, R. A. (1973). *In* "Methods in Cell Biology" (D. M. Prescott, ed.), Vol. 6, pp. 67–112. Academic Press, New York.

Tobey, R. A., and Crissman, H. A. (1972a). *Cancer Res.* **32**, 2726.

Tobey, R. A., and Crissman, H. A. (1972b). *Exp. Cell Res.* **75**, 460.

Tobey, R. A., and Crissman, H. A. (1974). *Cancer Res.* (in press).

Tobey, R. A., and Ley, K. D. (1971). *Cancer Res.* **31**, 46.

Tobey, R. A., Anderson, E. C., and Petersen, D. F. (1967). *J. Cell. Physiol.* **70**, 63.

Tobey, R. A., Crissman, H. A., and Kraemer, P. M. (1972). *J. Cell Biol.* **54**, 638.

Tolles, W. E., and Bostrom, R. C. (1956). *Ann. N. Y. Acad. Sci.* **63**, 1211–1218.

Trujillo, T. T., and Van Dilla, M. A. (1972). *Acta Cytol.* **16**, 26.

Udenfriend, S. (1969). "Fluorescence Assay in Biology and Medicine," Vol. 2, p. 660. Academic Press, New York.

Udenfriend, S., Stein, S., Böhlen, P., Paron, W., Leingruber, W., and Weigele, M. (1972). *Science* **178**, 871.

Vahouny, G. V., Wei, R., Starkweather, R., and Davis, C. (1970). *Science* **167**, 1616.

Van Dilla, M. A., Mullaney, P. F., and Coulter, J. R. (1967). Los Alamos Scientific Laboratory report LA-3848-MS, p. 100, National Technical Information Service, U.S. Department of Commerce, Springfield, Virginia.

Van Dilla, M. A., Trujillo, T. T., Mullaney, P. F., and Coulter, J. R. (1969). *Science* **163**, 1213.

Van Dilla, M. A., Steinmetz, L. L., Daws, D. T., Calvert, R. N., and Gray, J. W. (1974). *In* "Proceedings of the 1973 IEEE Nuclear Science Symposium." Vol. NS-21: 714–720.

Ward, D. C., Reich, E., and Goldberg, I. H. (1965). *Science* **149**, 1259.

West, S. S. (1969). *Phys. Tech. Biol. Res.* **3**, Part C, 253–320.

Wheelus, L. L., and Patten, S. F. (1973a). *Acta Cytol.* **17**, 333.

Wheelus, L. L., and Patten, S. F. (1973b). *Acta Cytol.* **17**, 391.

Wied, G. L., Bartels, P. A., Bahr, G. F., and Oldfield, D. G. (1968). *Acta Cytol.* **12**, 357.

Willighagen, R. G. J., van der Heul, R. O., and van Rijssel, T. G. (1963). *J. Pathol. Bacteriol.* **85**, 279.

Wyatt, P. J. (1968). *Appl. Opt.* **7**, 1879.

Chapter 13

Purification of Surface Membranes from Rat Brain Cells

KARI HEMMINKI

Department of Medical Chemistry, University of Helsinki, Helsinki, Finland

I. The Neuronal Surface

The uniqueness of neuronal plasma membranes[1] in maintaining excitability is a motive for their purification. Neurons may possess plasma membrane elements differentiated morphologically and functionally, such as synaptic, axonal, and dendritic plasma membranes, in addition to the perikaryal plasma membrane, which corresponds to plasma membranes of other cell types (Fig. 1A).

The complexity of neuronal plasma membranes has seriously impeded their characterization and limited the choice of starting material. Only a few techniques have been described for the isolation of surface membranes from separated cells. These include purification of plasma membranes from neuronal and glial cell-enriched fractions (Henn *et al.*, 1972) and cultured

---

[1] In this article membranes are operationally called plasma membranes when prepared from solid tissue and surface membranes when prepared from dissociated cells, irrespective of the terms used in the original papers. The terms may be used exchangeably to connote cell surface in general.

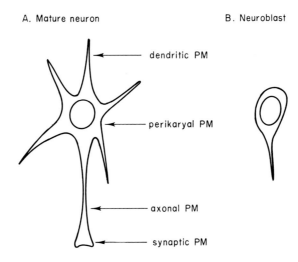

A. Mature neuron   B. Neuroblast

dendritic PM

perikaryal PM

axonal PM

synaptic PM

FIG. 1.   Plasma membranes of a mature (A) and an immature neuron (B).

neuroblastoma (Glick *et al.*, 1973). Preparations of plasma membranes from solid nervous tissue are more numerous. Usually these preparations are designated by the "typical" plasma membrane type found in electron micrographs. Preparations from brain tissue include synaptosomal plasma membranes (Whittaker *et al.*, 1964; DeRobertis *et al.*, 1966; Morgan *et al.*, 1971; Cotman and Matthews, 1971; Levitan *et al.*, 1972; Gurd *et al.*, 1974) and axonal plasma membranes (DeVries *et al.*, 1972). Although many of these preparations are reported to be markedly enriched in respect to plasma membrane markers, it frequently remains unestablished to what extent the preparations contain the particular type of neuronal plasma membrane indicated (cf. Fig. 1A). Other workers have used simpler starting material such as squid axons (Camejo *et al.*, 1969; Fischer *et al.*, 1970; Denburg, 1972; Barnola *et al.*, 1973) or eel electroplax (Bauman *et al.*, 1970) to be applied in the purification of receptors (Denburg, 1973; Barnola *et al.*, 1973; Meunier and Changeux, 1973).

This article reviews work carried out in our laboratory with surface membranes of immature brain cells. The selection of the material stems from our interests in the development of neuronal membranes, and particularly the control of their synthesis. Rat brain cortex appeared to have many advantages for such an approach. Multiplication of neuroblasts is thought to be terminated in rat cortex at the time of birth, while that of spongioblasts (glial stem cell) begins one or two weeks later (Caley and Maxwell, 1968; McIlwain and Bachelard, 1971; Burdman, 1972). This guarantees an enrichment of neuroblasts over spongioblasts. Moreover, immature neurons (Fig. 1B) lack most of their processes, making the dissociation of cells relatively easy, as has been shown (Hemminki, 1972a). Thus brain cells, intact by several parameters, can be isolated in high yields and maintained in cell cultures for some days. Surface membranes of the dissociated cells may be isolated as described below and the membrane prepared may not be as heterogeneous as that of adult neurons. Recently, we have been able to purify plasma membranes from solid brain tissue, which will also be described.

II. Isolation Methods

A. Dissociated Cells as Starting Material

The purification techniques for surface membranes from immature brain cells have been described in previous original articles and in a recent review (Hemminki, 1973a) and will be only shortly outlined here. In all cases rats 1 to 6 days old were used. Their cortices were dispersed into small pieces, which were incubated in the presence of 0.1% trypsin (Hemminki, 1972a). Dissociated cells recovered were washed by centrifugation and used for the isolation of surface membranes.

1. METHOD 1

The cellular pellet was resuspended in a hypotonic medium containing 90 mM sucrose, 1 mM MgCl$_2$, and 40 mM Tris buffer, pH 7.6 for 10 minutes and homogenized in a Teflon–glass homogenizer (Hemminki and Suovaniemi, 1973). The homogenate was filtered instantaneously through a nylon mesh (hole size 30 μm) under a vacuum, and osmolarity was restored by the addition of 1.0 M sucrose. The subsequent steps of purification are shown in Fig. 2A. An electron micrograph of the preparation shows membrane strands and vesicles without contamination with myelin or non-membranous elements (Fig. 3).

2. METHOD 2

The cells were resuspended, swollen, and homogenized in the hypotonic medium as described above. The difference from Method 1 is that the membranes were collected at the 4000 g pellet whereas Method 1 used the supernatant (Fig. 2). The sedimentation properties, electron microscopy, and incorporation kinetics, and the presence of microtubular protein in the lysed supernatant of surface membranes as prepared by Method 2, suggested that

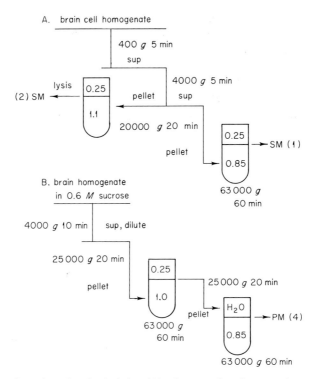

Fig. 2. A flow sheet for the isolation (A) of neuronal surface membranes (SM) and (B) of plasma membranes (PM). The figures in parentheses refer to the isolation methods described in the text. The figures depicted in the centrifuge tubes refer to molar concentrations of sucrose. Sup=supernatant.

the fraction is enriched in axonal surface membrane as compared to Method 1, yielding perikaryal membrane (Hemminki, 1973a).

3. Method 3

The cellular pellet was resuspended in 5 volumes of a mixture containing 1 mM ZnCl₂ and dimethyl sulfoxide (4:1, v/v). The suspension was left on ice for 10 minutes and centrifuged at 6000 g for 10 minutes. The supernatant was layered onto 1.0 M sucrose and centrifuged at 63,000 g for 60 minutes. The interface was collected and sedimented to be used as surface membranes (Hemminki, 1972b,1973c). This method is a modification of that described by Warren and Glick (1969). It has the largest recovery but, perhaps, the lowest purity (Hemminki, 1973a,c). Characterization of the fraction is hampered by inactivation of some marker enzymes by the ZnCl₂– dimethyl sulfoxide treatment. Sheets of surface membranes as produced by Method 3 are shown in Fig. 4.

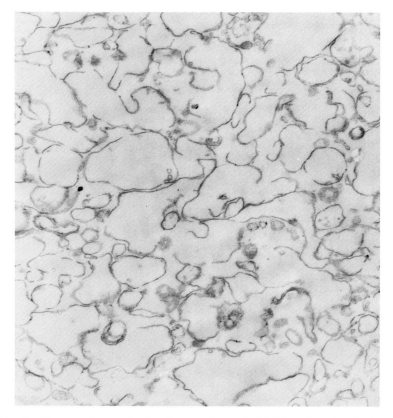

FIG. 3. Electron micrograph of surface membranes isolated from immature brain cells according to Method 1 (see text). × 11,000.

B. Brain Tissue as Starting Material

METHOD 4

We described recently a method for the preparation of plasma membranes from solid tissue (Hemminki, 1973d). In this procedure brain tissue was first swollen in 0.1 M Tris, pH 7.5 for 5 minutes. Addition of sucrose to 0.6 M was instantaneously followed by homogenization. The subsequent steps of the purification are described in Fig. 2B. The morphology of this fraction as studied by electron microscopy was largely simlar to that of Fig. 3 (Hemminki, 1973d).

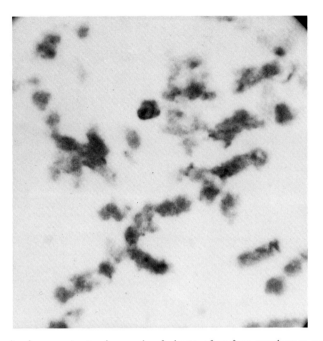

FIG. 4. A phase contrast micrograph of sheets of surface membranes prepared by Method 3 (see text) × 200.

III. Composition of Brain Plasma Membranes

The surface and plasma membrane fractions have been characterized by plasma membrane marker enyzmes as collected in Table I. Surface membranes purified according to Methods 1 and 2 yielded preparations where Na^+, K^+-activated ATPase, a plasma membrane marker, was concentrated 6-to 7-fold as compared to the homogenate. In plasma membranes, prepared from solid brain tissue, the concentration was somewhat higher, about 9-fold. These figures indicate purity comparable to that of the best preparations of neuronal plasma membranes (see Table I). 5'-Nucleotidase and acetylcholinesterase were also moderately concentrated in surface and plasma membrane preparations.

Protein/cholesterol and protein/phospholipid ratios of surface membranes were 7.2 and about 1.5, respectively. These figures differ quite extensively from plasma membranes (Method 4). By these ratios surface membranes resemble synaptic plasma membranes; by contrast, plasma membranes are more like the highly purified squid axolemma (Table I).

TABLE I

COMPARISON OF PLASMA MEMBRANES PREPARED FROM NERVOUS TISSUE

| | SM (Method 1) | SM (Method 2) | PM (Method 4) | Neuronal SM Henn et al. (1972) | Synaptic PM | | Axolemma | Myelin Agrawal et al. (1970) |
					Morgan et al. (1971)	Cotman et al. (1969) Cotman and Matthews (1971)	Cotman et al. (1969) Barnola et al. (1973)	
Total yield (%)	1.1	1.3	0.2	0.002[c]	0.1[c]	2[c]	0.4	2[c]
Na-K-ATPase over homogenate	7.1	6.3	9.2	5.2	11.5	2.6	11–17	—
5'-nucleotidase over homogenate	3.6	—	2.3	5.3	4.2	1.2	—	—
Acetylcholinesterase over homogenate	1.5	2.8	—	—	0.9	1.4	6.5	—
Protein/cholesterol	—	7.2	2.9	—	6.3	6.1	1.1	0.2[c]
Protein/phospholipid	1.3	1.7	0.89	—	1.4	1.4	0.5	0.1[c]
Phospholipids (%)								
P-choline	53.5	52.1	53.9	—	41.6	43.5	45.9	33[c]
P-ethanolamine	33.6	36.4	27.8	—	34.2	36.2	34.4	44
P-serine	8.4	6.8	12.1	—	13.2	15.2	8	8
P-inositol	4.5	4.7	3.8	—	3.5	—	—	—
Sphingomyelin	—	—	0.8	—	5.1	3.7	10.0	15
Others	—	—	1.6	—	2.4	1.0	—	—

[a] PM = plasma membrane; SM = surface membrane.
[b] Lipid analysis was published by Breckenridge et al. (1972).
[c] Recalculated.

Although the gross composition of surface membranes was like that of synaptic plasma membranes, the distribution of individual phospholipids was quite different (Table I). Phosphatidylcholine constituted more than 50% of all phospholipids whereas in other brain membranes it was at most 46%. Another major difference was the low level of sphingomyelin present in surface and plasma membranes isolated from newborn rat brain. This phospholipid class constituted no more than 1% of the total phospholipid. This may reflect developmental changes in brain membranes, as sphingomyelin appears relatively late in rat brain cortex. Sphingomyelin is a major phospholipid of myelin, and is abundantly deposited in rat brain during myelination (Ansell, 1971). Yet sphingomyelin is present in the homogenate and smooth endoplasmic reticulum of newborn rat brain (Hemminki, 1973e), and thus its low concentration is typical only of surface and plasma membranes described here. If contamination with myelin in the preparations of synaptic plasma membranes and axolemma can be excluded (Table I) the concentration of sphingomyelin clearly differentiates synaptic plasma membranes, squid axolemma, and brain axolemma (DeVries and Norton, 1974) from the membranes described in this article.

IV.   Synthesis of Proteins and Glycoproteins of Surface Membranes

Protein and glycoprotein composition of surface and plasma membranes has been previously studied by sodium dodecyl sulfate–polyacrylamide gel electrophoresis (Hemminki, 1972b, 1973b,c). Major polypeptides were usually detected in the 45,000 57,000, 72,000, and 94,000 dalton region, and the patterns appear to be largely similar to preparations of synaptic plasma membranes (Levitan et al., 1972; Gurd et al., 1974), neuronal plasma membranes (Karlsson et al., 1973), and olfactory axolemma (Grefrath and Reynolds, 1973). Differences were detected between well purified smooth endoplasmic reticulum and surface membranes (Hemminki, 1973e). This contrasts with the findings of Gurd et al. (1974) showing extensive similarities with synaptic plasma membranes and microsomes. Yet total microsomes as used by Gurd et al. (1974) contain plasma membranes. The most prominent glycoproteins were detected in the 90,000 dalton region (Hemminki, 1972b, 1973e). Surface and plasma membranes drastically differed from smooth endoplasmic reticulum, which was poor in glycoproteins.

The time course of incorporation of radioactive leucine and fucose into surface membranes is shown in Fig. 5. The precursors were administered

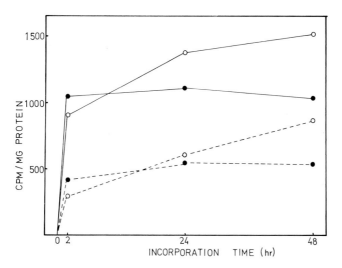

FIG. 5. Incorporation of leucine-³H (solid lines) and fucose-³H (dashed lines) into surface membrane (○) and homogenate (●) proteins and glycoproteins. Newborn rats (4–5 days of age) received 10 μCi of radioactive leucine or fucose intraperitoneally and were killed at times indicated. Brain cortices were dissociated into cells according to Hemminki (1972a) and surface membranes were isolated by Method 2 (see text).

intraperitoneally, the animals were killed, and the cells were dissociated from brain cortices followed by isolation of surface membranes by Method 2. Figure 5 demonstrates typical differences in labeling kinetics of surface membranes as compared to the total homogenate: the incorporation continues more slowly in plasma membranes indicating a delayed transport of newly synthesized plasma membrane constituents in agreement to previous observations on liver (Ray *et al.*, 1968) and brain (Hemminki, 1972b). After long labeling times plasma membranes attain a higher specific radioactivity than the total homogenate.

The labeling rates of individual polypeptides are studied in Fig. 6. Samples of surface membranes, labeled for 2 and 48 hours as in Fig. 5, are separated on polyacrylamide gels containing sodium dodecyl sulfate. The gels are sliced in 2-mm cylinders, whose radioactivity is determined (Hemminki, 1974). The peaks of radioactivity match with the major polypeptide bands. At 48 hours labeling time, the radioactivity also comigrated with the visible polypeptides. These data indicate that at least the most prominent polypeptides of surface membranes appear to label at fairly even rates. Elsewhere we provide additional evidence on this, using plasma membranes isolated from solid brain tissue and measuring the specific radioactivities (Hemminki, 1974).

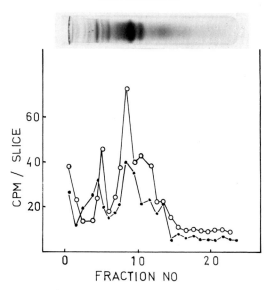

FIG. 6. Distribution of radioactivity along a sodium dodecyl sulfate–acrylamide gel containing surface membrane proteins. Newborn rats were labeled with leucine-³H and surface membranes were isolated as described in Fig. 5 after 2 hours (- ● -) and 48 hours (- ○ -) of incorporation. A gel containing surface membranes as stained with amido black is shown on top of the figure to indicate the distribution of surface membrane polypeptides.

ACKNOWLEDGMENTS

This study was supported by the Sigrid Jusélius Foundation and the National Research Council for Medical Sciences, Finland.

REFERENCES

Agrawal, H. C., Banik, N. L., Bone, A. H., Davison, A. N., Mitchell, R. F., and Spohn, M. (1970). *Biochem. J.* **120**, 635–642.

Ansell, G. B. (1971). *In* "Chemistry and Brain Development" (R. Paoletti and A. N. Davison, eds.), pp. 63–71. Plenum, New York.

Barnola, F. V., Villegas, R., and Camejo, G. (1973). *Biochim. Biophys. Acta* **298**, 84–94.

Bauman, A., Changeux, J.-P., and Benda, P. (1970). *FEBS (Fed. Eur. Biochem. Soc.) Lett.* **8**, 145–148.

Breckenridge, W. C., Combos, G., and Morgan, I. G. (1972). *Biochim. Biophys. Acta* **266**, 695–707.

Burdman, J. A. (1972). *J. Neurochem.* **19**, 1459–1469.

Caley, D. W., and Maxwell, D. S. (1968). *J. Comp. Neurol.* **113**, 1–16.

Camejo, G., Villegas, G., Barnola, F. V., and Villegas, R. (1969). *Biochim. Biophys. Acta* **193**, 247–259.

Cotman, C., Blank, M. L., Moehl, A., and Snyder, F. (1969). *Biochemistry* **8**, 4606–4612.

Cotman, C. W., and Matthews, D. A. (1971). *Biochim. Biophys. Acta* **249**, 380–394.

Denburg, J. L. (1972). *Biochim. Biophys. Acta* **282**, 453–458.

Denburg, J. L. (1973). *Biochim. Biophys. Acta* **298**, 967–972.

DeRobertis,, E., Alberici, M., Rodriguez de Lores Arnaiz, G., and Azcurra, J. M. (1966). *Life Sci.* **5**, 577–582.

DeVries, G. H., and Norton, W. T. (1974). *J. Neurochem.* **22**, 259–264.

DeVries, G. H., Norton, W. T., and Raine, C. S. (1972). *Science* **175**, 1370–1372.

Fischer, S., Cellino, M., Zambrano, F., Zampigni, G., Tellez Nagel, M., Marcus, D., and Canessa-Fischer, M. (1970). *Arch. Biochem. Biophys.* **138**, 1–15.

Glick, M. C., Kimhi, Y., and Littauer, U. Z. (1973). *Proc. Nat. Acad. Sci. U.S.* **70**, 1682–1687.

Grefrath, S. P., and Reynolds, J. A. (1973). *J. Biol. Chem.* **248**, 6091–6094.

Gurd, J. W., Jones, R. L., Mahler, H. R., and Moore, W. J. (1974). *J. Neurochem.* **22**, 281–290.

Hemminki, K. (1972a). *Exp. Cell Res.* **75**, 379–384.

Hemminki, K. (1972b). *Life Sci.* **11**, 1173–1179.

Hemminki, K. (1973a). *Comment. Biol.* **70**, 1–27.

Hemminki, K. (1973b). *Exp. Cell Res.* **82**, 31–38.

Hemminki, K. (1973c). *Int. J.* Neurosci. **5**, 81–85.

Hemminki, K. (1973d). *FEBS (Fed. Eur. Biochem. Soc.) Lett.* **38**, 79–82.

Hemminki, K. (1973e). *Biochim. Biophys. Acta* **298**, 810–816.

Hemminki, K. (1974). *Biochim. Biophys. Acta* **359**, 83–89.

Hemminki, K., and Suovaniemi, O. (1973). *Biochim. Biophys. Acta* **298**, 75–83.

Henn, F. A., Hansson, H., and Hamberger, A. (1972). *J. Cell Biol.* **53**, 654–661.

Karlsson, J.-O., Hamberger, A., and Henn, F. A. (1973). *Biochim. Biophys. Acta* **298**, 219–229.

Levitan, I. B., Mushynski, W. E., and Ramirez, G. (1972). *J. Biol. Chem.* **247**, 5376–5381.

McIlwain, H., and Bachelard, H. S. (1971). *In* "Biochemistry and the Central Nervous System," 4th ed., pp. 406–444. Churchill, London.

Meunier, J.-C., and Changeux, J.-P. (1973). *FEBS (Fed. Eur. Biochem. Soc.) Lett.* **32**, 143–148.

Morgan, I. G., Wolfe, L. S., Mandel, P., and Combos, G. (1971). *Biochim. Biophys. Acta* **241**, 737–751.

Ray, T. K., Lieberman, I., and Lansing, A. I. (1968). *Biochem. Biophys. Res. Commun.* **31**, 54–58.

Warren, L., and Glick, M. C. (1969). *In* "Fundamental Techniques in Virology" (K. Habel and N. P. Salzman, eds.), Vol. 1, pp. 66–71. Academic Press, New York.

Whittaker, V. P., Michaelson, I. A., and Kirkland, R. J. A. (1964). *Biochem. J.* **90**, 412–417.

Chapter 14

The Plasma Membrane of KB Cells; Isolation and Properties

F. C. CHARALAMPOUS AND N. K. GONATAS

*Departments of Biochemistry and Pathology,
University of Pennsylvania School of Medicine,
Philadelphia, Pennsylvania*

I. Introduction

Studies of the plasma membrane of animal cells have attracted the attention of an ever increasing number of investigators with the realization that knowledge of its structure and of the mechanisms involved in its assembly and functional regulation is of fundamental importance in understanding the molecular mechanisms underlying many biological processes; these include active transport, contact inhibition of movement, cell adhesiveness, density-dependent inhibition of growth, several critical aspects of malignancy, endocytosis, cellular secretion, role of receptors involved in the transduction of afferent and efferent messages, immunologic specificity, synaptic transmission, and information storage and retrieval.

One of the approaches employed in studies of the plasma membrane involves its isolation followed by characterization of its properties. Although a variety of preparative and analytical isolation procedures have been devised, it appears that no single method can be successfully employed for the isolation of the plasma membranes of the many different cell types. Apart from the considerable complexities inherent in working with membranes, many isolation methods do not meet rigorous criteria concerning the purity of the isolated membranes.

In assessing the purity of the plasma membranes it is necessary to employ both biochemical and morphologic criteria. Ideally, the minimal biochemical criteria should include: (1) Two or more markers which are present in the plasma membranes but absent in all other cell fractions. The degree of enrichment of each of these markers in the plasma membrane fraction must be the same. This is of particular value in those situations in which a marker is present not only in the plasma membranes but also in other cell compartments. (2) Two or more markers present in each of the other cell membranes or organelles but absent in the plasma membranes. The choice of these markers must be such as to exclude those that are easily inactivated or solubilized from their natural cell structures during the isolation procedure. Under these conditions the absence of these markers in the plasma membrane fraction will constitute unambiguous evidence against contamination of this fraction with other cell membranes. (3) The use of two different isolation methods yielding plasma membranes of similar purity and properties. This will minimize the possibility that the presence or the absence of a particular component or feature in the isolated plasma membranes is an artifact resulting from the particular conditions employed during the isolation.

Here we describe the isolation of the plasma membrane of KB cells by two different methods, namely, the latex bead ingestion and the $ZnCl_2$ methods. Optimal conditions for bead uptake and the isolation procedures employing discontinuous sucrose gradient centrifugation are presented. Both methods yield plasma membranes of high and comparable purity based on morphological and biochemical criteria. The concentrations of various enzymes in the plasma membranes are compared with those present in the mitochondria and microsomes of these cells. Most of these results have been published recently (Charalampous *et al.*, 1973).

II.   Culture Conditions and Various Media

The KB cells (certified line No. 17 of the American Type Culture Collection) were grown in suspension culture in Eagle's minimal essential medium with 10% horse serum (Charalampous *et al.*, 1961) at a cell density between 2×10^5 and 5×10^5 cells per milliliter.

In the experiments for bead uptake the cells were incubated in Krebs-Ringer bicarbonate buffer (KRB) of the following composition, in millimolar concentrations: KCl, 4.74; $CaCl_2$, 2.53; KH_2PO_4, 1.19; $MgSO_4$, 1.19; NaCl, 118.5; $NaHCO_3$, 24.9; and glucose, 2.0. The solution, equilibrated with a mixture of 95% O_2–5% CO_2, had a pH of 7.4.

The phosphate-buffered saline (PBS) contained, in millimolar concentrations: sucrose, 250; EDTA, 1.0; NaCl, 137; KCl, 2.7; Na_2HPO_4, 8.1 KH_2PO_4, 1.47; and 0.5 gm of bovine serum albumin per liter. The pH was 7.4.

The sucrose solutions used in the isolation of plasma membranes by the latex bead method were made in Tris-EDTA buffer (0.02 M Tris, pH 6.8, containing 1 mM EDTA), whereas those used in the $ZnCl_2$ method were made in distilled water. The concentrations are based on weight/weight.

III.   Enzymic Assays

The following enzymic activities were assayed according to the methods indicated: $Na^+ + K^+$-activated ATPase (Post and Sen, 1967), 5'-nucleotidase (Heppel and Hilmoe, 1955), NADPH–cytochrome c reductase (Masters *et al.*, 1967), NADH–cytochrome c reductase and rotenone-insensitive NADH–cytochrome c reductase (Mackler, 1967), succinate–cytochrome c reductase (Tisdale, 1967), cytochrome c oxidase (Chen and Charalampous, 1969), acid phosphatase using p-nitrophenylphosphate as substrate (Ostrowski and Tsugita, 1961), phosphoprotein phosphatase (Paigen, 1958), and cathepsins (Mycek, 1970).

The concentrations of cytochromes aa_3, b, c, and c_1 were calculated from difference spectra (reduced minus oxidized) according to the method of Williams (1964). Cytochrome b_5 was calculated from difference spectra (reduced minus oxidized) using an extinction coefficient of 20.0 mM^{-1} cm^{-1} at 556 minus 575 nm (Garfinkel, 1957). Cytochrome P-450 was calculated from difference spectra (reduced CO minus reduced) using an extinction coefficient of 91 mM^{-1} cm^{-1} at 450 minus 490 nm (Omura and Sato, 1967).

IV. Analytical Methods

Inorganic phosphate was determined by the method of Fiske and SubbaRow (1925), and protein by the method of Lowry *et al.* (1951). Radioactive samples were counted in a Packard Tri-Carb liquid scintillation spectrometer (Packard Instrument Co., Inc., Downers Grove, Illinois) using as scintillation fluid a mixture of 792 ml of dioxane, 30 ml of toluene, 4.5 gm of 2,5-diphenyloxazole (PPO), and 90 gm of naphthalene. The latex beads are completely soluble in this mixture and do not affect the counting efficiency.

V.   Isolation of Plasma Membranes by the Latex Bead Method

a. Preparation of Latex Beads.   Ten milliliters of a 10% suspension of polystyrene latex beads 1.01 μm in diameter (Dow Chemical, Midland, Michigan) were centrifuged at 6000 g for 15 minutes and suspended in 4.0 ml of KRB medium. Bead clumps were broken up by sonicating the suspension twice for 15 seconds each with a 30-second interval, using a Branson Model W-185C sonifier (Branson Instruments Co., Stamford, Connecticut) at an output setting of 4.

b. Uptake of beads by KB Cells.   The harvested cells (5 × 10^8) were washed with 20 volumes of ice-cold 0.15 M NaCl and were resuspended in 10 ml of KRB medium. To the cell suspension 4 ml of beads prepared as described above were added, and the mixture was incubated in a 50-ml Erlenmeyer flask for 30 minutes at 37°C with gentle agitation. At the end of this period 20 ml of 0.3 M sucrose were added, and the mixture was centrifuged at 400 g for 4 minutes. The excess beads were removed by washing the cells five times with 20 ml each of 0.3 M sucrose, followed by a final wash with 30 ml of PBS.

c. Isolation of Membrane-Enclosed Beads. The washed cells from the previous step were suspended in 10 ml of PBS and homogenized in a Dounce, type B, homogenizer employing 25 full strokes. The extent of cell breakage which was complete was monitored by phase microscopy. The cell homogenate was diluted with an equal volume of 60% sucrose, and 6–7 ml of the resulting mixture were placed at the bottom of each of three cellulose nitrate tubes of the SW-25.1 rotor (Beckman Instruments, Inc., Spinco Division, Palo Alto, California) and were then overlaid with approximately 6–7 ml each of 25, 20, and 10% sucrose. The tubes were centrifuged at 79,000 g for 90 minutes. The membrane-enclosed beads concentrated at the interface between the 10 and 20% sucrose layers were carefully collected after removing the overlying sucrose solution. The suspension of the membrane-enclosed beads was diluted with an equal volume of the Tris-EDTA buffer and was centrifuged at 30,000 g for 15 minutes. The resulting pellet was resuspended in 6–7 ml of 30% sucrose and was centrifuged once more in the discontinuous sucrose gradient described above. The band containing the membrane-enclosed beads was recovered from the gradient, and after dilution with an equal volume of the Tris-EDTA buffer it was centrifuged at 30,000 g for 15 minures. The pelleted material was washed with 20 ml of 10% sucrose and was finally suspended in 1–3 ml of 10% sucrose or 0.02 M Tris, pH 6.8.

d. Isolation of Membranes. The suspension of the membrane-coated beads from the preceding step was sonicated for twelve 10-second pulses in an ice bath in the Branson Sonifier (Model W-185C) at an output setting of 4 with 1-minute intervals between each pulse. The temperature of the suspension did not rise above 5°C. The sonicated sample was centrifuged at 15,000 g for 20 minutes to remove the latex beads. The membranes contained in the supernatant fluid were collected by centrifugation at 100,000 g for 1 hour in Spinco rotor No. 40.

VI. Isolation of Plasma Membranes by the ZnCl$_2$ Method

This procedure is a modification of the ZnCl$_2$ method described by Warren and Glick (1969). KB cells (5 × 10^8) were harvested by centrifugation at 400 g for 5 minutes, washed twice with 50 volumes of cold 0.15 M NaCl, and resuspended in 10 ml of cold NaCl solution. To 1-ml aliquots of this suspension 3 ml of 0.001 M ZnCl$_2$ were added and the mixtures were incubated at 25°C for exactly 6 minutes. The incubations were terminated by chilling in ice, and each cell suspension was homogenized manually in a Dounce homogenizer (type B) at 4°C using about 150 strokes. The cell breakage, monitored by phase microscopy, was 80–90%. The combined homogenate (40 ml) was mixed with an equal volume of 60% sucrose, and 40-ml portions of the resulting mixture were layered over discontinuous sucrose gradients consisting of 30 ml each of 55, 50, 48, 45, 43, 40, and 35% sucrose solutions in 250-ml polycarbonate centrifuge bottles. The gradients were centrifuged at 1400 g for 30 minutes in the Sorvall RC2-B centrifuge (Ivan Sorvall, Inc., Newton, Connecticut) using the HS-4 rotor. The membrane-enriched fraction recovered from the 43 and 45% sucrose layers of each gradient contained mainly whole membrane envelopes, large membrane fragments, and some unidentified particulate matter. These fractions were combined to give the "principal membrane fraction." In order to recover more membranes from the sucrose gradient the fractions recovered between 45 and 50% and between 40 and 43% sucrose were diluted with water to 35% sucrose and centrifuged separately at 5000 g for 12 minutes. The resulting pellets were suspended in 2 ml of 10% sucrose and each suspension was further purified on a discontinuous sucrose gradient consisting of 2 ml each of 45, 40, 35, and 30% sucrose. The gradients were centrifuged at 250 g for 30 minutes, and the membrane-enriched fractions recovered from the 35 and 40% sucrose layers were combined and centrifuged at 5000 g for 12 minutes. The pellet was suspended in 1 ml of 35% sucrose and was combined with the principal membrane fraction obtained from the first sucrose gradient (3.6 ml total volume). The combined membrane fraction was further purified by layering 0.6-ml aliquots over each of six similar sucrose gradients having the following composition: 0.5 ml of each of 65, 50, and 40% sucrose, and 1 ml each of 55, 50, and 45% sucrose. The gradients were centrifuged at 50,000 g for 1 hour in the SW 50-1 rotor (Beckman Instruments, Inc., Spinco Division). The membranes recovered from the 55% sucrose layer of the 6 gradients were diluted with water to 35% sucrose and were pelleted by centrifugation at 6000 g for 12 minutes. After the membranes were washed in 10 ml of 0.25 M sucrose they were resuspended in 3 ml of the same sucrose solution and were used promptly in the various assays.

VII.   Isolation of Mitochondria and Microsomes from KB Cells

The method is a modification of the centrifugal fractionation procedure of Schneider (1948) using 0.25 M sucrose. All operations were performed at 0–4°C. Solutions: 0.25 M sucrose; 0.25 M sucrose in 0.01 M triethanolamine, 1 mM EDTA, of pH 7.2 (STE); imidazole-glycylglycine buffer, 0.015 M, pH 7.4.

a. Preparation of Cell Homogenate. The cells (5×10^8) were harvested by centrifugation at 400 g for 4 minutes, washed with 0.15 M NaCl, and resuspended in 16 ml of 0.25 M sucrose. They were then homogenized in a Dounce homogenizer employing about 150 full strokes with a B pestle. Cell breakage was usually greater than 95%.

b. Isolation of Mitochondria. The homogenate was centrifuged twice at 700 g for 10 minutes to remove nuclei and unbroken cells. The supernatant fluid was centrifuged in the HB-4 Sorvall swinging bucket at 8000 g for 20 minutes, and the supernatant was centrifuged once more at this speed. The two pellets thus obtained are the crude mitochondrial fraction whereas the supernatant fluid is the postmitochondrial fraction containing the microsomes.

The two pellets comprising the crude mitochondrial fraction were overlaid with 2 ml of STE, and the "fluffy top layer" was removed by gentle agitation. The two pellets were combined by suspending them in 20 ml of STE and were centrifuged at 8000 g for 20 minutes. The pellet was washed twice more in the same manner and each time any fluffy layer was removed as described above. The washed pellet was suspended in 2–3 ml of imidazole–glycylglycine buffer to give the *final mitochondrial fraction.*

c. Isolation of Microsomes. The postmitochondrial fraction obtained in the previous step was centrifuged twice at 15,000 g for 15 minutes in the HB-4 rotor, and the small amount of precipitated material was discarded. The microsomes were precipitated by centrifuging the supernatant fluid at 100,000 g for 90 minutes in the No. 40 rotor of the Spinco centrifuge. The pelleted microsomes were resuspended in 20 ml of STE and were centrifuged again as above. The resulting pellet was suspended in 4–5 ml of imidazole glycylglycine buffer to give the *final microsomal fraction.*

VIII.   Electron Microscopy

The KB cells were washed in 0.15 M NaCl at 4°C and fixed in 5% glutaraldehyde in 0.1 M cacodylate buffer, pH 7.4, for 1–3 hours at 4°C. The cells were postfixed in 1–2% osmium tetroxide in 0.1 M cacodylate buffer for another hour at 4°C. After dehydration in ethanol, cells were embedded in Araldite. The treatment with propylene oxide was omitted because it extracts latex beads.

Sections, cut with diamond knives, were stained with uranyl acetate and lead citrate. Subcellular fractions were fixed and processed as cell suspensions. Packing of fractions was achieved after centrifugation in a Sorvall RC2-B centrifuge at 5000–10,000 rpm in a swinging-bucket rotor. Some membrane preparations were stained in block with 3% aqueous uranyl acetate. Electron micrographs were taken with an Elmiskop IA electron microscope.

IX.   Kinetics of Bead Uptake

In several preliminary experiments the optimal conditions for maximal bead uptake were determined with respect to the composition of the incubation medium, the time and temperature of incubation, the size of the latex beads, and the ratio of number of beads per cell. The amount of beads ingested was quantified as follows: At the end of the incubation, excess beads were removed as described in Section V, and the cell pellet was extracted with dioxane (2 ml/1 × 10^6 cells) for 4 hours at room temperature. This time was found sufficient for extracting all the ingested beads. After removal of insoluble material by centrifugation, the absorbance of the supernatant fluid at 259 nm was used to calculate the amount of polystyrene beads ingested (Weisman and Korn, 1967). Similar incubations without beads were used as controls.

For bead uptake two different incubation media were tried, the growth medium and Krebs-Ringer bicarbonate. Among varous size beads the following were tried: 0.795, 0.81, 1.01, and 1.10 μm in diameter. The ratio of number of beads per cell was varied from 500 to 4000. The results of these experiments showed that bead uptake was maximal when the cells were incubated with beads of 1.01 μm in diameter in KRB medium at a bead

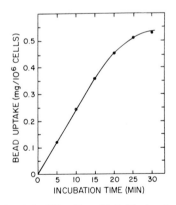

FIG. 1.   Uptake of latex beads by KB cells at 37°C. The incubation mixture contained, per milliliter, 1 × 10^7 cells and 16 mg of latex beads. Each point represents the average of triplicate incubations.

multiplicity of 2000–4000 beads per cell. The time course of bead uptake is shown in Fig. 1. The uptake of beads was linear with time for the first 15 minutes and was complete by 30 minutes. The rate of bead uptake was proportional to the cell concentration up to 1.8×10^7 cells per milliliter. Lowering the incubation temperature to 0°C abolished bead uptake.

X.   Morphological Studies

A.   Whole Cells

Intracytoplasmic latex beads, 0.7–1 μm in cross sectional diameter, were observed after incubation of KB cells with latex beads for 30 minutes. The beads were randomly scattered in the cytoplasm (Fig. 2). A single membrane

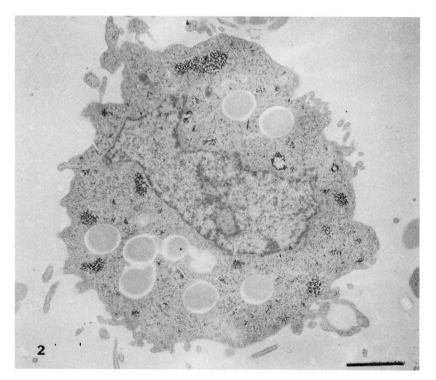

FIG. 2.   KB cell incubated for 30 minutes with latex beads. Note clusters of glycogen particles. Bar, 1 μm. × 14,000.

separated the bead from the cytoplasm; occasionally several beads had coalesced and were surrounded by a single membrane. There was no evidence of continuity or fusion of any cytomembrane (i.e., endoplasmic reticulum, Golgi apparatus, lysosome) with the membrane surrounding the bead. The cells which had ingested beads showed no unusual cytoplasmic or nuclear alterations except for prominence of clusters of glycogen particles in the vicinity of the intracytoplasmic beads. Cells in mitosis also had engulfed beads.

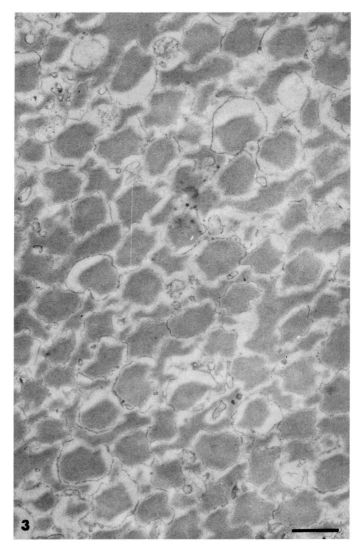

FIG. 3.   Isolated membrane-enclosed beads. Bar, 1 μm. ×18,000.

B.   Membrane-Enclosed Beads

The fraction of the isolated membrane-coated beads showed somewhat distorted beads surrounded by a single membrane (Fig. 3). Fragments of membranes or small vesicles were occasionally seen in this fraction but other constituents such as mitochondria or rough endoplasmic reticulum were not observed. The thickness of the membrane surrounding the bead was 75–100 Å wide (Fig. 4). The membrane was irregular with segments that appeared fuzzy, and other segments in which a triple-layered structure 45–50 Å wide was observed.

C.   Plasma Membranes

The isolated fraction of plasma membranes was extremely uniform and consisted of round, oval, or tubular profiles (Fig. 5). The diameter of these structures varied from 0.1 to 0.3 μm, with occasional larger (0.5 μm) diameters. Occasionally, the membrane of the vesicles could be resolved to a trilaminar structure, 75–85 Å thick. Mitochondria, free ribosomes or polysomes, glycogen granules, or other identifiable structures were not observed in this fraction.

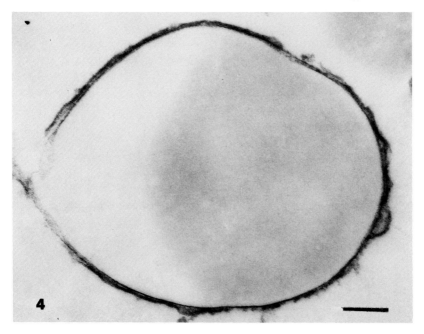

Fig. 4.   Isolated beads with surrounding membrane, which occasionally shows trilaminar structure. Bar, 0.1 μm. $\times$128,000.

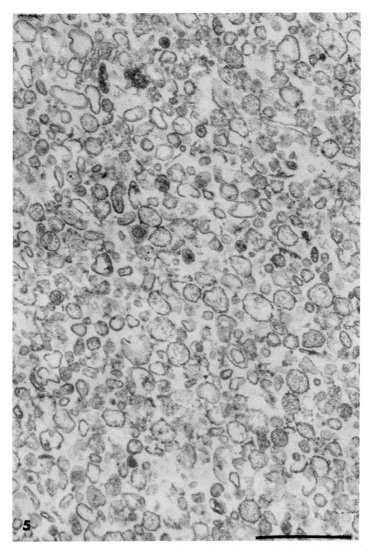

FIG. 5.   Plasma membrane preparation isolated by the latex bead method. Note homogeneity of fraction. Bar, 1 μm . ×30,000.

D.   Microsomes

This preparation differed from the plasma membrane fraction in two aspects: (1) the vesicles were larger (0.3–0.6 μm) and more uniform, round

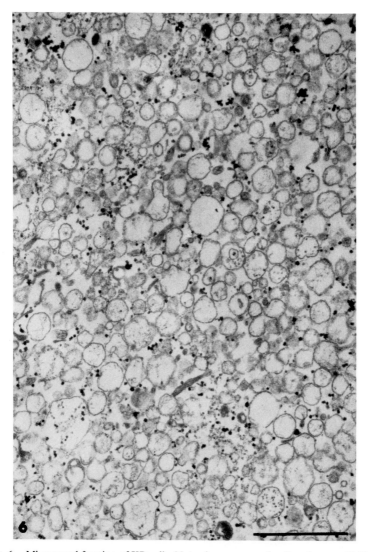

FIG. 6.   Microsomal fraction of KB cells. Note glycogen granules. Bar, 1 μm. ×30,000.

or oval; (2) numerous glycogen granules were observed in this fraction, whereas glycogen was never observed in the plasma membrane fraction (Fig. 6).

XI. Biochemical Studies

Concentration of Various Enzyme Activities in Mitochondria, Microsomes, and the Plasma Membrane of KB Cells

Table I summarizes the specific activities of various enzymes in these fractions and in the unfractionated cell homogenate. The values represent the averages from three to five separate experiments. Maximal deviation of the individual values from the average was $\pm 10\%$.

1. Na$^+$ + K$^+$-Activated ATPase and 5′-Nucleotidase

The specific activities of these enzymes were highest in the plasma membranes where they were concentrated 21–23-fold over the concentra-

TABLE I

Subcellular Distribution of Certain Enzyme Activities in KB Cells[a]

Enzyme	Homogenate	Mitochondria	Microsomes	Plasma membranes	
				Beads[b]	ZnCl$_2$[b]
Na$^+$ + K$^+$-activated ATPase	4.3	1.6	0.0	100.0	98.0
5′-Nucleotidase	65.0	87.0	77.0	1400.0	1375.0
Succinate–cytochrome c reductase	9.4	216.0	0.4	3.6	7.0
Cytochrome oxidase	2.1	19.6	0.0	0.0	0.005
NADPH–cytochrome c reductase	10.0	0.9	31.2	50.4	51.0
Rotenone-insensitive NADH–cytochrome c reductase	ND[d]	11.5	103.0	1.7	1.9
Cytochrome b_5	ND	ND	0.04	0.08	0.076
Cytochrome P-450	ND	ND	0.003	<0.0003[c]	<0.0003[c]
Acid phosphatase	20.5	22.4	20.8	30.0	ND

[a]The values represent specific activities which in all cases, except for cytochrome oxidase and cytochromes b_5 and P-450, are expressed as nanomoles of substrate used or product formed per minute per milligram of protein. Cytochrome oxidase activities are expressed as $k \times 10^{-3}$ sec^{-1} per milligram of protein, where k is the first-order rate constant for the oxidation of ferrocytochrome c. The values for cytochromes b_5 and P-450 are in nanomoles per milligram of protein. The concentration of rotenone in the assay was 0.1 mM. All fractions were sonicated for 30 seconds at 0°C before their use in the assays since this treatment gave higher activities for cytochrome oxidase and succinate–cytochrome c reductase, without affecting the other enzyme activities.

[b]This refers to the method of isolation of the membranes.

[c]Lowest detectable level.

[d]ND not determined.

tion in cell homogenates. Microsomes were devoid of this ATPase, and only trace amounts of this activity were present in the mitochondria, reflecting, very likely, marginal contamination by the plasma membranes. The specific activity of 5'-nucleotidase in the microsomal and mitochondrial fractions was not very different from that of cell homogenates. It is concluded, therefore, that these two enzymes are unique components of the plasma membrane of KB cells.

2. NADPH-Cytochrome c Reductase and Cytochrome b_5

Both the reductase considered to be a microsomal system (Williams and Kamin, 1962), and cytochrome b_5 reported to be present in both microsomes (Strittmatter, 1963) and the outer mitochondrial membrane of rat liver cells (Sottocassa et al., 1967) were also present in the plasma membrane. The specific activities in the latter fraction are, respectively, 1.6 and 2.0 times greater than the corresponding values of the microsomes. Vassiletz et al. (1967) also reported the presence of NADPH–cytochrome c reductase and cytochrome b_5 in the plasma membranes of liver cells. However, in this case the specific activities in the plasma membrane were 40% and 10%, respectively, of the corresponding values of the microsomal fraction.

3. Cytochrome P-450

This microsomal cytochrome was not detected in the plasma membranes, in agreement with similar findings with the plasma membranes of liver cells (Vassiletz et al., 1967).

4. Rotenone-Insensitive NADH-Cytochrome c Reductase

The highest specific activity of this system was found in microsomes as compared with the mitochondria. It was virtually absent in the plasma membranes (1.7% of the specific activity found in microsomes). This activity has been also found in the outer mitochondrial membrane of rat liver cells (Sottocassa et al., 1967).

5. Cytochromes aa_3, b, c, and c_1

These mitochondrial enzymes as well as cytochrome oxidase activity were not found in the microsomes or the plasma membranes isolated by the latex bead method. The presence of these cytochromes was determined by difference spectrophotometry (reduced minus oxidized spectra) at the temperature of liquid nitrogen under conditions which could have detected amounts equivalent to 1% of the corresponding specific activities of the mitochondrial fraction. Succinate–cytochrome c reductase was detected in these membranes, but its specific activity was 1.7% of that of the mitochondrial fraction.

The plasma membranes isolated by the $ZnCl_2$ method exhibited traces of cytochrome oxidase activity and low levels of succinate–cytochrome c reductase the specific activity of which was 3.2% of that of the mitochondrial fraction. They also contained trace amounts of cytochrome c_1 at a concentration equivalent to 1% of that of mitochondria, but they were free of cytochromes b and c.

6. Lysosomal Enzymes

Acid phosphatase activity was found in all the fractions at approximately similar specific activity which was comparable to that of the unfractionated homogenate. Cathepsins A, B, C, and D, and phosphoprotein phosphatases could not be detected in the homogenates or any of the other fractions.

XII.   Kinetics of Transport of α-Aminoisobutyric Acid (AIB) in KB Cells After Ingestion of Latex Beads

In these studies the kinetics of uptake of $1\text{-}^{14}C$-labeled α-aminoisobutyric acid (Calbiochem, San Diego, California) were investigated with cells after ingestion of latex beads. It was reasoned that if the plasma membrane sites involved in AIB transport are interiorized during phagocytosis of latex beads such cells will exhibit decreased rate of AIB uptake and lower V_{max} for AIB influx as compared with cells not exposed to beads. If, on the other hand, the kinetics of AIB transport are unaffected by bead ingestion, it would indicate that the transport sites of the plasma membrane are either not interiorized or that they are regenerated following their interiorization. The results are shown in Fig. 7. From the Lineweaver-Burk plots of the data the V_{max} (nanomoles of AIB taken up per minute per 10^6 cells) and K_m (mM) values were derived. Cells not exposed to beads have a V_{max} of 18.0 as compared with 7.9 for cells after bead ingestion. The K_m was 1.0 mM for both types of cells. These results show that the number of AIB transport sites of the plasma membrane is decreased after bead ingestion as a consequence of plasma membrane interiorization accompanying phagocytosis.

XIII.   Purity of the Plasma Membranes

1. With regard to the plasma membranes prepared by the latex bead method, the results presented in Table I and in the text, as well as the electron micrographs, show that this fraction satisfies the criteria of purity mentioned

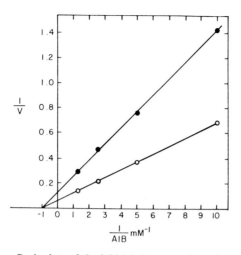

FIG. 7. Lineweaver-Burk plots of the initial influx rate of α-aminoisobutyric acid (AIB) versus AIB concentration. The cells were first incubated for 20 minutes with (● ——— ●) or without (○——○) latex beads in Krebs-Ringer bicarbonate buffer; after removal of excess beads, they were incubated at 37°C with different concentrations of ^{14}C-labeled AIB in the same buffer for 1 minute. Velocity is expressed as nanomoles of AIB uptake per 10^6 cells per minute.

in the introduction. The plasma membranes contain $Na^+ + K^+$-activated ATPase and 5′-nucleotidase, the specific activities of which increase by the same factor (22–23-fold) over the corresponding values in the unfractionated homogenate. Furthermore, this ATPase is not detectable in the other fractions, and the 5′-nucleotidase is present at very low levels in both mitochondria and microsomes. The latter levels do not differ significantly from the level in the whole homogenate.

Of the five mitochondrial enzyme markers, i.e., succinate–cytochrome c reductase, cytochrome oxidase, and cytochromes b, c, and c_1, four were absent in the plasma membrane fraction, and only the presence of trace amounts of succinate–cytochrome c reductase activity would suggest a maximal contamination of 1.7% by mitochondrial components.

Of the microsomal markers cytochrome P-450 was not detectable in the plasma membranes, and the specific activity of the rotenone-insensitive NADH-cytochrome c reductase of the plasma membranes was 1.7% of that of the microsomes, indicating a maximal possible contamination by microsomes of 1.7%. It is interesting to note that with respect to the NADPH–cytochrome c reductase and cytochrome b_5, both of which have been considered primarily as microsomal components, their specific activities are significantly higher in the plasma membrane of KB cells than in the microsomes. It must be concluded, therefore, that these enzymes are present in both the plasma membrane and the microsomes of the KB cells.

With regard to the acid phosphatase the profile of its specific activity among the various fractions indicates no preferential concentration in any particular fraction, suggesting adsorption of this enzyme on the various fractions during their isolation.

The electron micrographs show the morphological homogeneity of the plasma membranes (Fig. 5) and of the membrane-enclosed latex beads (Fig. 3) from which the plasma membrane fraction was derived.

2. With respect to the membranes isolated by the $ZnCl_2$ method, the data of Table I show that the specific activities or concentrations of the various enzymes are practically identical to those of the latex bead membranes. An exception is the succinate–cytochrome c reductase, the specific activity of which was twice as high in the $ZnCl_2$ membranes. This together with the presence of trace amounts of cytochromes aa_3 and c_1 in the latter membranes indicates that the degree of contamination of the $ZnCl_2$ membranes by mitochondrial enzymes, although small (3.2%), is twice that of the latex bead membranes.

On the basis of these results it can be concluded that both isolation methods yield plasma membranes of comparable purity. However, the latex bead method offers the advantage of studying the inner (cytoplasmic) surface of the plasma membrane by labeling isolated membrane-coated beads with various nonpermeant probes, e.g., iodination with lactoperoxidase (Phillips and Morrison, 1971; Hubbard and Cohn, 1972), protein labeling with pyridoxal phosphate in the presence of sodium borotritide (Rifkin et al., 1972), and lipid labeling with 2,4,6-trinitrobenzenesulfonate (Bonsall and Hunt, 1971).

XIV.   Validity of the Latex Bead Method

The isolation of plasma membranes by two independent methods is very useful not only in assessing the purity of the membranes, but also in evaluating the validity of the latex bead method. One question that can be raised conerns the possibility that the membranes isolated by this method represent a specialized portion of the plasma membrane engaged in phagocytosis rather than a representative sample of the whole membrane. That this is unlikely in the case of KB cells is evident from the excellent agreement between the specific activities of the four enzymes studied with plasma membranes which were isolated by the two methods, i.e., the $Na^+ + K^+$- activated ATPase, the 5′-nucleotidase, the NADPH-cytochrome c reductase, and cytochrome b_5. Such agreement is improbable if the latex bead method was selecting specialized areas of the cell surface membrane. Additional evidence supporting the above conclusion comes from the kinetics of transport of α-aminoisobutyric acid. The results of Fig. 7 show that the number of transport sites (V_{max}) of the plasma membrane was considerably decreased following phagocytosis of latex beads, a finding that is expected if these sites were interiorized. Hence, the plasma membrane interiorized as a result of phagocytosis carried with it amino acid transport sites and the specific enzyme markers of the plasma membrane. These findings agree with those of Lutton (1973), who observed decreased binding of concanavalin A by the plasma membrane receptors of peritoneal macrophages following bead ingestion. Our results are not in agreement with those of Tsan and Berlin (1971), who showed that in polymorphonuclear leukocytes and alveolar marcophages the rate of transport of various nonelectrolytes was the same before and after bead ingestion. They interpreted these results as indicating that the carrier sites involved in these transport systems were not interiorized during the internalization of plasma membrane accompanying phagocytosis. The differences between our results and those of Tsan and Berlin may result from differences in the structure of the plasma membranes of the cells used in these studies or they may reflect different mechanisms of phagocytosis operative in different cells.

Another aspect concerning the validity of the latex bead method is the possibility that the plasma membrane surrounding the ingested beads may undergo modification following interiorization (Werb and Cohn, 1972). Such modification could result from fusion of the ingested beads with lysosomes on prolonged incubation following phagocytosis. In the present method this possibility is unlikely since the isolation of the membrane-coated beads is carried out immediately after phagocytosis of the beads (after 20–30 minutes of incubation with latex beads). Under these conditions fusion of the beads with lysosomes or among themselves is not detectable unless the incubation with beads is prolonged beyond 1 hour. This conclusion is further supported by the excellent agreement of the specific activities of the four plasma membrane markers of membranes isolated by the latex bead and the $ZnCl_2$ methods.

XV. Damage of the Plasma Membrane by Tris Buffer

Tris buffer causes a breakdown of the isolated plasma membranes of KB cells, and KB cells exposed to isoosmotic Tris buffer (phH 7.0–7.4) lose progressively their viability and various plasma membrane functions, such as concentrative amino acid transport and active translocation of Na$^+$ and K$^+$. These effects have been observed with Tris from various sources even after repeated crystallizations. The presence of sucrose (10% or higher) in the Tris buffer protects the isolated plasma membranes against the Tris effect. For these reasons, methods employing Tris in the isolation of plasma membranes (Warren and Glick, 1969; Bosmann *et al.*, 1968; Werb and Cohn, 1972) are not suitable for use with KB cells.

XVI. Applicability of the Latex Bead Method to Other Cells

In principle the latex bead method can be employed with any cell type provided the proper conditions for bead ingestion are optimized (size and multiplicity of beads, incubation medium, time and temperature of incubation, and absence of fusion betweeen phagosomes and lysosomes). The method has been used for the isolation of phagosomes from *Acanthamoeba* (Wetzel and Korn, 1969), L cells (Heine and Schnaitman, 1971), and polymorphonuclear leukocytes and macrophages (Tsan and Berlin, 1971; Smolen *et al.*, 1971; Werb and Cohn, 1972). In our laboratories we have developed optimal conditions for the ingestion of latex beads by mouse neuroblastoma cells (clone Neuro-2A) in monolayer (differentiated state) and suspension cultures (undifferentiated form). In addition we have observed that only lymph node macrophages ingest latex beads (1.01 μm in diameter) after intradermal-subcutaneous injection of a 10% aqueous suspension of beads (Fig. 8). Thus the isolation of plasma membranes by the latex bead procedure described here can be accomplished with cells grown in culture as well as with certain cells from tissues after *in vivo* injection of the beads.

ACKNOWLEDGMENTS

This work was supported by National Science Foundation grant, GB-35378X and U.S. Public Health Service grant, NS-05572-09.

REFERENCES

Bonsall, R. W., and Hunt, S. (1971). Biochim. Biophys. Acta **249**, 281.
Bosmann, B. H., Hagopian, A., and Eylar, E. H. (1968). *Arch. Biochem. Biophys.* **128**, 51.
Charalampous, F. C., Wahl, M., and Ferguson, L. (1961). *J. Biol. Chem.* **236**, 2552.
Charalampous, F. C., Gonatas, N. K., and Melbourne, A. D. (1973). *J. Cell Biol.* **59**, 421.
Chen, W. I., and Charalampous, F. C. (1969). *J. Biol. Chem.* **244**, 2767.
Fiske, C. H., and SubbaRow, Y. (1925). *J. Biol. Chem.* **66**, 375.
Garfinkel, D. (1957). *Arch. Biochem. Biophys.* **71**, 111.
Gordesky, S. E., and Marinetti, G. V. (1973). *Biochem. Biophys. Res. Commun.* **50**, 1027.

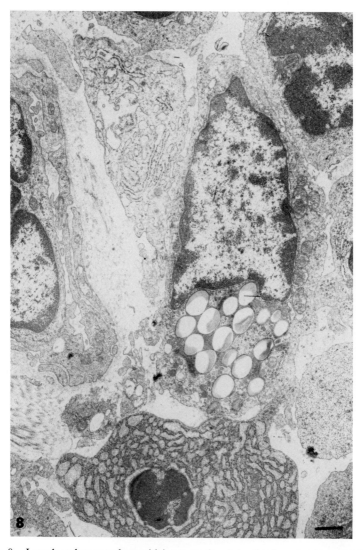

FIG. 8.   Lymph node macrophage with intracytoplasmic beads after subcutaneous injection of latex beads. Bar, 1 μm. ×15,000.

Heine, J. W., and Schnaitman, C. A. (1971). *J. Cell Biol.* **48**, 703.
Heppel, L. A., and Hilmoe, R. J. (1955). *In* "Methods in Enzymology" (S. P. Colowick and N. O. Kaplan, eds.), Vol. 2, p. 546. Academic Press, New York.
Hubbard, A. L., and Cohn, Z. A. (1972). *J. Cell Biol.* **55**, 390.
Lowry, O. H., Rosebrough, N. J., Farr, A. L., and Randall, R. J. (1951). *J. Biol. Chem.* **193**, 265.

Lutton, J. D. (1973). *J. Cell Biol.* **56**, 611.

Mackler, B. (1967). *In* "Methods in Enzymology" (R. W. Estabrook and M. E. Pullman, eds.), Vol. 10, p. 551. Academic Press, New York.

Masters, B. S. S., Williams, C. H., and Kamin, H. (1967). *In* "Methods in Enzymology" (R. W. Estabrook and M. E. Pullman, eds.), Vol. 10, p. 565. Academic Press, New York.

Mycek, M. J. (1970). *In* "Methods in Enzymology" (G. Perlman and L. Lorand, eds.), Vol. 19, p. 286. Academic Press, New York.

Omura, T., and Sato, R. (1967). *In* "Methods in Enzymology" (R. W. Estabrook and M. E. Pullman, eds.), Vol. 10, p. 556. Academic Press, New York.

Ostrowski, W., and Tsugita, A. (1961). *Arch. Biochem. Biophys.* **94**, 68.

Paigen, K. (1958). *J. Biol. Chem.* **233**, 388.

Phillips, D. R., and Morrison, M. (1971). *Biochemistry* **10**, 1766.

Post, R. L., and Sen, A. K. (1967) *In* "Methods in Enzymology" (R. W. Estabrook and M. E. Pullman, eds.), Vol. 10, p. 762. Academic Press, New York.

Rifkin, D. B., Compans, R. W., and Reich, E. (1972). *J. Biol. Chem.* **247**, 6432.

Schneider, W. C. (1948). *J. Biol. Chem.* **176**, 259.

Smolen, J. E., Shahet, S. B., Cohen, P., Boehner, R. L., and Karnovsky, M. J. (1971). *J. Clin. Invest.* **50**, 87a.

Sottocassa, G. L., Kuylenstierna, B., Ernster, L., and Bergstrand, A. (1967). *In* "Methods in Enzymology" (R. W. Estabrook and M. E. Pullman, eds.), Vol. 10, p. 448. Academic Press, New York.

Strittmatter, P. (1963). In "The Enzymes" (P. D. Boyer, H. Lardy, and K. Myrbäck, eds.), Vol. 8, p. 113. Academic Press, New York.

Tisdale, H. D. (1967). *In* "Methods in Enzymology" (R. W. Estabrook and M. E. Pullman, eds.), Vol. 10, p. 213. Academic Press, New York.

Tsan, M. F., and Berlin, R. D. (1971). *J. Exp. Med.* **134**, 1016.

Vassiletz, I. M., Derkatchev, E. F., and Neifakh, S. A. (1967). *Exp. Cell Res.* **46**, 419.

Warren, L., and Glick, M. (1969). *In* "Fundamental Techniques in Virology" (K. Habel and N. P. Salzman, eds.), Vol. 1, p. 66. Academic Press, New York.

Weisman, R. A., and Korn, E. D. (1967). *Biochemistry* **6**, 485.

Werb, Z., and Cohn, Z. A. (1972). *J. Biol. Chem.* **247**, 2439.

Wetzel, M. G., and Korn, E. D. (1969). *J. Cell Biol.* **43**, 90.

Williams, C. H., and Kamin, H. (1962). *J. Biol. Chem.* **237**, 587.

Williams, J. N. (1964). *Arch. Biochem. Biophys.* **107**, 537.

Chapter 15

The Isolation of Nuclei from Paramecium aurelia

DONALD J. CUMMINGS AND ANDREW TAIT[1]

Department of Microbiology,
University of Colorado Medical Center,
Denver, Colorado

I. Introduction

The isolation of nuclei from ciliate protozoa, in particular *Paramecium aurelia*, offers several unique challenges. First, like most ciliates, paramecia contain two types of nuclei: a micronucleus which is diploid and is responsible for maintaining the genotype of the cell; and a macronucleus which contains approximately 400 times the genetic material of the micronucleus and is necessary for the growth, fission and phenotypic expression of the cell (see Sonneborn, 1970; Beale, 1954). These two types of nuclei maintain close contact within the cell, and, in the case of *Tetrahymena*, the nuclear membranes (Elliott and Kennedy, 1973) of the macronucleus and micro-

---

[1] Present address: Institute of Animal Genetics, University of Edinburgh, Scotland.

nucleus can be continuous. Structurally, these two nuclei differ both in morphology and in size (Cummings, 1972; Jurand and Selman, 1970); the macronucleus measures between 15 and 35 μm and is composed of two sizes of subnuclear particles, which are distributed throughout the nucleus, connected by a matrixlike material; the micronucleus is vesicular with a centrally located chromatin body and measures only about 2 μm in diameter. The separation and isolation of these intimately associated nuclei from one another as well as from other cellular substructures (cilia, trichocysts, mitochondria, pellicular material, etc.) represents a formidable technical problem. In addition, the huge size of the macronucleus from *Paramecium* precludes certain methods for disrupting the cells, such as harsh homogenization in a Waring Blendor. Recently, Muramatsu (1970) discussed many of these technical problems involved in isolating nuclei from many different organisms, and the reader is referred there for a more extensive discussion of suggested methods for disruption.

In this chapter, we present our rationale for the disruption of cells, and the isolation of macronuclei will be explored in detail. Since methods for the isolation of micronuclei are still developing, these will be discussed at a preliminary level with emphasis on the variety of methods attempted. The procedures for the isolation of macro- and micronuclei from *Tetrahymena* is described by Gorovsky *et al.* in this volume (Chapter 16), and the reader is encouraged to seek out the similarities and dissimilarities in the methods described for these two ciliate protozoa.

II.   Isolation of Macronuclei

A.   Early Work and Rationale

In our original method for isolating macronuclei (Cummings, 1972), our aim was to use conditions that (1) caused cell disruption but left the nuclei intact, (2) allowed isolation of at least 50% of the nuclei with minimal contamination from other subcellular structures, (3) protected the nuclei so that macromolecular compositional studies could be performed and, (4) resulted in a method that was reproducible and applicable to a variety of *Paramecium* syngens as well as to other ciliates.

We approached the problem for cell disruption using several different nonionic detergents (Triton X-100, Tween 80, sarkosyl, Nonidet P40) and sodium deoxycholate. All of these detergents led to either incomplete disruption of the cells or disruption of the nuclei as well. In agreement with the results of others (Blobel and Potter, 1967; Evenson and Prescott, 1970; Gorovsky, 1970; Isaacks *et al.*, 1969; Kumar, 1970; Muramatsu, 1970;

Stevenson, 1967), varying amounts of divalent cations and/or spermidine protected the nuclei from disruption. It must be emphasized here that the basis of any procedure is the combined disruption of the cell and the simultaneous protection of the nuclei and that these are essentially antagonistic processes. Increasing the concentration of nuclear protective agents prevented disruption of the cells; consequently, a balance between these opposing factors is necessary. Of the two best nonionic detergents for disruption of the cells, Nonidet P40 appeared to be more gentle than Triton X-100 and was consequently chosen as being both effective and relatively mild in its action. Also in agreement with other workers (see Muramatsu, 1970), it was found that Ca^{2+} protected the nuclei to a greater extent than did Mg^{2+}. We also found that spermidine (or spermine) protected the nuclei from disruption by nonionic detergents or by deoxycholate. The combination of Ca^{2+} and spermidine permitted the use of lower concentrations of either reagent used independently, and thus alleviated certain problems such as the possibility that concentrations of Ca^{2+} greater than 5 mM can lead to excessive "hardness" of the nuclei and make compositional studies awkward (see Muramatsu, 1970). The concentration of Nonidet necessary to disrupt the cells completely was about 2%.

In order to reduce the amounts of reagents necessary for the protection of nuclei, we studied the effect of using deoxycholate in conjunction with Nonidet. Deoxycholate was chosen as the synergistic agent because (a) nonionic detergents other than Nonidet were not much different from Nonidet itself, and (b) the nuclei obtained ultimately were contaminated to a lesser extent with cytoplasmic fragments. It was found that 0.33% Nonidet, 0.22% deoxycholate in the presence of 0.17 M sucrose, 4.8 mM CaCl₂, and 150 μg of spermidine per milliliter caused paramecia to round up, swell, and become immotile. The macronuclei were clearly visible within the cells, and their appearance did not alter, as judged by phase microscopy with or without staining with methyl green, during any step in the procedure. Some disruption of the cells was apparent, but the major changes that occurred in the presence of the detergents and protective reagents involved excessive enlargement, swelling, and apparent increased fragility of the cell. The increased fragility was a necessary feature of the method, for then and only then could the cells be disrupted by mild agitation. One or two applications on a vortex mixer were sufficient to rupture these cells; for uniformity, however, we routinely homogenized the suspension by three strokes of a loosely fitting Teflon pestle attached to a Tri-R homogenizer at setting −3. This minimal homogenization served to fracture the fragile cells, and released the intact nuclei without apparent change in their size or morphology. All these operations were performed at 0–4°C.

Our first attempts to isolate the macronuclei from the homogenization mixture using one-step sucrose gradients resulted in a low yield of nuclei, while differential centrifugation resulted in contamination by other sub-structures. Blobel and Potter (1967) showed that decreased yields of nuclei on one-step sucrose gradients resulted from the rapid accumulation of cellular debris at the interface of sample and the sucrose layer and that this debris screen trapped the nuclei, preventing them from entering the sucrose layer. This was circumvented by increasing the sucrose concentration of the applied sample. Therefore, after homogenization of the cells in the Tri-R stirrer, the suspension containing nuclei, cell debris, etc., was mixed 1:1 with a solution containing 2.35 M sucrose, 4.8 mM CaCl2, 0.02 M NaCl, and 100 μg of spermidine per milliliter, to yield a final sucrose concentration of 1.26 M. Mixing with this solution also reduced the concentrations of the detergents, and the Ca^{2+} and spermidine were included for continued protection of the nuclei. Rapid harvesting of the macronuclei was accomplished by layering this final suspension onto an equal volume of a solution containing 2.1 M sucrose, 3 mM CaCl$_2$, 100 μg/ml spermidine, and 0.02 M NaCl and centrifuging in a SW rotor for 10 seconds at a force of 10,000 g, decelerating without brake. This brief centrifugation was sufficient to quantitatively sediment the macronuclei. These macronuclear pellets were reproducibly free of cellular debris; the only contamination consisted of unidentified crystals in some of the homogenates, and even these could be avoided by sedimenting the macronuclei onto a D_2O-sucrose cushion rather than to the bottom of the tube. The purity of these macronuclei preparations is illustrated in Fig. 1A and the integrity of the nuclear membrane and internal morphology is presented in Fig. 1B. Another feature of this brief centrifugation procedure was that while sedimentation of micronuclei into the sucrose cushion occurred, sufficient numbers remained in the homogenate to allow subsequent isolation of micronuclei (see Section III, A).

This procedure for isolating macronuclei was successful on three different syngens as well as one stock of *Euplotes minuta*; in general it was a reliable reproducible method. The major disadvantage of the procedure was that it used detergents at 0.55% total concentrations. There is some indication that certain enzymes, for example, DNA polymerase (Chang and Bollum, 1971), can be lost from isolated nuclei in the presence of 0.5% detergents. For this reason, we embarked on an extensive modification of the macronuclei procedure, keeping in mind the criteria demanded earlier.

B. Recent Methods

Our present method for isolating macronuclei was developed by attempting a wide range of procedures and can be conveniently considered under three main headings.

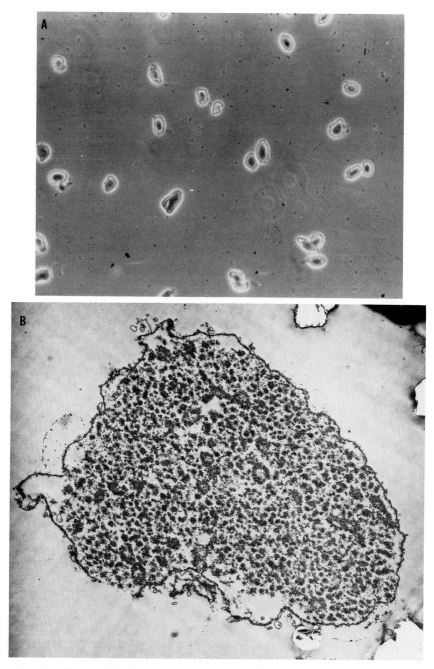

FIG. 1.　Macronuclei isolated using older methods (Cummings, 1972). (A) Low magnification micrograph; (B) electron micrograph illustrating general morphology.

1. Methods for Cell Disruption

Primarily, two methods for disruption were used: the Parr Bomb and homogenization in a glass–Teflon homogenizer. In both these methods, the aim was to achieve a high percentage of cell breakage, while minimizing damage to the nuclei. This can be effected by varying the conditions for breakage as well as altering the two antagonists in the suspension medium, i.e., protecting agents, such as Ca^{2+}, polyamines, and sugars, and disrupting agents, such as detergents. Our aim here was to determine methods that would lead to disruption of the cells without introducing the need for increasing the concentrations of reagents necessary for protecting the nuclei. The work will be presented somewhat as a progress report in order to adequately describe the multitude of conditions attempted and their results.

The rationale behind the use of the Parr Bomb (see Section III, C) is to allow nitrogen gas to diffuse rapidly into the cells under pressure; when the pressure is released the N_2 forms bubbles within the cell, which rupture the cell wall (Hunter and Commerford, 1961). Should this procedure be successful, an obvious advantage would be that no detergents would be necessary. Cells were resuspended in 0.05 M NaCl, 2 mM phosphate pH 7.1, 3 mM $CaCl_2$, and 0.15 M sucrose and subjected to different pressures of N_2. After release of the pressure, the ejected homogenate was examined for macronuclei by light microscopy. The nuclei were found to be distorted in shape, i.e., greatly elongated, and were attached to large fragments of cytoplasmic debris. When sucrose was omitted from the suspension medium, total disruption of the nuclei occurred upon ejection. These observations are summarized in Table I. Not all the suspension medium was ejected through the nozzle of the Parr Bomb. About 0.2–0.4 ml remained in the container within the reaction vessel. This fluid was semi-frozen and after thawing, examination by light microscopy revealed that the cells in this suspension were still intact and motile.

With these observations in mind, we deduced that disruption of the cells occurred by the shear produced on ejection from the nozzle and that possibly the N_2 was not entering the cells. Furthermore, sucrose, which was deemed necessary in most nuclei isolation procedures (see Muramatsu, 1970), appeared to be necessary to protect the nuclei after disruption of the cell but did not appear to be needed prior to disruption.

The next series of experiments was designed to increase the diffusion of nitrogen into the cells and produce disruption without ejection through the nozzle. In order to facilitate diffusion of N_2, low concentrations of ionic and nonionic detergents or dimethyl sulfoxide (DMSO) were used to "soften" the cell wall. Using low concentrations of deoxycholate (DOC)

TABLE I

PRELIMINARY FINDINGS WITH PARR BOMB

| | Conditions | | | |
Medium	Pressure (psi)	Time (min)	Cells after ejection	Macronuclei after ejection
NaCl, 0.05 M;	550	10	Some whole cells	Attached to
Sucrose, 0.15 M;	550	20	Some whole cells	cytoplasmic debris
Phosphate, 2 mM;	700	20	Breakage	and distorted
CaCl$_2$, 3 mM,;	880	20	Breakage	
pH 7.1	965	20	Breakage	
NaCl, 0.05 M;	750	20	Breakage	Very few
Phosphate, 2 mM;	600	15	A few whole cells	A few, but attached
CaCl$_2$, 3 mM;				to cytoplasmic
Sucrose, 0.1 M;				debris
pH 7.1	550	12	More whole cells	More macronuclei,
				but still associated
				debris
	600	12	Some whole cells	Some nuclei but
				long and distorted
NaCl, 0.05 M;	750	20	Breakage	Smashed
Phosphate, 2 mM;	600	10	Some whole cells	Smashed
CaCl$_2$, 3 mM;	450	10	Whole cells	None
pH 7.1				

and Nonidet (NP40), total concentration 0.25%, complete disruption of both the cells and the nuclei was observed after decompression. In order to protect the nuclei, spermine was added to the suspension medium at 1 mg/ml. The spermine was indeed found to protect the nuclei while still allowing a high percentage of cell breakage. If the spermine concentration was lowered to 100 μg/ml, the nuclei were again totally disrupted.

The ability of DMSO to facilitate N$_2$ diffusion into the cells was also tested, and the results from this set of experiments are summarized in Table II. In control preparations (i.e., identical conditions at 0° but not subjected to N$_2$ pressurization), considerable swelling and, in some cases, lysis occurred. This raised the immediate question of whether pressurization in the Parr Bomb was required at all. This procedure involved equilibration of the N$_2$ gas and cells for 15–20 minutes at 0°C. Clearly, the cell wall and membrane were being conditioned by exposure to the detergents alone during this time, and so gentle mechanical shear of these "softened" cells was attempted. Homogenization of such cell suspensions by 8–10 strokes on the Tri-R homogenizer at low speed resulted in almost identical nuclei

TABLE II

FINDINGS WITH PARR BOMB WITHOUT EJECTION

Medium	DOC (%)	NP40 (%)	DMSO (%)	Spermine (μg/ml)	Conditions Pressure (Psi)	Time (min)	Cells	Nuclei
NaCl, 0.05 M; Phosphate, 2 mM; CaCl$_2$, 3 mM; pH 7.1	0.1	0.15	—	—	650	20	None	None
	0.05	0.075	—	100	650	15	A few whole cells, some partially broken	A few, but attached to debris
	0.07	0.075	—	100	700	15	None	Many, but broken and much debris
	—	—	10	100	800	15	70% Whole	Swollen
	0.05	0.075	5	100	800	15	60% Whole	Swollen
	0.07	0.075	—	1000	800	15	A few whole	Many, but some damaged
	0.07	0.075	—	1000	Tri-R homogenization		A few	Many, only a few damaged
NaCl, 0.05 M; Phosphate, 2 mM; pH 7.1	0.07	0.075	—	200	Tri-R homogenization		Very few	Many, but some swollen

and cell breakage as that occurring in the Parr Bomb (Table II). In view of the greater simplicity of homogenization compared with N_2 pressurization, the Parr Bomb was abandoned as a reliable method for obtaining macronuclei. However, the time spent in examining its usefulness led to two important conclusions: (a) sucrose, or any sugar, was required to stabilize the nuclei and in fact interfered with the "softening" of the cells. (b) This conclusion enabled us to untimately greatly reduce the concentration of detergents in the incubation mixture while still achieving sufficient conditioning for mild cell disruption.

2. CONDITIONS FOR DISRUPTION

This series of experiments was aimed at reducing the detergent concentrations used for "softening" the cells and involved testing the effect of varying the concentration of protective agents versus the concentration of detergents. In all these experiments, the cell breakage and the condition of the nuclei relative to those in the whole cells were monitored. Two main types of experiments were undertaken: (a) No Ca^2, low concentrations of detergents, varying spermine concentrations; (b) varying Ca^{2+} concentrations, low concentrations of detergents, 250–1000 μg/ml spermine.

The range of conditions used in these experiments is indicated in Table III. It was found that 0.145% total detergents, 3 mM $CaCl_2$, and 1 mg of spermine per milliliter or 0.04% total detergents, 1.0 mM $CaCl_2$, and 250 μg of spermine per milliliter yielded the best preserved nuclei, although the latter conditions gave slightly less effective cell disruption. No differences were noted when spermine was substituted for by spermidine or putrescine. In addition, little difference in the condition of the nuclei was noted when homogenization was carried out by hand or at a low setting on the Tri-R stirrer.

3. METHODS FOR PURIFICATION

Having achieved conditions for cell breakage with minimum damage to the nuclei, further experiments were initiated to determine the best conditions for purification of the nuclei from the cell homogenate. Since this placed more stringent requirements on the intactness and stability of the nuclei, it was found necessary to further study and modify the conditions for cell lysis, while still maintaining essentially the original optimal conditions.

The first set of experiments in this series showed that the detergent concentration had to be raised to avoid contamination by pellicle debris in the nuclear preparations. In order to do this without damaging the nuclei, the concentration of protective agents was raised. As the protective agents also protect the cells from lysis, it was found to be advantageous to increase

TABLE III

EFFECTS OF LOW CONCENTRATIONS OF DETERGENTS VS VARYING CONCENTRATIONS OF PROTECTIVE AGENTS[a]

DOC (%)	NP40 (%)	CaCl$_2$ (mM)	Spermine (μg/ml)	Cell breakage (%)	Macronuclei	Treatment in Tri-R[b]
0.007	0.0075	—	50	50–60	Few, long and thin	Setting 3; 6 strokes
0.005	0.0045	—	100	60	More, but many long and thin	Setting 5; 6 strokes
0.01	0.015	—	250	80–90	Few, long and thin	Setting 2; 6 strokes
0.01	0.01	—	175	80	More, but damaged	Setting 3; 6 strokes
0.01	0.01	—	300	70	Many, good condition	Setting 3; 6 strokes
0.01	0.01	—	250	80	Many, some damaged	Hand homogenization in pestle; 6 strokes
0.02	0.02	1.0	250	95	Many, some rounded	Hand homogenization
0.07	0.075	3	1000	100	Many, some swollen	Hand homogenization

[a] All experiments were performed by incubating cells at 0°C for 15–20 minutes in phosphate buffer described in Table II (without sucrose).
[b] Lowest speed of Tri-R homogenizer is at a setting 1 with the highest speed at a setting 10.

TABLE IV

Range of Variation in Lysis Conditions

$CaCl_2$ (mM)	Detergents (total %)	Spermine (μg/ml)	Additions[b]	Cell breakage (%)	Cushion[a]	Macronuclei
1.0	0.04	250	None	95	2.1 M Sucrose	Extensive damage, plus pellicle contamination
1.5	0.04	250	CaCl₂ to 4.5 mM after homogenization	93	2.1 M Sucrose	Some damage, plus pellicle contamination
3.0	0.08	500	CaCl₂ to 4.5 mM after homogenization	100	2.1 M Sucrose	Less damage and less pellicle
3.0	0.06	500	CaCl₂ to 4.5 mM before homogenization	99	75% Conray	None
3.0	0.05	500	CaCl₂ to 4.5 mM before homogenization	97	2.1 M Sucrose	Good nuclei, but some crystals
3.0	0.05	500	CaCl₂ to 4.5 mM 0.1 M raffinose prior to homogenization	97	15% Glycerol, 2.1 M sucrose	Very good, plus crystals
3.0	0.04	500	CaCl₂ to 4.5 mM, 0.1 M raffinose prior to homogenization	97	20% glycerol, 2.1 M sucrose	Excellent, plus 5–10% crystals

[a] All sucrose solutions contain 0.02 M NaCl, 3 mM $CaCl_2$, 100 μg/ml spermine tetrahydrochloride.

[b] All solutions for lysis contain 0.05 M NaCl, 2 mM phosphate, pH 7.1. Lysis was done by incubating at 0°C for 20 minutes with detergents and then homogenizing 8–10 strokes at setting 2.5, followed by a 1:1 dilution of homogenate by cushion layer. The homogenate was layered on a 10-ml cushion and centrifuged for 1 min at 14,000 rpm, decelerating without brake. It was especially important to regulate the temperature at 1–3°C during centrifugation in order to reproducibly control the viscosity of these sucrose layers.

291

the concentration of these agents after softening the cells and prior to homogenization. In all these experiments the homogenate was diluted 1:1 with 2.1 M sucrose, layered on a 2.1 M sucrose cushion, centrifuged at 10,000 g for 1 minute and the macronuclear pellet was then examined by light microscopy. A second set of experiments concerned alterations in the constituents of this cushion to see whether further improvements could be made. Optimal conditions found previously (1 mM CaCl$_2$, 0.04% detergents, 250 mg of spermine per milliliter) were used as a basis for comparison and a compilation of the most notable variations are listed in Table IV.

It was found that 15% glycerol/2.1 M sucrose reduced the breakage of nuclei when compared with 2.1 M sucrose, while still giving 5–10% crystals in the preparation. The number of crystals could be reduced considerably by including a 5-ml cushion of glycerol/sucrose in D_2O. The use of other constituents in various combinations (see Bhargava and Halvorson, 1971; Gorovsky, 1970) in the cushion were tested: sorbitol, raffinose, glycerol/D_2O, PVP-10, PVP-40, Conray, and raffinose/D_2O. Although some of these conditions appeared to give marginally reduced damage to the nuclei, this was, in all cases, accompanied by an increase in contamination by food vacuoles, crystals, cytoplasmic debris, and other cell constituents. The main finding which improved the nuclear preparation, without significantly affecting the efficiency of cell disruption, was the addition of raffinose and CaCl$_2$ prior to the homogenization.

C.   Final Scheme for Isolating Macronuclei

After this extensive investigation of the different conditions, the following conclusions were drawn:

1. Cells left at 0° for 10 minutes prior to the addition of detergents plus protective reagents led to more reproducible swelling of cells.

2. Addition of raffinose/CaCl$_2$ after swelling but prior to homogenization minimized damage to nuclei.

3. The optimal conditions in terms of minimum breakage of nuclei and minimum contamination by other cell constituents was found to include the use of 20% glycerol/2.1 M sucrose as a cushion. Contamination by crystals amounted to 5–10% which could be reduced significantly by including a glycerol/sucrose/D_2O cushion. The scheme of macronuclear preparation illustrated in Fig. 2 was therefore devised as that method which minimized nuclear damage and contamination by other cell constituents and at the same time yielded a maximum of cell breakage concomitant with a good yield of nuclei. The purity and integrity of these macronuclei are illustrated in Fig. 3; Fig. 3A is at low magnification indicating that mainly crystals contaminate the preparation; Fig. 3B was selected to illustrate that

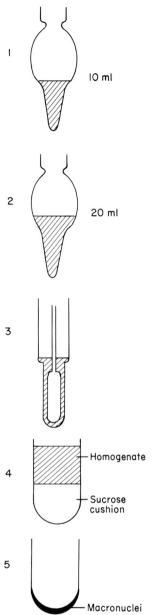

1   10 ml   Packed cells are incubated for 10 minutes at 0°C.

2   20 ml   Packed cells are incubated in equal volume of homogenization buffer for 20–25 minutes at 0°C.

3   Homogenization mixture plus one-third volume of raffinose–CaCl₂ solution is homogenized 10 times at 2.5 setting in Tri-R stirrer.

4   ⌐ Homogenate   Homogenized paramecia preparation is mixed with an equal volume of 2.1 M sucrose-glycerol solution, layered over the same solution, and centrifuged for 1 minute at 14,000 rpm in SW 25 Beckman rotor, without brake.

             Sucrose cushion

5   Homogenization mixture is withdrawn, interface material is removed, and the surface of the sucrose layer is washed 3 times with deionized H₂O. The sucrose cushion is then withdrawn, and the macronuclei pellet is resuspended in the appropriate solution.

             Macronuclei

Fɪɢ. 2.   Recent macronuclear isolation procedure.
Key: (Part 2) *Homogenization buffer*: 0.04% w/v sodium deoxycholate; 0.04% v/v Nonidet P40; 6 mM CaCl₂; 37.5 mM NaCl; 1.5 mM sodium phosphate; 1 mg/ml spermine tetrahydrochloride; pH 7.2.   (Part 3) *Raffinose–CaCl₂ solution*: 6mM CaCl₂; 0.28 M raffinose.   (Part 4) *Sucrose–glycerol solution*: 2.1 M sucrose, 3 mM CaCl₂; 20% glycerol; 20 mM NaCl; 100 μg/ml spermine tetrahydrochloride.

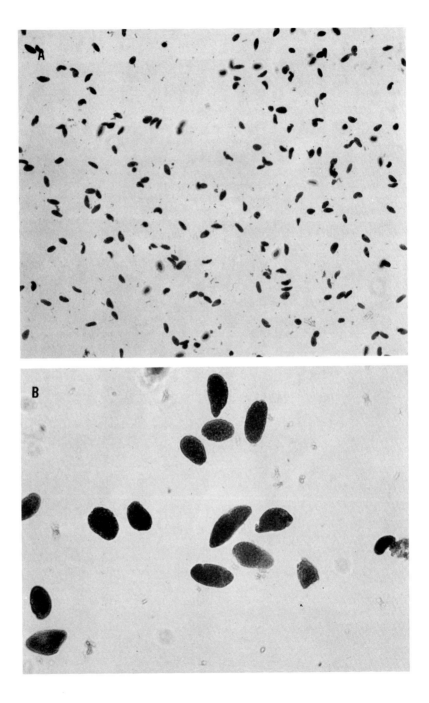

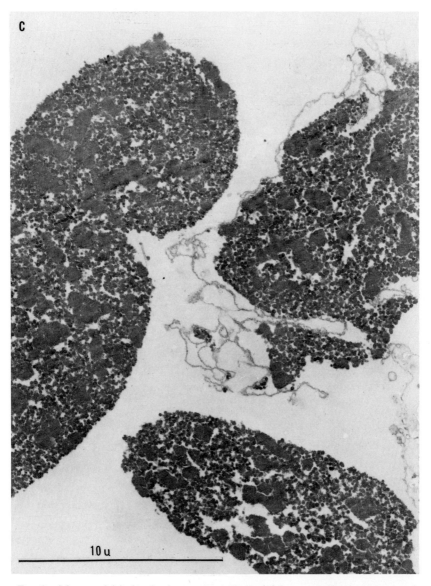

FIG. 3. Macronuclei isolated using recent methods. (A) Low magnification micrograph; (B) low magnification micrograph selected to illustrate occasional presence of micronuclei; (C) electron micrograph illustrating the presence of a nuclear membrane; this micrograph of macronuclei from balanced growth cells also illustrates the enlargement of large particles. Bar, 10 μm.

a few micronuclei sometimes cosediment with the macronuclei; Fig. 3C shows that the nuclear membrane is intact. Figure 3C is a section of macronuclei isolated from logarithmically grown cells; note that the large particles are more diffuse and enlarged, indicating more rapid production of ribosomes, as one would expect at this growth stage.

D.   Assessment of Method

In Section II,A we stipulated that there were several criteria for any procedure that purported to isolate macronuclei. These involved breakage

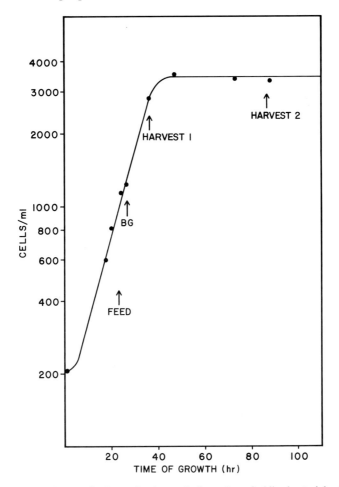

Fig. 4.   Growth curve for *P. aurelia*. Arrows indicate time of adding bacterial pellets (feed), balanced growth (BG), and harvest time for compositional studies.

of the cells, yield of nuclei, integrity of the nuclei necessary for compositional studies, reproducibility, and applicability. We have already indicated that the final scheme presented in Fig. 2 satisfied the criteria for breakage of the cells as well as yielding good amounts of nuclei. With regard to reproducibility, the only time this method has failed was when the temperature during centrifugation was inadvertently allowed to drift below 0° C. This prevented the nuclei from sedimenting through the highly viscous glycerol/sucrose layer in the prescribed time. In our earlier report (Cummings, 1972), we stated that the method for isolating macronuclei used there was applicable to three syngens of paramecia and one stock of *Euplotes minuta*. We have now used the present method on those same three syngens plus two more, as well as two additional stocks within these five syngens. In addition, the present method worked very well in isolating macronuclei from one stock of *Paramecium caudatum*. Consequently, we conclude that this method is both reproducible and applicable.

Finally, we come to compositional studies. Earlier, we reported that the macronuclei consisted of 22% DNA, 10% RNA, and 68% protein. These values agreed very well with the composition of macronuclei isolated from *Tetrahymena* (Lee and Scherbaum, 1966), where a range of 15 to 23% DNA was found. The range in values for our DNA measurements was 18 to 26%. With the present method, compositional studies were done on cells harvested in late logarithmic and stationary growth (see Fig. 4), and these values are listed in Table V. No difference was observed in the percent of DNA in macronuclei isolated in logarithmic or stationary growth. Whether the actual mass of the nuclei isolated from these different growth states varied was not determined. It is noteworthy that the percent DNA composition of whole cells was only 1.2% compared with the 22% found in macronuclei. Such a great increase in percent DNA would be expected since it has been reported that macronuclei contain more than 90% of the total DNA in the cell (Allen and Gibson, 1972).

This use of the percent DNA in the macronucleus as an indication of the integrity of the nucleus can be illustrated best by consideration of values obtained with another method. Isaacks *et al.* (1973) recently reported a method for isolating macronuclei which involved homogenization of paramecia in 1% Triton X using a Waring Blendor. Micrographs of their nuclei indicated that many of the macronuclei were swollen and fragmented. Significantly, these authors found that macronuclei contained only 9% DNA and that the average DNA content of an isolated nucleus was only 30% that of an intact cell. Moreover, only 21% of the total nuclei were recovered in the final preparation. Data for comparison with our procedure are contained in Table V. The DNA content of the intact cell measured by us was the same as that measured by other workers (Isaacks *et al.*, 1973;

TABLE V

COMPOSITION AND YIELD OF MACRONUCLEI

	Cells/ml[a]	Packed volume	%Breakage	Number of nuclei/ml[a]	Recovery (%)
Late log cells	2.13×10^7	2.4 ml	97	1.23×10^7	58
Stationary cells	1.42×10^7	1.7 ml	>98.5	8.17×10^6	60

	Mg protein (%)	Mg RNA (%)	Mg DNA (%)	Total Mg
Late log macro-nuclei	11.5 (68)	1.86 (11)	3.59 (21)	16.92
Stationary macronuclei	8.8 (66)	1.29 (10)	3.2 (24)	13.28
Late log cell lysate	314 (82)	64.9 (17)	4.64 (1.2)	384
Stationary macronuclei [b]	— (68)	— (10)	— (22)	—

[a] Number of cells was calculated by counting 20-μl sample of known volume of cell suspension. Nuclei were counted by resuspending in exact volume, fixing 5 μl of the suspension in 1% osmic acid, staining with lacto-orcein, and then counting.

[b] From Cummings (1972) for comparison with earlier results.

Stevenson, 1967). From another viewpoint, compositional data should reflect the physiological changes which occur during the growth cycle. We have harvested macronuclei from paramecia during balanced, late logarithmic, and stationary growth and found that the specific activity of DNA polymerase progressively decreased 9-fold during this cycle (Tait and Cummings, 1974).

From a variety of viewpoints, therefore, we have demonstrated that the scheme presented in Fig. 2 for preparing macronuclei is reliable and can be used for studying many features of macronuclear composition and development.

III.   Isolation of Micronuclei

A.   Early Work

It is clear from the discussion of the methodology involved in developing a procedure for the isolation of macronuclei that their size and morphology were both an advantage and a disadvantage. The extreme care required for the rupturing of the cells was brought about from fear of doing damage to the nuclei. However, it was extremely easy to determine when damage was inflicted; the nuclei, and nuclear fragments could be readily seen in the light microscope and an assessment could be made as to their integrity, distortion, etc. With the micronuclei, the situation is entirely different. It is difficult to examine these nuclei *in situ* (Jurand and Selman, 1970) and their size and morphology makes difficult any rationale for deciding their state during isolation.

Our starting point for the isolation of micronuclei was primarily the early method for obtaining macronuclei (Cummings, 1972). After rupture of the cells with detergents in the presence of protective reagents and removal of the macronuclei by brief centrifugation through a sucrose cushion, the homogenized supernatant was withdrawn from the centrifuge tubes and relayered onto fresh sucrose solutions, and then the micronuclei were collected by centrifugation for 1 hour at 40,000 g. These micronuclear pellets were further purified by differential centrifugation and their purity and morphology is illustrated in Fig. 5. This procedure, when successful, yielded good preparations of micronuclei, uncontaminated by macronuclear fragments and most cellular debris but containing variable quantities of bacteria. The disadvantages were that (1) reproducibility was fair at best; often few if any, micronuclei were recovered; (2) even when successful, only fair yields were obtained; the causes for this may be that the nuclei disintegrated during preparation, which could also account for the poor reproducibility, or that micronuclei cosedimented in association with macronuclei. The photograph in Fig. 3B was selected to show some micronuclei in the field of macronuclei.

B. Recent Developments

The method previously described (Cummings, 1972) was used as a starting point for our attempts to develop a method which gave high yields of micronuclei with improved purification and reproducibility. From these attempts, several methods were devised, but on further use of each of these methods, great variability in yield was encountered. Subsequent modification, adjustments, etc., resulted in some improvement, but there still remains a wide variability in the yield of micronuclei. This variability appears to result from the breakage of the micronuclei in the initial homogenization, and ways of circumventing this are currently under investigation. The work involved can be divided into three sections.

1. ATTEMPTS TO IMPROVE PURIFICATION USING THE MACRONUCLEAR FREE SUPERNATANT

During the testing of different conditions for the purification of macronuclei, the effect of these conditions on the preparation of micronuclei

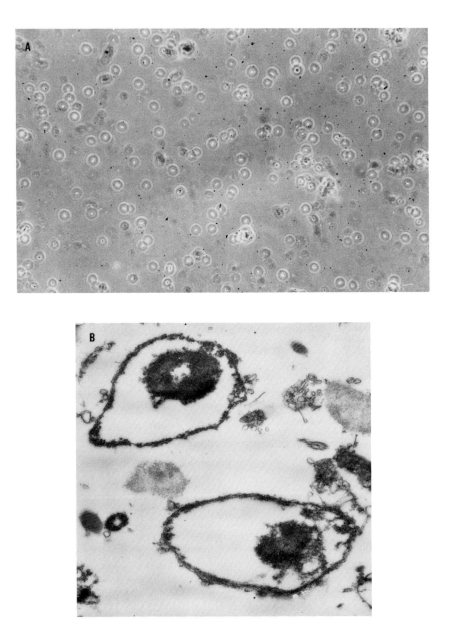

FIG. 5. Micronuclei isolated using old procedure (Cummings, 1972). (A) Low magnification micrograph; (B) electron micrograph illustrating morphology.

was examined. The method involved layering the "macronuclei-free supernatant" (i.e., the homogenate remaining on top of the sucrose cushion after removal of the macronuclei by centrifugation), onto a 2.1 M sucrose cushion and centrifuging for 1 hour at 18,000 rpm. The pellets obtained were then examined, after staining with lacto-orcein, by light microscopy. These preparations were highly contaminated with macronuclear fragments, crystals, and some food vacuoles.

With this method of preparing a macronuclear supernatant, a variety of different density gradients and density cushions were used in an attempt to either band the micronuclei while the macronuclear fragments and crystals formed a pellet at the bottom of the tube or to raise the density of the cushion to a point at which the micronuclei would not sediment whereas the contaminants would. Although a wide range of conditions were tested, no good purification was achieved. The most useful observation from these experiments was the ease with which micronuclei are held up at any interface between two solutions of differing density. Improved yields of micronuclei could be obtained by ensuring that the difference in density between a layer containing micronuclei and the cushion was at a minimum. For example, the macronuclei-free supernatant was mixed with sucrose to a concentration of 1.8 M before centrifuging onto a cushion of 2.1 M sucrose.

2. NEW METHODS FOR CELL BREAKAGE

A new method for isolating micronuclei was developed after the observation of large numbers of micronuclei in homogenates produced by using a cream homogenizer (Suyama and Preer, 1965; Beale *et al.*, 1972). This procedure had two advantages; first, it used no detergents, second, the macronuclei were totally disrupted, except for some fragments, thus, reducing the contamination of the micronuclear preparations.

The procedure consisted of homogenizing a suspension of cells in an equal volume of cells and 0.02 M Tris-HCl 7.2, 0.25 M sucrose, 10 mM MgCl$_2$, 0.5 mM dithiothreitol, 5 times through a cream homogenizer (see Fig. 6). The homogenate was then centrifuged at 6000 rpm for 10 minutes, and the pellet was resuspended (by gentle homogenization in a Teflon–glass homogenizer) in 1.8 M sucrose, 0.02 M NaCl, 3 mM CaCl$_2$, 100 μg/ml spermine. This suspension was then layered on a cushion of 2.1 M sucrose (in the same solution) and centrifuged for 30 minutes at 18,000 rpm in a Spinco SW 25.2 rotor. The pellets obtained contained large numbers of micronuclei together with crystals, trichocysts, and macronuclear fragments. The optimal pH for homogenization was found to be pH 7.2; all subsequent experiments used this homogenization buffer and the other constituents in the concentrated sucrose solutions.

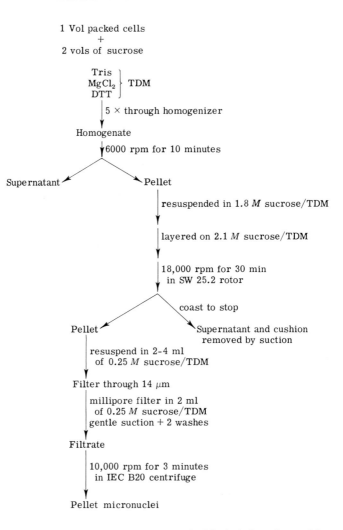

1 Vol packed cells
+
2 vols of sucrose

Tris ⎫
MgCl₂ ⎬ TDM
DTT ⎭

5 × through homogenizer

Homogenate

6000 rpm for 10 minutes

Supernatant ← → Pellet

resuspended in 1.8 *M* sucrose/TDM

layered on 2.1 *M* sucrose/TDM

18,000 rpm for 30 min
in SW 25.2 rotor

coast to stop

Pellet ←   → Supernatant and cushion
                      removed by suction

resuspend in 2-4 ml
of 0.25 *M* sucrose/TDM

Filter through 14 μm

millipore filter in 2 ml
of 0.25 *M* sucrose/TDM
gentle suction + 2 washes

Filtrate

10,000 rpm for 3 minutes
in IEC B20 centrifuge

Pellet micronuclei

FIG. 6.   Flow chart for one recent method for isolating micronuclei.

3.   PURIFICATION OF MICRONUCLEI

A variety of methods were used to remove the contaminants from the micronuclear preparations: (a) treatment with DNase and centrifugation at 2000 rpm in a bench centrifuge; (b) treatment with 0.04% detergents, followed by centrifugation; (c) layering onto solutions of different sucrose concentrations (0.5 *M*, 0.75 *M*, 1.0 *M*) and centrifugation for 1 minute at

2000 rpm to sediment crystals and macronuclear fragments, followed by centrifugation of the supernatant to sediment the micronuclei. (d) omitting sucrose from the initial homogenization buffer.

Most of these methods increased the purity of the preparation, but only by a small factor. Examination of such preparations by negative staining and electron microscopy, showed that pieces of mitochondria, clumps of crystal, crystals, and trichocysts were usually present. However, the exclusion of sucrose from the homogenization buffer appeared to result in the removal of contaminating mitochondrial fragments. It should also be noted that these methods resulted in considerable reduction in the yield of micronuclei.

These methods used to remove crystals, macronuclear fragments, and trichocysts by differential sedimentation did not seem to be very effective. The other criterion by which these contaminants were known to differ from micronuclei was shape and size, and so simple filtration was used in an attempt to further purify the micronuclei. Four sizes and types of Millipore filters were used, and the results are summarized in Table VI.

Clearly, the 14 μm filters gave the best results and electron microscopy (negative staining) of such preparations showed very little contamination, except some trichocysts and small crystals. Recently, Lipps *et al.* (1974) used 7 μm mesh nylon gauze to purify micronuclei from *Stylonychia mytilus*. As mentioned earlier, if the sucrose was omitted from the homogenization buffer, lower yields of micronuclei were obtained; therefore, 0.125 M sucrose was used, which will, while increasing the yield of micronuclei, still preclude the fragments of mitochondria.

C. Current Procedures for Isolating Micronuclei

Three general schemes are in use and are presented in the form of (1) flow chart (Fig. 6) and (2) schematic chart (Fig. 7).

TABLE VI

Filter	Pore size (μm)	Filtrate composition	Top of filter
Nylon	3	No micronuclei	Micronuclei, macronuclear fragments, crystals
Nylon	7	20% of micronuclei; a few trichocysts	Micronuclei, macronuclear fragments, crystals
Teflon	10	No micronuclei	Micronuclei, macronuclear fragments, crystals
Nylon	14	80% of micronuclei; some trichocysts	A few micronuclei, macronuclear fragments, crystals

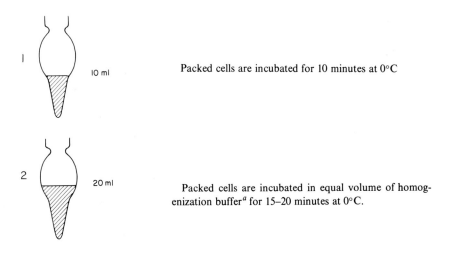

1   10 ml   Packed cells are incubated for 10 minutes at 0°C

2   20 ml   Packed cells are incubated in equal volume of homogenization buffer[a] for 15–20 minutes at 0°C.

3 Homogenization mixture plus one-third volume of raffinose, ±CaCl₂ solution, is then homogenized at 0°–4°C by either of two methods: (a) Hand homogenizer, and (b) Parr Bomb.

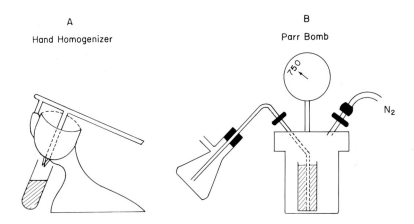

A
Hand Homogenizer

B
Parr Bomb

4 Homogenized solution is centrifuged for 1 minute at 2300 rpm to remove debris. Supernatant solution mixed with an equal volume of 3.0 M sucrose, 0.02 M Tris, pH 7.2, 3 mM $CaCl_2$, 100 μg/ml spermine tetrahydrochloride.

Fig. 7.   Schematic diagram for recent methods for isolating micronuclei. Fig. 7 continued on facing page.

Key: (Part 2) *Homogenization buffer*: 6 mM $CaCl_2$; 37.5 mM NaCl; 1.5 mM sodium phosphate; 500 μg/ml spermine tetrahydrochloride; pH 7.2. Addition of 0.02 to 0.04% detergents led to more complete disruption of cells, but did not appreciably affect the yield of micronuclei.

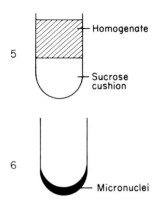

Homogenate

5

Sucrose cushion

Homogenized supernatant-sucrose solution is layered over 2.1 M sucrose, 0.02 M Tris, pH 7.2, 3 mM CaCl$_2$, 100 μg/ml spermine tetrahydrochloride, and centrifuged for 30–40 minutes at 18,000 rpm in SW 25 rotor, without brake.

6

Micronuclei

Homogenized supernatant is withdrawn, interface material is removed, and the sucrose layer surface is washed 3 times with deionized water and then withdrawn. Micronuclei pellet was resuspended in 37.5 mM NaCl, 1.5 mM sodium phosphate, pH 7, 150 μg/ml spermine tetrahydrochloride, 4 mM CaCl$_2$, 0.12 M sucrose.

Micronuclear suspension is centrifuged for 20–30 seconds in IEC SW rotor at 2000 rpm, and the supernatant solution is filtered through a Millipore filter, 14-μm pore size (Type NC), with mouth-applied suction. The filter was washed once with resuspension buffer described in Step 6.

Micronuclear filtrate is centrifuged for 3 minutes at 10,000 rpm in IEC B20 centrifuge, and the micronuclear pellet is resuspended in the appropriate buffer.

An indication of the yields and purity of micronuclei obtained using the method in the flow chart is indicated in Fig. 8; and the morphology of the nuclei from the cream homogenization procedure in the schematic chart (Fig. 7) is presented in Fig. 9. The Parr Bomb appeared to give good preparations of micronuclei. Its main disadvantage was controlling the rate of ejection; too rapid ejection yielded much disruption of other cellular components but resulted in loss of homogenate; controlled ejection avoided this volume loss, but the micronuclei obtained were no better in quantity or condition than those obtained using the cream homogenizer.

D.   Final Considerations for Isolation of Micronuclei

As mentioned initially, considerable variation in the yield was observed and further studies were undertaken to examine the reasons for this. While some factors were elucidated and also the probable cause was identified, a method circumventing these difficulties has not yet been developed.

During the course of these experiments, our initial homogenizer was broken and only two more good preparations were achieved with a new homogenizer. At this point the method of feeding the cells was changed, so that they were growing continuously and were harvested in late logarithmic growth routinely. From this point on, few really good preparations were achieved, and in some instances as few as 1–2 micronuclei per field were observed by light microscopy using oil immersion (good preparations had 10–40 micronuclei per field, suspension volume = 2 ml). One obvious

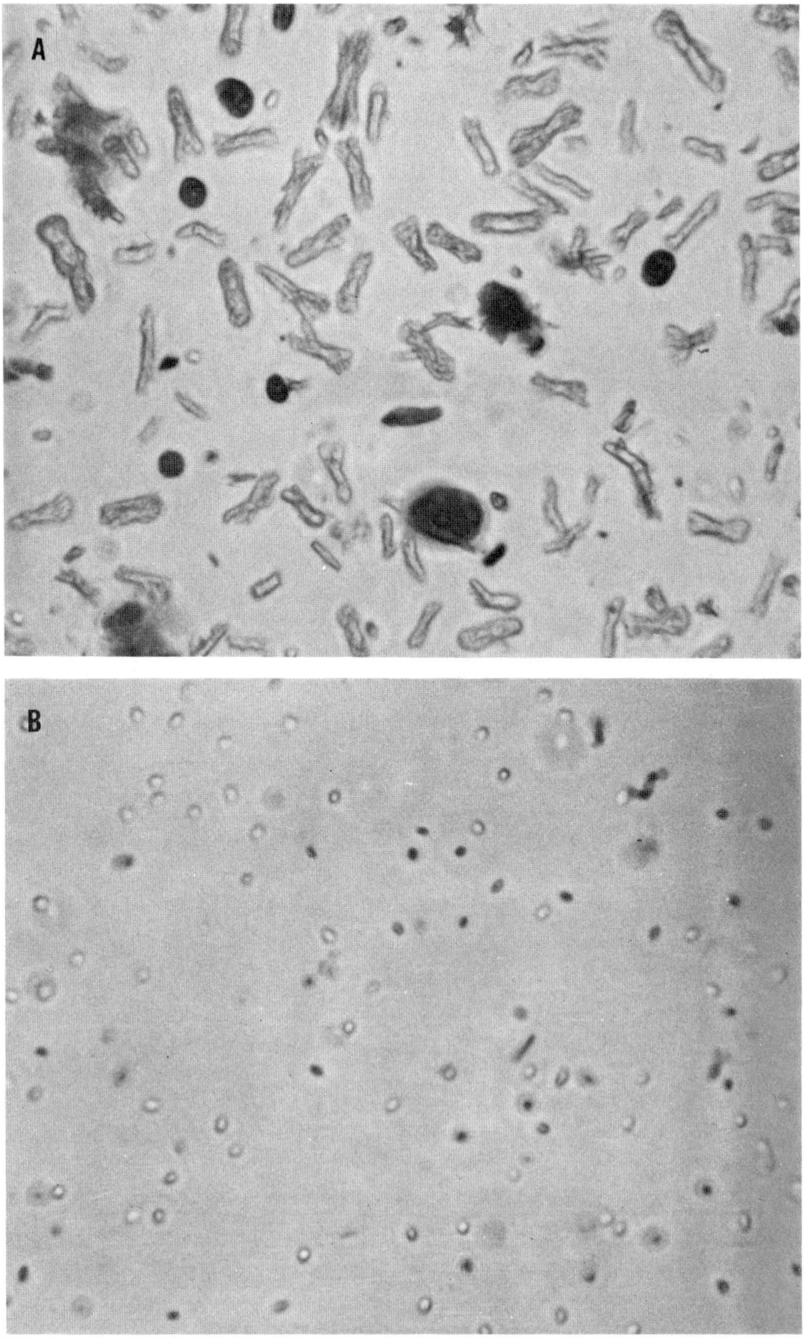

FIG. 8. Low magnification micrographs of micronuclei obtained from scheme in Fig. 6. (A) Resuspended micronuclear preparation before filtration showing the presence of micronuclei, macronuclear fragments, crystals, etc.; (B) micronuclei filtrate after filtration through 14-μm filter and centrifugation.

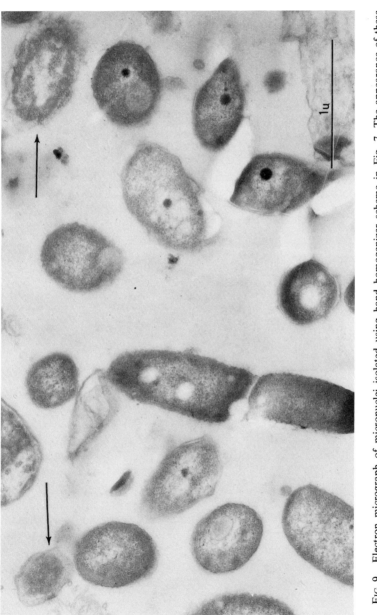

FIG. 9. Electron micrograph of micronuclei isolated using hand homogenizer scheme in Fig. 7. The appearance of these micronuclei was variable; some had the usual condensation of chromatin (arrows) while in others, the chromatin material was dispersed; also, occasionally doublets were present, indicating that division was taking place.

possibility was that the micronuclei were more fragile in late log compared to stationary phase, and although it was found that slightly higher yields were obtained with stationary phase cells, these yields were still not good. In the preparations where low yields were obtained, the other fractions in the purification procedure were examined and very few micronuclei were observed. This suggested that the nuclei were being broken during the preparation, probably at the homogenization step. To counteract these possibilities, agents known to protect macronuclei were added to the homogenization buffer. Homogenization in $0.12\,M$ sucrose, $0.02\,M$ Tris-HCl pH, 7.2, $250\,\mu g/ml$ spermine, $3\,mM$ $CaCl_2$ was used and found to increase the yields from 2–5 micronuclei per field to 10–20 micronuclei per field, but this was still not as good as had been found previously, and here too, a considerable variation in yield was observed. Finally, we noted that the tension on the spring-loaded cream homogenizer had changed over the months of experiments. Better and more reproducible yields were achieved using this homogenizer at about one-half to two-thirds maximum tension. As illustrated in Fig. 8, when these methods worked, substantial numbers of micronuclei were obtained. In addition, electron micrographs of these preparations indicated that the isolated micronuclei were in good condition.

Clearly, further studies are necessary on this problem, and a good starting point would seem to be the present methods and the assumption that low yields are due to breakage of the micronuclei on homogenization.

IV. Perspectives

Our aim in this chapter was to present a comprehensive description, including failures, of methods for isolating macro- and micronuclei from *Paramecium aurelia*. With regard to macronuclei, it is our opinion that within the framework of the method, few changes can be imagined that would lead to further improvements. The method was successful for many stocks of *P. aurelia* as well as for two other protozoa. The method for obtaining micronuclei is quite another story. Our attempts here resulted in more failures than successes, but it is hoped that this presentation will be of some use to other workers attempting the purification of micronuclei from ciliate protozoa.

ACKNOWLEDGMENT

The authors thank V. A. Chapman and S. S. DeLong for their assistance with the electron microscopy and J. Goddard for her critical readings of the manuscript. Most of the recent work reported here was supported by a Grant GB 35383 from the National Science Foundation.

REFERENCES

Allen, S., and Gibson, I. (1972). *Biochem. Genet.* **6**, 293–313.
Beale, G. H. (1954). "The Genetics of *Paramecium aurelia*." Cambridge Univ. Press, London and New York.
Beale, G. H., Knowles, J. K. C., and Tait, A. (1972). *Nature (London)*, **235**, 396–397.
Bhargava, M. M., and Halvorson, H. O. (1971). *J. Cell Biol.* **49**, 423–429.
Blobel, G., and Potter, V. R. (1967). *Science* **154**, 1662–1665.
Chang, L. M. S., and Bollum, F. J. (1971). *J. Biol. Chem.* **246**, 5835–5837.
Cummings, D. J. (1972). *J. Cell Biol.* **53**, 105–115.
Elliott, A. M., and Kennedy, J. R. (1973). *In* "Biology of Tetrahymena" (A. M. Elliott, ed.), pp. 57–87. Dowden Hutchinson & Ross, Stroudsburg, Pennsylvania.
Evenson, D. P., and Prescott, D. M. (1970). *Exp. Cell Res.* **61**, 71–78.
Gorovsky, M. A. (1970). *J. Cell Biol.* **47**, 619–630.
Hunter, M. J., and Commerford, S. L. (1961). *Biochim. Biophys. Acta* **47**, 580–586.
Isaacks, R. E., Santos, B. G., and Van Wagtendonk, W. J. (1969). *Biochim. Biophys. Acta* **195**, 268–275.
Isaacks, R. E., Santos, B. G., and Musil, G. (1973). *J. Protozool.* **20**, 477–481.
Jurand, A., and Selman, G. G. (1970). *J. Gen. Microbiol.* **60**, 357–364.
Kumar, A. (1970). *J. Cell Biol.* **45**, 623–634.
Lee, Y. C., and Scherbaum, O. H. (1966). *Biochemistry* **5**, 2067–2075.
Lipps, H. J., Sapra, G. R., and Ammermann, D. (1974). *Chromosoma* **45**, 273–280.
Muramatsu, M. (1970). *In* "Methods in Cell Physiology" (D. M. Prescott, ed.), Vol. 4, pp. 195–230. Academic Press, New York.
Sonneborn, T. M. (1970). *In* "Methods in Cell Physiology" (D. M. Prescott, ed.), Vol. 4, pp. 241–339. Academic Press, New York.
Stevenson, I. (1967). Ph.D. Thesis, University of Edinburgh, Scotland.
Suyama, Y., and Preer, J. R. (1965). *Genetics* **52**, 1051–1058.
Tait, A., and Cummings, D. J. (1974). *Biochim. Biophys. Acta*, (in press).

Chapter 16

Isolation of Micro- and Macronuclei of Tetrahymena pyriformis

MARTIN A. GOROVSKY, MENG-CHAO YAO, JOSEPHINE
BOWEN KEEVERT, AND GLORIA LORICK PLEGER

Department of Biology,
The University of Rochester,
Rochester, New York

I. Introduction

The physiological properties that make the ciliated protozoan *Tetrahymena pyriformis* useful as a research tool have been reviewed recently (Elliott, 1973; Everhart, 1972; Hill, 1972). In brief, *Tetrahymena* can be cultured axenically with generation times as short as 2–2.5 hours, and population densities as high as 1 to 2×10^6 cells/ml can be obtained in rich media. Owing to its absolute dietary requirement for a number of amino acids, both a purine and a pyrimidine, isotopic labeling techniques are readily employed with *Tetrahymena*. Thus, high specific activities of particular classes of macromolecules can be obtained in defined media. *Tetrahymena* is also susceptible to most of the metabolic inhibitors commonly used in cell biology. Finally, recent advances in techniques for mutagenesis, conjugation, and mutant selection (Bruns and Brussard, 1974; Orias and Flacks, 1973; Orias and Bruns, 1975) offer the possibility that the power of genetic analysis will soon be added to these highly favorable growth properties, further enhancing the attractiveness of *Tetrahymena* for studies in cell and molecular biology.

We have been particularly interested in various aspects of nuclear structure and function in *Tetrahymena pyriformis*. With the exception of a few amicronucleate or partially amicronucleate strains of *Tetrahymena* each cell contains two types of nuclei, a smaller micronucleus and a larger macronucleus. Although these two nuclei are formed from daughter nuclei of a single mitotic division (the second postzygotic division) during conjugation, they differ in most of their structural and functional properties in vegetative cells. Micronuclei divide mitotically prior to cell division, undergo DNA synthesis early in the cell cycle, do not contain nucleoli or heterogeneous ribonucleoprotein particles, and are genetically inactive. Macronuclei divide amitotically when the cell divides, synthesize DNA in the middle third of the cell cycle, have large numbers of RNA-containing inclusions, and are responsible for the genetic activity which controls the phenotype of the vegetative cell. Micro- and macronuclei thus have a common origin during conjugation and reside in the same cytoplasm during vegetative growth, yet differ strikingly in structure and function. Therefore, we have suggested that these two nuclei may be used as a model system to explore the mechanisms whereby the same (or closely related) genetic information is maintained in different structural and functional states in eukaryotic cells (Gorovsky, 1973; this reference also contains a review of the properties of macro- and micronuclei).

Prior to an analysis of the molecular mechanisms underlying the differences between macro- and micronuclei, it was necessary to develop methods to isolate and purify them. In this report, we describe the methods currently in use in our laboratory for isolating macro- and micronuclei.

It should be noted that methods other than the one reported here have been used to isolate macronuclei from *Tetrahymena* (Byfield and Lee, 1970; Lee and Scherbaum, 1965; Mita *et al.*, 1966; Prescott *et al.*, 1966; Ringertz *et al.*, 1967; Rosenbaum and Holz, 1966). The relative merits of a few of these techniques for isolating macronuclei have been reviewed recently (Everhart, 1972). However, none of these methods describes the simultaneous isolation of macro and micronuclei, and so they will not be discussed further. To our knowledge, only two methods have been described for isolating both macro- and micronuclei from *Tetrahymena*. One of these (Muramatsu, 1970) was reported only briefly, and subsequent work utilizing this method has not been forthcoming. The second (Gorovsky, 1970) is superseded by the procedures described here.

II.   Cell Culture

There are probably almost as many different methods for culturing *Tetrahymena* as there are laboratories studying the organism. Since growth conditions have a considerable effect on the ease with which nuclei can be isolated (see Section III,E), the particular methods used in our laboratory will be described briefly. More extensive treatments of nutrition and culturing techniques can be found in recent reviews (Elliott, 1973; Everhart, 1972; Hill, 1972).

A.   Media

The culture medium (Dr. F. Child, personal communication) consists of: 2.0% proteose peptone (Difco), 0.2% dextrose, 0.1% yeast extract (Difco), 0.003% Sequestrine (Geigy Chemical Corp., Ardsley, New York).

This medium is made up at 10 × the above concentrations and is then cleared by centrifugation at 7000 g for 30 minutes. This removes insoluble material which, if not removed, contaminates the nuclear preparations. The concentrate is stored frozen ($-20°$C) in appropriate-sized aliquots, and is diluted and autoclaved prior to use.

In experiments in which long-term labeling of nucleic acids by radioactive precursors is described, cells are grown in the above medium minus the yeast extract.

B.   Maintenance Cultures

Stock cultures are maintained in 5 ml of the enriched proteose peptone medium in 15-ml screw-cap culture tubes kept upright at room temperature. These cultures are transferred as necessary or every 2–4 weeks if the cells are not being used. Viable transfers can be made from such tube cultures of many strains (Syngen I, HSM, GL) even after 3–4 months; however, some strains (W) may need to be transferred more frequently.

C.   Flask Cultures

Culture medium, 50 ml, in 250-ml or 300-ml Erlenmeyer flasks is used for small-scale nucleus isolations or as "feeder" cultures for larger cultures (See Section III,D). Fifty-milliliter cultures are inoculated simply by sterile transfer of the entire contents of a 5-ml tube culture into the flask. Depending upon the particular use to which they will be put, flask cultures are grown for various periods of time at appropriate temperatures in incubator shakers with moderate shaking (100–150 cycles/minute). For experiments in which cell numbers or particular culture stages are critical, we have found it convenient to grow cells in 300 ml nepheloculture flasks (Bellco Glass Corp., Vineland, New Jersey). Cell growth is then monitored simply by placing the side arm of the flask into a Bausch and Lomb Spectronic 20 spectrophotometer. Under the conditions used in our laboratory, an absorbance of 1.0 at 550 nm equals approximately 1×10^6 cells/ml. With practice, it is possible to manipulate culture conditions (age of initial inoculum, temperature, etc.) to provide cultures in a particular growth state (lag growth, log growth, deceleratory growth, etc.) as required.

For use as feeder cultures for larger cultures, a flask culture is usually initiated with a 2- to 7-day-old tube culture and then grown for 24–48 hours at 28°–30° C prior to use.

D.   Large Cultures

We routinely grow 1-liter and 3-liter cultures of cells in 2-liter and 5-liter diphtheria toxin bottles (VWR Scientific, Rochester, New York), respectively. The bottles are fitted with stoppers consisting of a 20-cm length of glass tubing (8 mm, OD) around which cotton and cheesecloth have been wrapped. Gas dispersion tubes (VWR Scientific) are attached to the glass tubing by a short length of rubber tubing. Another short length of cotton-plugged glass tubing is attached to the other end of the stopper. A small piece of rubber tubing is attached to the open end of the second tube, and both lengths of rubber tubing are closed with a pinch clamp during autoclaving to prevent siphoning of the medium. Two drops of Antifoam SAG 471 (Union Carbide Corporation, New York) per liter are added to large cultures prior to autoclaving to prevent foaming during aeration.

Large cultures are inoculated simply by sterile transfer of the contents of a 1- to 2-day-old flask culture (Section III,C) into the bottle. We have found it most convenient to grow these large cultures in a 5 or 10 gallon aquarium outfitted with a Bronwill constant temperature circulator (Model 20, VWR Scientific). The bottles are simply placed in the aquarium in an upright position, and the glass tubing of the stopper is connected to a source of compressed air. The flow rate is adjusted to maintain a thin layer of foam at the top of culture. For routine use (strain WH-6), cultures containing 1 liter of medium are grown at 28°C for 16–18 hours. Cultures containing 3 liters are started about 24 hours before they are needed. These conditions yield cultures containing approximately 5 to 7 × 10^5 cells/ml. Slight adjustments in these growth conditions may be necessary when other strains are used.

III.   Isolation of Nuclei

A.   Solutions

1.   MEDIUM A

The composition of Medium A is given in Table I. Gum arabic is prepared as a 20% stock solution and is stored frozen (−20°) in appropriate sized aliquots. The stock gum arabic solution should be cleared by centrifugation at 16,000 g for 10 minutes to remove particulate material, which would

TABLE I

NUCLEUS ISOLATION MEDIA

---

1. Medium A[a]
    0.1 *M* Sucrose
    4.0% Gum arabic
    0.1% Spermidine-HCl
    0.002 *M* MgCl$_2$
    pH to 6.75 with NaOH
2. Medium B[a]
    0.63% (v/v) *N*-octyl alcohol in Medium A
3. Microsomal medium
    0.25 *M* Sucrose
    0.10 *M* KCl
    0.01 *M* Tris, pH 7.2
    0.006 *M* MgCl$_2$
    0.001 *M* Dithiothreitol
    0.01% Spermidine-HCl

---

[a]Media A and B are slight modifications of the media devised by Kuehl (1964) for isolation of nuclei from plant cells.

otherwise contaminate the isolated nuclear preparations. The pH of Medium A is adjusted to 6.75 with 1 N NaOH prior to use.

2. MEDIUM B

Medium B consists of 0.63 ml of *N*-octyl alcohol per 100 ml of Medium A. It is made immediately prior to use and care must be taken during the nucleus isolation procedure to keep the octanol uniformly suspended by vigorous shaking before each use.

3. MICROSOMAL MEDIUM

The composition of microsomal medium is listed in Table I. This medium is not actually necessary for isolating nuclei; however, it is useful for washing nuclei prior to metabolic studies (see Section V,C).

4. METHYL GREEN SOLUTION

All steps in the nucleus isolation procedure are monitored by light microscopy. A small drop of a suspension of the appropriate cell fraction is mixed on a slide with a small drop of a solution of 0.4% methyl green in 6.0% (v/v) acetic acid containing 2×10^{-3} M CaCl$_2$ (Kuehl, 1964). Both macro- and micronuclei are rapidly stained green by this procedure. Macronuclei are easily visible at low power even in a crude homogenate. Micronuclei can be seen easily at higher magnification.

B. Collection of Cells

Cells are collected by centrifugation in 50-ml plastic graduated centrifuge tubes (IEC No. 2809) at 2000 g for 3 minutes. We have found it convenient to use a light, 16-place, swinging-bucket rotor (the Sorvall HL-8) for this purpose because its rapid acceleration and deceleration rates minimize the time spent at this step.

Cells can be washed in distilled water or in Medium A and pelleted again as described above. However, we have found that this step can be eliminated if care is taken to remove all the culture medium from the pelleted cells. A Dispo-pipet connected to an aspirator is used for this purpose.

C. Isolation of Macronuclei

1. BLENDING

Cells are resuspended in Medium B at a ratio of approximately 50 ml of medium per milliliter of packed cells. The cells are then blended for 15–30 seconds at high speed in a Waring Blendor (Model 5010) using either the semimicro cup (for volumes up to 100 ml) or the standard cup (for volumes from 100 to 300 ml; volumes greater than 300 ml are divided into smaller aliquots to prevent leakage during blending).

2. COLLECTING MACRONUCLEI

Macronuclei are pelleted by centrifugation at 6500 g for 5 minutes. For larger isolations it is convenient to use a large-volume, swinging-bucket rotor (the Sorvall HG-4) for this step. This keeps the number of tubes that must be handled small (≤ 4) and also ensures that the nuclear pellets are centered at the bottom of the tube, facilitating their removal. For 1-liter isolations (the size most frequently used in our work), collection of macronuclei usually is carried out in 4 Corex 150-ml screw-cap bottles (Sorvall No. 158). For smaller isolations, centrifugations at 2250 g for 5 minutes in a smaller-capacity swinging-bucket head (the Sorvall HL-8 or HB-4) can be used.

After centrifugation there should be a small tightly packed white pellet, which is rich in macronuclei. The materials which do not pellet (some macronuclei, micronuclei, nonnuclear cell components) are found either in the cloudy supernatant or in a "skin" at the top of the tube. The size and tightness of the skin varies with culture age, degree of blending, octanol concentration, etc.

The supernatants and skins are poured off and blended again for 10 seconds at low speed, and macronuclei are again pelleted. This procedure is repeated (usually a total of 4–5 times) until microscopic analysis indicates that the pellets contain approximately as many micronuclei as macronuclei. The importance of removing most of the macronuclei from this homogenate cannot be overemphasized since any macronuclei remaining in the homogenate will contaminate the micronuclear preparations.

3. WASHING MACRONUCLEI

The crude macronuclear pellets are resuspended in Medium B (one-tenth the volume of the original homogenate is sufficient) by vigorously pipetting with a Dispo-pipet. They are then pooled and are pelleted by centrifugation at 2000 g for 5 minutes. A second wash with medium B is usually followed by two washes with either Medium A or with microsomal medium to remove the octanol and most of the remaining cytoplasmic contamination.

D. Isolation of Micronuclei

1. COLLECTING MICRONUCLEI

The postmacronuclear supernatants are pooled and blended for 15 seconds at high speed, and micronuclei are pelleted by centrifugation at

16,000 *g* for 10 minutes. The resultant pellet contains micronuclei usually contaminated by small numbers of macronuclei and by significant amounts of cytoplasmic material.

2. NUCLEOPORE FILTRATION AND WASHING

The crude micronuclear pellet is resuspended in 10–20 ml of Medium A. The nuclei are then pelleted at 2000 *g* for 10 minutes. For reasons that are not completely clear, the micronuclei that remained either in suspension or in the skin when the original Medium B homogenate was centrifuged at low speeds pellet at these same speeds when resuspended in Medium A.

The washed nuclei are resuspended in 10–20 ml of Medium A and filtered under mild suction, first through a 47 mm nucleopore filter (VWR Scientific) with 8-μm diameter pores and then through a filter with 5-μm diameter pores. The filtrate is then centrifuged at 2000 *g* for 10 minutes. The micronuclear pellet is resuspended in 10–20 ml of Medium A and pelleted again by centrifugation at 2000 *g* for 10 minutes.

E. Critical Variables

1. CULTURE CONDITIONS

Nuclei can be isolated by the methods described above from cells in log phase or in the decelerating stage of culture growth. For older cultures (stationary phase) it is often necessary to add additional amounts of octanol or to extend blending times to prepare nuclei. We have had little success to date in trying to isolate nuclei from starved cells or from poorly aerated cultures.

2. CELL DENSITY DURING HOMOGENIZATION

Lower ratios of Medium B to cells than that suggested here (50:1) can be used, but usually additional centrifugations are required to collect all the nuclei. Higher ratios can be used without deleterious effect and may even result in purer nuclear preparations.

3. COMPONENTS OF NUCLEUS ISOLATION MEDIA

a. Octanol. The mechanism of action of octanol is not known. Cell lysis can be obtained with less octanol than used here, but at these lower concentrations the macronuclear pellet may be heavily contaminated by clumped cytoplasmic components. Nuclear blebbing resulting in the nuclei having a "bumpy" appearance is frequently observed in Medium B, but can be reversed by resuspending nuclei in octanol-free solutions (Gorovsky, 1970).

b. Divalent Cations. The presence or absence of divalent cations seems to have little effect on the isolation of nuclei (Gorovsky, 1970), and successful isolations have also been obtained in 0.01 *M* EDTA.

c. pH. The effects of pH and of different buffers have not been extensively examined. In isolations without spermidine, phosphate buffer was found to be inferior to Tris or unbuffered solutions, and no differences were found between pH 6.5 and 7.0 (Gorovsky, 1970). In spermidine-containing media, macronuclei have been isolated at pHs as low as 5.0, but increased difficulty in breaking cells is observed at pHs below 6.0.

d. Spermidine. To our knowledge, the first use of spermidine for the isolation of *Tetrahymena* nuclei was reported by Rosenbaum (1963; Rosenbaum and Holz, 1966). Prescott *et al.* (1966) also described the use of spermidine in conjunction with the detergent Triton X-100 for isolating ciliate nuclei. In agreement with these authors, we have found that spermidine has a remarkable stabilizing effect, particularly on macronuclei. In the absence of spermidine, the blending conditions used here disrupt macronuclei, making it necessary to use shorter blending times to isolate macronuclei and resulting in poorer yields than those obtained by this procedure (Gorovsky, 1970) (see Section VI,A). The mechanism of this stabilizing effect is unknown. In addition, we have found spermidine at a concentration of 0.1% to be a good inhibitor of endogenous histone proteolysis in isolated macronuclei (M. A. Gorovsky and J. B. Keevert, unpublished observations).

e. Gum Arabic. If gum arabic is left out of Medium A, both the yield and the condition of isolated macronuclei are worsened considerably. Recently, we have found that 4% dextran sulfate can substitute for gum arabic, but extensive studies on yields, purity, and metabolic properties of nuclei isolated using dextran have not yet been performed.

It is essential to note that gum arabic can contaminate both nucleic acid and histones extracted from isolated nuclei, making it necessary to wash nuclei in appropriate gum arabic-free solutions (SSC, distilled water, microsomal medium, etc.) before attempting further chemical analysis of the isolated nuclei.

F. Applicability

The method described here has been used to isolate macronuclei from a number of strains and syngens of *Tetrahymena pyriformis* with only slight changes in blending times and/or in the number of centrifugations necessary to completely pellet the nuclei (Table II). This method has also been reported to be useful in isolating macronuclei from *Paramecium aurelia* (Soldo and Godoy, 1972).

To date, micronuclei have been isolated by these methods only from Syngen I and from strain HSM (Table II) of *Tetrahymena*.

TABLE II

STRANS FROM WHICH NUCLEI HAVE BEEN ISOLATED[a–c]

Cell type[d]	Macronuclei	Micronuclei
Syngen 1		
WH-6-I	+	+
LWA-III	+	NA
B-IV	+	+
UM221-VI	+	NA
B-VII	+	+
Syngen 2		
DDH-58-7-5-I	+	NA
HAM-3-IX	+	NA
Syngen 5		
I	+	NA
UM30-II	+	NA
Syngen 6		
UM1060-I	+	NA
UM1091-II	+	NA
UM1147-III	+	NA
Syngen 9		
TC105-I	+	NA
TC3-III	+	NA
TC160-IV	+	NA
Syngen 11		
Au-50-1-II	+	NA
Strain GL[e]	+	–
Strain HSM	+	+

[a]Data for all cells other than strains WH-6, B-IV, B-VII, GL, and HSM represent unpublished experiments of C. A. Johmann.

[b]We are indebted to Dr. G. Holz for strain WH-6, Dr. P. J. Bruns for strains B-IV and B-VII, Dr. J. Nilsson for strain GL, Dr. I. Cameron for strain HSM, and Dr. D. Nanney for all of the other strains listed in this table.

[c]Symbols used: + = successful isolation; NA = isolation not attempted.

[d]Roman numerals indicate mating types; letters and arabic numerals indicate particular strain.

[e]An amicronucleate strain.

IV.   Yields and Purity of Isolated Nuclei

A.   Yields of Macronuclei

Table III illustrates the yields obtained by one of us (J.B.K.) in the ten most recent isolations performed over a period of approximately 4 months. It is clear that isolation of macronuclei is an efficient, highly reproducible process.

TABLE III
EFFICIENCY OF ISOLATION OF MACRONUCLEI FROM 1-LITER CULTURES

Isolation number	Total cell number $\times 10^8$	No. of macronuclei recovered $\times 10^8$	% Efficiency[a]
1	5.9	5.0	85
2	6.0	3.1	52
3	6.2	6.4	103
4	6.3	4.7	75
5	6.0	4.4	73
6	6.7	5.8	87
7	5.4	4.1	76
8	5.2	4.6	88
9	6.2	5.8	94
10	5.1	4.4	86

[a]Computed as (No. of macronuclei recovered)/(total cell No.) $\times$ 100.

B. Purity of Macronuclei

Evidence has been presented previously attesting to the purity of the isolated macronuclear preparation (Gorovsky, 1970). Light and electron microscopic examinations reveal that the macronuclear preparation does contain some micronuclei (1–2 micronuclei for every 10 macronuclei in the few cases where these have actually been counted), however the large differences in size, DNA content, histone content, RNA content, etc., between macro- and micronuclei suggest that, for most purposes, the properties of this fraction are essentially macronuclear. For example, macronuclei contain, on the average, about 20 times as much DNA as micronuclei. Therefore, DNA extracted from the macronuclear fraction should contain 100–200 times more macronuclear DNA than micronuclear DNA.

Nonnuclear contamination of the macronuclear fraction is slight, consisting of small numbers of ribosomes bound to the nuclear envelope and of occasional ciliary, pellicular, or membrane fragments. These can be removed by detergent washes or by pelleting the macronuclei through dense sucrose or Ficoll solutions. In practice however, it is usually unnecessary to remove these components since their amounts are negligible. If chromatin is to be prepared from macronuclei, most of these components are removed in the normal course of chromatin purification.

C. Yields of Micronuclei

Table IV lists the yields of micronuclei from 13 isolations performed by one of us (M. C. Y.) over a one-year period. It can be seen that recovery of

TABLE IV
YIELDS AND PURITY OF MICRONUCLEI ISOLATED FROM 1-LITER CULTURES

Isolation number	Total cell number $\times 10^8$	No. of micronuclei recovered $\times 10^8$	% Efficiency[a]	Micronuclei[b] / macronuclei
1	7.8	4.1	53	150
2	9.0	4.4	49	300
3	6.5	4.9	75	500
4	7.0	6.0	86	450
5	7.0	7.6	109	500
6	7.1	3.6	51	900
7	5.6	2.0	36	200
8	5.4	3.6	67	200
9	7.4	7.2	97	200
10	6.4	6.4	100	200
11	6.3	4.4	70	3950
12	5.2	2.0	38	1400
13	6.1	3.0	49	250

[a]Computed as (No. of micronuclei recovered)/(total cell No.) $\times$ 100.
[b]Based on number of macronuclei and micronuclei present in the micronuclear fraction. Determined by hemacytometer counting of nuclei after staining with methyl green.

micronuclei is considerably more variable than that of macronuclei (compare Table IV with Table III). On the average, approximately 6 micronuclei are recovered per 10 cells. However, there is frequently more than 1 micronucleus per cell in the strain of *Tetrahymena* on which these studies were done (M. C. Yao, unpublished observations) so that the actual yield of micronuclei is somewhat lower than 60%.

D. Purity of Micronuclei

1. MICROSCOPIC ANALYSIS OF CONTAMINATION BY MACRONUCLEI

Microscopic analysis of the micronuclear fraction indicates that it is contaminated by variable numbers of macronuclei. An estimate of the macronuclear contamination can be made by counting the numbers of macro- and micronuclei at high magnification using a hemacytometer. A typical experiment is presented in Table V, in which this method was used to assess both the yield and the purity of micronuclei at various steps during the isolation. In this experiment, a final ratio of 440 micronuclei per macronucleus was achieved. Since macronuclei contain about twenty times more DNA than micronuclei, we can estimate that micronuclear DNA (or histones) isolated from this preparation would be contaminated by approximately 5% macronuclear DNA. If we had purified DNA from the 16,000 g

TABLE V

RECOVERY AND PURITY OF MICRONUCLEI ANALYZED BY COUNTING STAINED NUCLEI
USING A HEMACYTOMETER.

Stage of isolation	No. of Micronuclei	% Efficiency[a]	Micronuclei / macronuclei
16,000 g Pellet	1.1×10^9	147	240
8 μm Filtrate	9.2×10^8	123	350
5 μm Filtrate	5.8×10^8	77	440

[a]Total cell number in this isolation was 7.5×10^8.

pellet without filtration of micronuclei, the yield would have approximately doubled, but the contamination by macronuclear DNA would have risen to about 10%. The ratios of micronuclei observed in 13 isolations are in Table IV. As with micronuclear recovery, considerable variability is seen in the amount of macronuclear contamination. However, the lowest ratio obtained was 150 to 1, so that even this preparation is estimated by microscopic analysis to consist of about 85% micronuclear components and only 15% macronuclear contamination.

It should be pointed out that these visual assessments provide only a crude estimate of macronuclear contamination. Macronuclear fragments which are too small to be recognized as such are not counted, and may even be mistaken occasionally for micronuclei.

2. ISOTOPIC ANALYSIS OF CONTAMINATION BY MACRONUCLEI

Visual assessment of macronuclear contamination is difficult, time consuming, and subject to inaccuracies due to misidentification of macronuclear fragments. Therefore, we have developed an isotopic method for measuring contamination which is considerably easier. This method takes advantage of the existence of amicronucleate strains of *Tetrahymena*. Cells from an amicronucleate strain (GL) are grown in the presence of a ^{14}C-labeled precursor of the chemical component to be studied (thymidine-^{14}C for DNA; lysine-^{14}C for histone, etc.). The ^{14}C-labeled amicronucleate cells are then pelleted, resuspended in Medium B, and mixed with the micronucleate cells from which macro- and micronuclei are to be prepared. The micronucleate cells either can be unlabeled or can be labeled with a tritiated version of the precursor used to label the amicronucleate cells (thymidine-^{3}H, lysine-^{3}H, etc.). Macro- and micronuclei are then isolated in the usual manner, and the chemical component being studied (DNA, histone, etc.) is extracted and purified by standard techniques. Since the ^{14}C-labeled cells were amicronucleate, any ^{14}C which appears in the micronuclear fraction must be due to contamination. A comparison of

TABLE VI

RECOVERY AND PURITY OF MICRONUCLEI ANALYZED BY DUAL ISOTOPE
SCINTILLATION COUNTING[a]

Stage of isolation	% ³H recovered[b]	% ¹⁴C recovered[b]	% Macronuclei in micronuclei
Homogenate	97.6	96.9	—
Purified macronuclei	67.6	70.0	—
16,000 g Pellet	4.8	1.2	24.0[d]
Purified micronuclei[c]	3.1	0.58	18.0[d]
Purified macronuclear DNA	45.7	55.2	—
Purified micronucelar DNA	1.8	0.27	12.4[e]

[a]Thymidine-¹⁴C-labeled cells from amicronucleate strain GL added to thymidine-³H-labeled cells of strain WH-6.

[b]Percentages based on amount of isotope incorporated into whole cells counted prior to the isolation.

[c]Purified by a single passage through a nucleopore filter with 5 μm pore diameter.

[d]Computed by comparing the ratio of ¹⁴C to ³H with that in purified macronuclei.

[e]Computed by comparing the ratio of ¹⁴C to ³H with that in purified macronuclear DNA. The difference between this value and that for sample 4 (the purified micronuclei from which the DNA was isolated) probably is due to the contamination of the micronuclear fraction by ¹⁴C incorporated into non-DNA-containing cytoplasmic components.

the specific activity (¹⁴C counts per minute per unit DNA or per unit protein) or the ¹⁴C to ³H ratio of the material extracted from micronuclei with that extracted from macronuclei yields a simple estimate of the degree of contamination of the micronuclear material by macronuclei. Table VI illustrates the results obtained during the course of isolation of nuclei from such a mixture of thymidine-¹⁴C-labeled amicronucleate cells (GL) with thymidine-³H-labeled micronucleate cells (Syngen I, strain WH-6).

This isotopic method yields slightly higher estimates of contamination of micronuclear components by macronuclei than does the microscopic counting method (in three experiments of the type illustrated in Table VI, estimates of 6.8%, 8%, and 12.4% contamination were obtained), but the discrepancy is not great. Clearly, this method relies on the assumption that macronuclear components from the amicronucleate cell behave identically to those of the micronucleate cell. While we have not yet thought of any direct way to test this assumption, the good agreement between the isotopic estimates of contamination, the optical method, and a number of independent chemical assessments of micronuclear contamination (see Section IV,D3) suggests that this assumption is not unreasonable.

3. CHEMICAL ANALYSIS OF CONTAMINATION BY MACRONUCLEI

We have found that a number of chemical components of macronuclei, such as methylated adenine (Gorovsky *et al.*, 1973a), acetylated subfractions of histone F2A1 (Gorovsky *et al.*, 1973b), and amplified cistrons for ribosomal genes (Yao *et al.*, 1974), are either absent or present in greatly reduced amounts in micronuclei. If one assumes that these components are, in fact, absent from micronuclei *in vivo*, their amounts in the isolated micronuclear preparation can serve as another estimate of the purity of the isolated micronuclei. Estimates of contamination of micronuclei by macronuclear components using this method are in good agreement with the visual and isotopic estimates described above and suggest that approximately 0–20% of the DNA or histone isolated from micronuclear preparations is from contaminating macronuclei. Clearly, if any of these components are actually present in small amounts in micronuclei, this method would overestimate the amount of contamination.

4. NONNUCLEAR CONTAMINATION

The micronuclear preparation is more heavily contaminated by cytoplasmic components than the macronuclear fraction. However, since most of our studies involved subsequent purification of easily identifiable nuclear components (chromatin, DNA, histones), we have not routinely attempted to remove these remaining contaminants. They can be removed, if necessary, by pelleting micronuclei through dense sucrose solutions or by equilibrium sedimentation in Metrizamide (Birnie *et al.*, 1973).

V. Structural, Chemical, and Metabolic Studies of Isolated Nuclei

A. Structural Studies

Electron microscopic observations have indicated that damage to the macronuclear envelope and difficulty in identifying nuclear pores in both macro and micronuclear envelopes constitute the most marked structural artifacts caused by the isolation procedure. Otherwise, the major morphological entities typical of ciliate macro- and micronuclei appear to be present in the isolated nuclei (Gorovsky, 1970).

B. Chemical Studies

Chromatin, DNA, and histones have been purified from isolated macro- and micronuclei in good yield. These components have been characterized by standard analytical techniques and have been found to be comparable to components isolated from the nuclei of higher organisms. To date, differences have been found in the amounts of methylated adenine in DNA (Gorovsky *et al.*, 1973a) in the amounts of moderately repetitive DNA sequences (M. -C. Yao and M. A. Gorovsky, 1974), in the degree of repetitiveness of the genes coding for 25 S and 17 S ribosomal RNA (Yao *et al.*, 1974), and in the particular types of histones in the two nuclei (Gorovsky, 1970; Gorovsky *et al.*, 1973b).

High molecular weight precursors of 25 S and 17 S ribosomal RNA have been extracted from macronuclei purified by these techniques (Kumar, 1970), and methods have also been devised for removing the ribosomes that remain bound to the nuclear envelope of isolated macronuclei (Gorovsky, 1969; M. A. Gorovsky and J. B. Keevert, unpublished observations). To date, we have had little success isolating the other major morphological components (nucleoli, interchromatin granules) contained in macronuclei.

C.   Metabolic Studies

We have only studied a few of the metabolic properties of isolated macronuclei, and have not yet attempted similar studies with micronuclei since they are much more difficult to obtain in large amounts. Also the small but variable levels of macronuclear contamination probably would make it difficult to ascertain whether low levels of metabolic activity were actually due to micronuclei or to contaminating macronuclei.

It is essential to note that nuclei isolated by the methods described here must be washed thoroughly to remove any traces of octanol and gum arabic, since these have deleterious effects on the function of the isolated nuclei *in vitro*. Microsomal medium (Table I) has been found to be useful for this purpose and also has served as a good medium for maintaining nuclear integrity when nuclei are resuspended in it for metabolic studies (Gorovsky, 1969; Gorovsky *et al.*, 1973b).

To date, we have found that isolated macronuclei can transfer radioactive acetate from acetyl coenzyme A to histones in a fashion similar to that observed *in vivo* (Gorovsky *et al.*, 1973b). Isolated macronuclei are also able to incorporate radioactive ribonucleoside triphosphates into RNA and tritiated thymidine triphosphate into DNA (A. R. Kimmel and M. A. Gorovsky, unpublished observation). Finally, the ribosomes that are bound to the outer envelope of the macronucleus can incorporate radioactive amino acids into peptides, which are then transferred into the macronucleus (Gorovsky, 1969; M. A. Gorovsky and J. B. Keevert, unpublished observations). The RNA, DNA, and peptides synthesized by isolated macronuclei have not yet been characterized.

ACKNOWLEDGMENT

This work was supported by National Science Foundation Grant No. GB-33716.

REFERENCES

Birnie, G. D., Rickwood, D., and Hell, A. (1973). *Biochim. Biophys. Acta* **331**, 283–294.

Bruns, P. J., and Brussard, T. B. (1974). *J. Exp. Zool.* **188**, 337–344.

Byfield, J. E., and Lee, Y. C. (1970). *J. Protozool.* **17**, 445–453.

Elliott, A. M. (1973). "Biology of *Tetrahymena*." Dowden, Hutchinson & Ross, Stroudsburg, Pennsylvania.

Everhart, L. P., Jr. (1972). *In* "Methods in Cell Physiology" (D. M. Prescott, ed.), Vol. 5, pp. 219–287. Academic Press, New York.

Gorovsky, M. A. (1969). *J. Cell Biol.* **43**, 46A.

Gorovsky, M. A. (1970). *J. Cell Biol.* **47**, 619–630.

Gorovsky, M. A. (1973). *J. Protozool.* **20**, 19–25.

Gorovsky, M. A. Hattman, S., and Pleger, G. L. (1973a). *J. Cell Biol.* **56**, 697–701.

Gorovsky, M. A., Pleger, G. L., Keevert, J. B., and Johmann, C. A. (1973b). *J. Cell Biol.* **57**, 773–781.

Hill, D. L. (1972). "Biochemistry and Physiology of *Tetrahymena*." Academic Press, New York.

Kuehl, L. (1964). *Z. Naturforsch. B* **19**, 525–532.

Kumar, A. (1970). *J. Cell Biol.* **45**, 623–634.

Lee, Y. C., and Scherbaum, O. H. (1965). *Nature (London)* **208**, 1350–1351.

Mita, T., Shiomi, H., and Iwai, K. (1966). *Exp. Cell Res.* **43**, 696–699.

Muramatsu, M. (1970). *In* "Methods in Cell Physiology" (D. M. Prescott, ed.), Vol. 4, pp. 195–230. Academic Press, New York.

Orias, E., and Bruns, P. J. (1975). *In* "Methods in Cell Biology" (D. M. Prescott, ed.) (in preparation).

Orias, E., and Flacks, M. (1973). *Genetics* **73**, 543–559.

Prescott, D. M., Rao, M. V. N., Evenson, D. P., Stone, G. E., and Thrasher, J. D. (1966). *In* "Methods in Cell Physiology" (D. M. Prescott, ed.), Vol. 2, pp. 131–142. Academic Press, New York.

Ringertz, N. R., Bolund, L., and DeBault, L. E. (1967). *Exp. Cell Res.* **45**, 519–532.

Rosenbaum, J. L. (1963). Ph.D. Thesis, Syracuse University, Syracuse, New York.

Rosenbaum, J. L., and Holz, G. G., Jr. (1966). *J. Protozool.* **13**, 115–123.

Soldo, A. T., and Godoy, G. A. (1972). *J. Protozool.* **19**, 673–678.

Yao, M.-C., and Gorovsky, M. A. (1974). *Chromosoma* **48**, 1–18.

Yao, M.-C., Kimmel, A. R., and Gorovsky, M. A. (1974). *Proc. Nat. Acad. Sci. U.S.* **71**, 3082–3086.

Chapter 17

Manipulations with Tetrahymena pyriformis on Solid Medium

ENORE GARDONIO, MICHAEL CRERAR,[1]
AND RONALD E. PEARLMAN

Department of Biology,
York University, Toronto,
Ontario, Canada

I. Introduction

Over a number of years, the ciliate protozoan *Tetrahymena pyriformis* has been an organism much studied by many biologists [for recent discussions on the biology and on the biochemistry and physiology of *Tetrahymena*, see Hill, (1972) and Elliott, (1973)]. Some of the characteristics of *Tetrahymena* that make it interesting for biochemical and molecular biological studies are that it is a unicellular eukaryotic organism that can be readily cultured axenically with a reasonably short generation time in liquid culture including completely defined medium and can be manipulated for the most part with standard microbiological techniques.

Correlation of a biochemical activity *in vitro* with function *in vivo* can often be made by isolating mutants of an organism with missing or altered

---

[1]*Present address:* Department of Medical Genetics, University of Toronto, Toronto, Ontario M5S 1A8 Canada.

enzyme activity. The usefulness of this approach has probably been the major reason for the recent rapid accumulation of knowledge in biochemistry and molecular biology. Most molecular biological studies to date have been carried out with prokaryotic organisms, particularly *Escherichia coli* and its viruses. Many researchers are now interested in extending these studies to eukaryotic organisms and in using the "one gene, one enzyme" correlation to study problems such as development, morphogenesis, and the control of these processes. Because of their ease of handling by standard microbiological techniques, some unicellular eukaryotic organisms are of great interest and importance in such studies. *Tetrahymena* is such an organism.

Mutant isolation and genetic manipulation in *Tetrahymena* has not progressed as rapidly as research in other areas. This is due in part to the complexity of gene transfer and expression in *Tetrahymena*, to the diffficulty in isolating clones of *Tetrahymena* because of the inability of this organism to divide on solid medium without coalescing, and to the fact that methods for "replica plating" are not readily available. Recently, however, drug-resistant mutants of *Tetrahymena* have been obtained carrying out all operations in liquid culture (Roberts and Orias, 1973a,b). A discussion of induction and isolation of mutants in *Tetrahymena* is in preparation for a future volume of this series (E. Orias and P. J. Bruns, in preparation).

Growth of *Tetrahymena* on agar medium has been reported (West *et al.*, 1962). Colonies observed in this study were probably not clones because the problem of restricting the organisms adequately without inhibiting growth was not completely solved for *Tetrahymena* by these workers. We have developed two methods for growing and cloning *Tetrahymena* on solid medium on petri plates. The first method involves the use of Sephadex beads on the plates to aid in restricting the movement of cells while still allowing them to remain motile and divide. This method has been described previously (Gardonio *et al.*, 1973). The second method involves growth of *Tetrahymena* in "micro-holes" punched in the surface of agar medium. Cells can be grown, cloned, and replica plated using this method. Thus these methods should be very valuable in some of the steps in mutant isolation and genetic manipulation in *Tetrahymena*.

II.   Growth and Cloning of *Tetrahymena pyriformis* on Solid Medium

A.   Plates Containing Sephadex

The first method we have developed for growth and cloning of *Tetrahymena* on solid medium uses agar medium sprinkled with Sephadex in petri plates. The Sephadex beads on the plates aid in restricting the movement of cells while still allowing them to remain motile and divide. Most of the liquid in which cells are added to the plate is taken up by the agar and by the Sephadex, and spaces between Sephadex beads remain dry. Thus *Tetrahymena* cannot migrate between Sephadex beads and single cells are localized around beads in a liquid environment with sufficient nutrients to allow for motility, growth, and division.

1. Preparation of Medium

Solid medium is prepared in sterile, plastic petri dishes (100 × 15 mm) containing medium in 1.5% agar (Difco) topped with 2–3 ml of medium in 0.3% agar. The medium we generally use is 2% proteose peptone (Difco) and 0.1% liver extract L (Nutritional Biochemical Co.) (complex medium). Defined media can also be used. Experiments reported here used the defined medium of Elliott *et al.* (1954) generally supplemented with 0.04% proteose peptone. To minimize chance of contamination, plates can be supplemented with 250 µg each of penicillin (Sigma) and streptomycin sulfate (Sigma) per milliliter.

Sephadex G-25, fine (Pharmacia), is applied to the plates from a glass jar closed with a metal cap of diameter 4.5 cm. The cap contains small holes, approximately 0.5 mm drilled through the cap concentrically from the center. After autoclaving, the jar containing the Sephadex is inverted and clamped at a height to give optimal spreading of the Sephadex on a plate. This height is determined by trial before each experiment. In most of our experiments, the jar is clamped 36 cm above the plate. The Sephadex is released by gently tapping the bottom of the jar.

2. Growth and Visualization of the Cells

In order for *Tetrahymena* to grow and divide on solid medium, the cells must remain in a liquid environment where they are motile. If trapped on top of dry agar, cells are nonmotile and they become round and do not divide. If the medium is too wet, the *Tetrahymena* swim over a large area of the plate and colonies observed probably do not arise from single cells. Thus single cells must be kept in an environment where they remain motile enough to allow division yet are trapped so that they cannot migrate across the plate. This is accomplished by sprinkling Sephadex G-25 on top of solid medium on petri plates.

Plates containing either complex or defined medium are prepared and are then dried for 2 days at 37°C or at room temperature for at least a week. Medium, 0.5 ml, containing between 100 and 150 *Tetrahymena* for maximum plating efficiency is pipetted on top of the medium and distributed uniformly by tilting the plate. Before the medium is absorbed by the dry agar and before the *Tetrahymena* are trapped on the dry surface of the agar, Sephadex is sprinkled evenly on top of the plate. Plates are placed in a plastic bag to minimize evaporation, inverted and incubated at the appropriate temperature. After incubation for approximately 5 days, colonies can be observed and counted with a dissecting microscope before staining (Fig. 1) or after staining with Giemsa stain (Fig. 2). The staining procedure is essentially that of Chu (1971). Methanol (3 ml) is layered on top of the agar for 15 minutes. The methanol is carefully removed and 3 ml of stain [2.5 ml of Giemsa stain (Fisher Scientific Co.), 3.0 ml of methanol and 94.5 ml of water] is carefully poured onto the agar. Colonies are observed within 5 minutes and after 10 minutes the dye is removed. The stained colonies are well separated from each other and can be counted with a colony counter (American Optical Corp.).

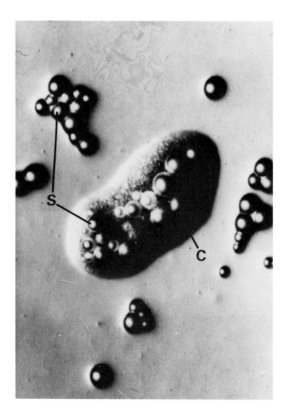

FIG. 1.   Photomicrograph of a colony of *Tetrahymena pyriformis* amicronucleate strain GL viewed under a dissecting microscope (SMXXB, Carl Zeiss, Jena). Approximately 100 cells were plated on top of agar in complex medium and incubated for 6 days at 28°C. S, Sephadex beads; C, cells in a colony surrounding Sephadex beads. From Gardonio *et al.* (1973). Approximately × 200.

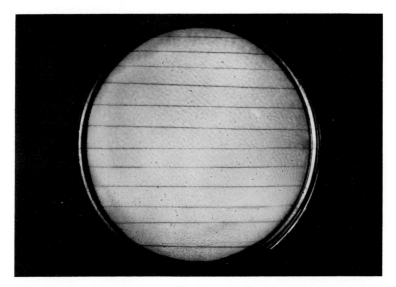

FIG. 2. Colonies of *Tetrahymena pyriformis* strain GL viewed after being stained with Giemsa as described in the text. Approximately 116 cells of *T. pyriformis* GL were plated onto agar in complex medium, incubated for 5 days, and then stained with Giemsa. The small grains (lighter in the middle and gradually darker toward the outside of the plate) are Sephadex beads. Colonies can be seen as mostly irregularly shaped spots, larger and darker than the Sephadex beads. Parallel lines were drawn in pencil on the back of the plate to aid in counting. From Gardonio *et al.* (1973).

3. CHARACTERISTICS OF GROWTH

a. Efficiency of Plating.   Table I shows the plating efficiencies of wild-type *T. pyriformis* amicronucleate strain GL-C, and micronucleate strains HSM and B-7 (kindly supplied by Dr. M. Gorovsky, University of Rochester) on complex medium plates. Data for strain GL-C on defined medium plates supplemented with proteose peptone are also presented. To determine plating efficiency, a known number of cells (determined electronically with a Coulter Counter B, Coulter Electronics Co.) are added to the plates. Colonies are counted after 5 days of growth either before or after staining. Plating efficiency does not depend on whether colonies are counted before or after staining.

b. Viability of Colonies.   Colonies can be picked from unstained plates with a wire loop, wire probe, or sterile toothpick and transferred to liquid medium for growth. Data for colonies picked at random with a wire loop from plates of a number of strains of *Tetrahymena* are presented in Table I. Care must be taken in picking colonies for transfer to liquid medium since the operation is performed while viewing the plate with a microscope.

TABLE I

EFFICIENCY OF PLATING *Tetrahymena pyriformis* ON SOLID MEDIA CONTAINING SEPHADEX[a]

Strain	Temp (°C)	Agar medium	Dissecting microscope				Giemsa staining				No. of colonies picked from plates	No. of colonies which grew in liquid complex medium
			Colonies/plate	Mean	Standard deviation	Plating efficiency (%)	Colonies/plate	Mean	Standard deviation	Plating efficiency (%)		
GL	28	Complex	88, 65, 57, 90, 105, 79, 41, 120	81	± 25	70	95, 56, 51, 81, 92	81	± 24	70	10	9
GL	28	Defined plus proteose peptone (0.04%)	45, 38, 36	40	± 5	29	45, 37, 35	39	± 5	29	10	6
HSM	28	Complex	32, 70, 46, 53, 37, 53	49	± 14	53	27, 62, 37, 45, 32, 45	41	± 12	44	10	10
B-7[b]	32	Complex	64, 55, 61, 80, 87	69	± 14	70	62, 56, 61, 78, 82	68	± 11	69	10	9

[a]From Gardonio et al. (1973).

[b]This strain shows some tendency for cells to migrate from colonies surrounding Sephadex beads. Some colonies have been observed not surrounding beads. To prevent this apparent migration, extra care in drying the plates must be used. Also, Sephadex should be spread very sparsely so that the distance between beads is greater than with the other strains.

TABLE II

Demonstration That Colonies on Sephadex Plates Arise from Single Cells[a,b]

Strain	No. of colonies picked	No. of colonies that grew in drug-free liquid complex medium	No. of colonies that grew in liquid complex medium containing cycloheximide (10 μg/ml)	No. of colonies that grew in liquid complex medium containing chloramphenicol (250 μg/ml)	No. of colonies that grew in liquid complex medium containing cycloheximide (10 μg/ml) and in liquid complex medium containing chloramphenicol (250 μg/ml)
CA-103 plus CHX F-3[c]	40[d]	30	21	8	1
CHX F-3	5[e]	4	4	0	0
CA-103	5[f]	5	0	5	0

[a]From Gardonio et al. (1973).

[b]All experiments were done with complex medium at 28°C.

[c]Approximately 46 CHX F-3 and 41 CA-103 were plated on each of 5 plates.

[d]Colonies were picked at random from three different plates.

[e]Colonies were picked at random from plates onto which approximately 93 cells/plate had been added.

[f]Colonies were picked at random from plates onto which approximately 103 cells/plate had been added.

Unless a micromanipulator is used some "mispicking" can occur. Mispicking rather than the existence of nonviable colonies probably is the reason for low apparent viability in some experiments.

   c. Colonies are Clones.   Colonies observed on plates are clones derived from single cells. To demonstrate this, we plated a mixture of drug-resistant mutants of syngen 1 of Tetrahymena. The mutants used were strain CA-103 (Roberts and Orias, 1973a) resistant to 250 μg of chloramphenicol per milliliter and strain CHX F-3 (Roberts and Orias, 1973b) resistant to 10 μg of cycloheximide per milliliter. As a control, plates containing only one of the above mutants were included in the experiments. After incubation for 5 days, colonies were picked at random from the plates before being stained and incubated in 2 ml of liquid complex medium. Samples from tubes that showed growth after 48 hours were transferred to liquid complex medium containing either 250 μg chloramphenicol or 10 μg cycloheximide per milliliter. After incubation for 2 days, the drug-containing media were scored for growth. Table II shows the results of one such experiment. Of 30 colonies that grew in drug-free liquid medium, only one grew in both cycloheximide- and chloramphenicol-containing medium. The other 29 colonies grew in either medium plus chloramphenicol or in medium plus cycloheximide, but not in both. This suggests that these colonies are clones arising from single mutant cells. The reason for the preponderance of cycloheximide-resistant colonies picked may simply be attributable to random picking or may be because of the slow growth rate of CA-103 (Roberts and Orias, 1973a). Colonies picked from plates containing a single mutant cell type, not a mixture of the two mutants, grew only in medium containing the drug to which they were resistant.

   d. Selection on Plates.   To demonstrate that this technique for plating and cloning can be used in selective procedures for the isolation of mutants, we observed the growth of various mutant strains of Tetrahymena on selective medium. The mutant strains used had been selected in liquid culture (Roberts and Orias, 1973a,b). Strain CA-103, resistant to 250 μg of chloramphenicol per milliliter, was plated on solid complex medium containing 250 μg of chloramphenicol per milliliter and on plates containing 10 μg cycloheximide per milliliter. Drugs were generally included in the agar and were added to the agar by filtration through sterile membrane filters (Sartorius, 0.2 μm). Table III shows that strain CA-103 grows on plates containing chloramphenicol but not on plates containing cycloheximide. A similar experiment done with strain CHX F-3, a mutant resistant to 10 μg of cycloheximide per milliliter, shows that growth occurs on plates containing cycloheximide, but not on plates containing chloramphenicol (Table III). Wild-type cells do not grow on plates in the presence of either chloramphenicol (250 μg/ml) or cycloheximide (10 μg/ml). This experiment demonstrates that strains

TABLE III

GROWTH ON SEPHADEX PLATES CONTAINING CYCLOHEXIMIDE OR CHLORAMPHENICOL[a]

Strain	Medium	Temperature (°C)	No. of colonies on plates containing cycloheximide (10 μg/ml)	Mean	No. of colonies on plates containing chloramphenicol (250 μg/ml)	Mean
CHX F-3[b]	Complex	28	95	79 ± 11	0	
			73		0	
			69		0	
			80		0	
CA-103[c]	Complex	28	0		70	78 ± 11
			0		93	
			0		75	
			0		73	
			0		80	

[a]From Gardonio *et al.* (1973).

[b]Approximately 105 *Tetrahymena* (cell number electronically determined) added to each plate.

[c]Approximately 100 *Tetrahymena* (cell number electronically determined) added to each plate.

carrying genetic markers involved with drug resistance behave the same when grown on solid or liquid medium. We conclude that isolating clones on solid medium containing selective agents can be used in selection procedures for the isolation of mutant strains.

B. Plates with "Micro-Holes"

We have developed a second method for manipulating *Tetrahymena* on solid medium. This method involves making a large number of holes in the surface of agar medium in a petri dish. Cells are spread over the plate and become localized in holes. Conditions are suitable for growth and division and after approximately 48 hours, colonies can be observed in holes.

1. APPARATUS FOR MAKING HOLES

The apparatus (Fig. 3) consists of a stainless steel box 15 cm long by 15 cm wide by 10.7 cm high, one side of which has a removable plexiglass window. A circular plate of radius 4.13 cm having 1000 probes attached to it (Fig. 4) is attached inside the box to a plunger by a pin. This arrangement allows easy removal of the plate for cleaning. The plunger has an adjustable screw

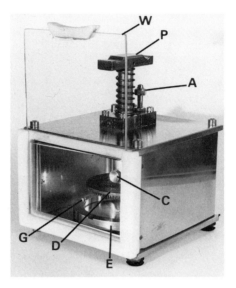

Fig. 3. Apparatus for making holes in solid medium on petri plates. W, removable plexiglass window; P, plunger; A, adjusting screw; C, pin holding plate to plunger; D, plate with probes; E, stainless steel ring which holds petri plate securely below probes; G, pin to orient replicating ring below probes.

Fig. 4. Some components of the apparatus for making holes in solid medium on petri plates and for replica plating. R, aluminum replicating ring; S, screw used to hold and orient petri plates on the replicating ring; D, plate with probes; H, petri plate holder, used as an aid in orienting the petri plate and replicating ring (it has the same dimensions as the stainless steel ring (E, Fig. 3) which is welded below the probes in the hole making apparatus); F, holes on replicating ring which match the pin G, on the petri plate holder.

which determines the depth the probes will go into the agar plates. The probes are made from 19-gauge needles which are cut to a length of 6 cm.

Beneath the plate a stainless steel ring (diameter 8.8 cm) is welded to the bottom of the box. The ring holds the petri plates securely under the probes (Fig. 3). The ring has two pins approximately 170° apart, which are used to orient the replicating ring in the proper position. The replicating ring is an aluminum ring, beveled on the inside, which holds the petri plates (Fig. 4). It has two holes which are also 170° apart, exactly the same distance apart as the pins on the stainless steel ring at the bottom of the box.

To punch the holes in the agar, petri plates containing the appropriate medium are placed in the beveled aluminum ring, which is then aligned on the ring below the probes. Proper alignment is obtained by inserting the holes in the replicating ring into the pins in the stainless steel ring. The probes are then lowered into the agar with the plunger, which has been adjusted with the adjusting screw to depress a known distance.

The entire apparatus is sterilized by autoclaving prior to each use.

2. PREPARATION OF PLATES

A prime requirement for plates is that they all have approximately the same depth of medium. We routinely use a volume of 25 ml of medium in sterile 100- by 15-mm petri plates. The same volume of medium in each plate ensures that the depth of the holes is the same in each plate. Once the plates are poured, they must solidify on a level surface, otherwise, holes of different depth will occur in different areas of the plate.

The volume of medium per plate as well as the concentration of agar to be used in the medium was determined empirically. Also the depth that the holes had to be punched was determined empirically. Agar concentrations of from 0.5% to 1.5% were tried, and we finally decided on a 1% concentration of agar in the medium. Either complex or defined medium can be used. This concentration of agar appears to minimize the occurrence of hairline cracks between the holes.

Originally we planned to make wells in the agar with the hollow probes. To do this, however, we had to punch very deep into the agar so as to be able to pull out the central plug formed by the hollow probe. This appears to cause stress on the agar since hairline cracks occur between adjacent wells. These cracks are big enough to allow the *Tetrahymena* to migrate from one well to another once a high enough concentration of cells has been reached in a well. To prevent this we adjust the plunger so that the probes make shallower holes. We now set the probes to penetrate deep enough into the agar medium to form a cylindrical depression around a central plug. Because of the shallow depth that the probes go into the agar, the plug is not removed (Fig. 5). We find that this procedure eliminates the formation

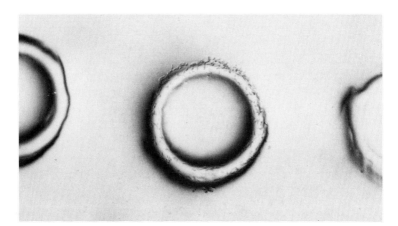

FIG. 5.   Photomicrograph of a colony of *Tetrahymena pyriformis* mutant strain CHX F-3 in a single hole in solid complex medium on a petri plate viewed under a dissecting microscope (SMXXB, Carl Zeiss JENA). Approximately 70 cells were added to the plate in a volume of 1 ml of complex medium. Plates were dried open in an incubator at 28° C for approximately 1 hour. The plates were then covered, placed in plastic bags, and incubated at 28°C for 2 days. Cells are observed in a shallow depression around the outside of a central plug in the agar. Approximately × 200.

between holes of hairline cracks that might form channels for migration of cells.

3.   PROCEDURE FOR REPLICA PLATING

To obtain a replicate of a master plate, both the master and replica plates must be aligned and oriented in exactly the same positions beneath the probes. This exact orientation is necessary since replication is done from a preholed master plate to another preholed plate, and the holes must match. To obtain proper alignment, each petri plate must be oriented in exactly the same position in the replicating ring. Once this is accomplished, the replicating ring is always oriented in the same position relative to the probes because of the holes in the replicating ring.

Each replicating ring has a screw with a sharp point which is used to keep the petri plate tightly inside the replicating ring (Fig. 4). To orient the petri plate on the replicating ring, an indentation is made on the petri plate at the exact height that the screw on the replicating ring will tighten on the petri plate. The indentation on the petri plate is made with a hand drill with the aid of a replicating ring which has a hole at the exact position of the screw in the regular replicating rings. To orient any plate properly, all that has to be done is to indent the master and replica plates with a hand drill and then place them in the replicating ring in such a position that the tip of the screw fits into the indentation on the petri plate.

4. CHARACTERISTICS OF GROWTH

a. Efficiency of Plating and Viability of Colonies. Cells, usually in 1 ml of liquid medium, are applied to plates that have had holes punched in the medium. The plates are tilted back and forth a few times so cells are spread evenly over the surface of the plate. After the cells are spread, excess medium on the surface is removed as quickly as possible in order to trap the cells in the holes. Excess medium is removed in approximately 1 hour by placing the plates upright with the lid off in an incubator at 28°C. This procedure ensures that a cell in one hole will not be able to migrate to another hole. Bacterial and fungal contamination of plates is eliminated by sterilizing the incubator with an ultraviolet lamp between uses and by including 250 μg penicillin and 250 μg streptomycin per milliliter of medium.

After drying, the plates are covered, placed in a plastic bag, and incubated at the appropriate temperature for at least 2 days. At this time, colonies are easily seen when viewed with a stereo dissecting microscope (Fig. 5). If the

FIG. 6. Colonies of *Tetrahymena pyriformis* mutant strain CHX F-3 in solid complex medium on a petri plate viewed without staining. Approximately 70 cells were added to the plate in 1 ml of complex medium. The plate was placed at 28°C with the lid off for approximately 1 hour. The plate was then placed in a plastic bag and incubated at 28°C until colonies were visible, a minimum of 3 days. The colonies are seen as dark rings at the edges of holes in the center of the plate and can be easily counted with the aid of a colony counter (American Optical). In this photograph, colonies are not distinguished at the edges of the plate owing to technical problems with refraction of light during photography.

plates are incubated for 3 days or longer, the colonies appear as rings around the edges of holes (Fig. 6) and can be observed and counted using a colony counter.

To determine the plating efficiency, a known number of cells (determined with an electronic cell counter) are added to a plate. After 2 days at the appropriate temperature, the number of holes containing cells are counted. Plating efficiency can be determined from these data assuming that each colony in a hole arises from a single cell in that hole (see next section). Table IV shows data obtained with two micronucleate and one amicronucleate strain of *Tetrahymena* growing at 28°C in complex medium. Plating efficiency up to 70% can be obtained.

Cells in holes on the plates are viable and appear healthy. The cells are seen to be motile in the confines of the holes. Cells can be picked from the holes using a sterile wooden applicator or toothpick. These cells can be transferred into liquid medium. Table IV shows that 100% of colonies picked from a plate grow in liquid medium. Although we have not made an accurate determination of the growth rate of *Tetrahymena* growing on plates, the doubling time is certainly near normal.

TABLE IV

EFFICIENCY OF PLATING *Tetrahymena pyriformis* ON SOLID MEDIUM
CONTAINING "MICRO-HOLES"

			Dissecting microscope			
Strain	Temperature (°C)	Agar medium	Colonies/ plate	Mean	Plating efficiency (%)	% of colonies picked from plates which grew in liquid complex medium
GL	28	Complex	43	47	70	100
			32			
			63			
			52			
			46			
CA-103	28	Complex	43	41	64	100
			51			
			37			
			34			
CHX F-3	28	Complex	44	40	62	100
			40			
			43			
			34			
			40			

b. Colonies are Clones.   It is expected that if enough holes are present and if few enough cells are added to each plate, no hole should contain more than a single cell. If 50 cells are added to a plate containing 1000 holes, the probability that any hole will contain two cells is approximately 0.001, the probability that any hole will contain one cell is approximately 0.048 and the probability that any hole will contain no cells is approximately 0.95.

Based on these statistical expectations, we have chosen to add approximately 70 *Tetrahymena* per plate which at a plating efficiency of approximately 70% represents approximately 50 colonies per plate. Under these conditions, colonies observed in holes arise from single cells; i.e., colonies are clones. This was demonstrated in essentially the same manner as described previously for the Sephadex plates. Equal numbers of cells of two mutant strains, CA-103 resistant to 250 μg of chloramphenicol per milliliter and CHX F-3 resistant to 10 μg of cycloheximide per milliliter, were added to a number of plates. For an appropriate control, each mutant strain was also added separately to a number of plates. Plates were incubated at 28°C for 72 hours. A total of sixty colonies were picked from four different plates containing the mixture of strains, and 10 colonies were picked from plates containing only the single strains. Colonies that were picked were incubated at 28°C for 48 hours in 2 ml of liquid complex medium. Data in Table V show that only two colonies did not grow in liquid complex medium. Aliquots of each colony in liquid medium were transferred to 2 ml of liquid complex medium containing either 250 μg of chloramphenicol per milliliter of 10 μg of cycloheximide per milliliter. After 48 hours at 28°C, no colonies from the mixed plates grew in both medium containing cycloheximide and in medium containing chloramphenicol (Table V). This suggests that no hole on plates containing a mixture of the two mutant strains of *Tetrahymena* contains cells of more than one mutant strain. It is very unlikely that under our conditions any hole initially contains more than a single cell. Therefore we conclude that colonies are clones.

c. Replica Plating.   Prior to pouring, all plates, both master and those which will be replicas, are indented as described previously. The master plates are poured, holes are punched, and the plates are inoculated with *Tetrahymena*. After approximately 24 hours of incubation at the appropriate temperature, the master plates are ready to be replicated.

To replica plate, the master plate is oriented in the replicating ring by the screw in the ring and the indentation on the plate. The plate is placed under the probes, and the probes are lowered onto the plate. The probes fit exactly into holes already on the plate because the orientation of the plate relative to the probes is exactly the same as when the holes were originally punched in the plate. The probes in the holes pick up some cells from the colonies. The master plate is removed, and a second plate with holes already punched

TABLE V

DEMONSTRATION THAT COLONIES ON PLATES WITH "MICRO-HOLES" ARISE FROM SINGLE CELLS

Strain	No. of Colonies picked	No. of colonies that grew in drug-free liquid complex medium	No. of colonies that grew in liquid complex medium containing cycloheximide (10 μg/ml)	No. of colonies that grew in liquid complex medium containing chloramphenicol (250 μg/ml)	No. of colonies that grew in liquid complex medium containing cycloheximide (10 μg/ml) and in liquid complex medium containing chloramphenicol (250 μg/ml)
CA-103 plus CHX F-3[a]	60[c]	58	26	32	0
CHX F-3[b]	10	10	10	0	0
CA-103[b]	10	10	0	10	0

[a] Approximately 35 cells of CA-103 and 35 cells of CHX F-3 were added to each plate.
[b] Approximately 65 cells of the designated strain were added to each plate.
[c] Twenty colonies were picked at random from each of two plates, and 10 colonies were picked at random from each of another two plates.

is aligned under the probes. Again, because of the indentation on the plate, the plate is oriented relative to the probes exactly as it was when the holes were punched and exactly as the master plate. When the probes are lowered, *Tetrahymena* from the master plate are transferred to corresponding holes in the replica plate. Master and replica plates are rapidly dried at 28°C, placed in plastic bags, and incubated until enough growth has occurred for the plates to be scanned for colonies. As is shown in Table VI, at least 90% of the colonies are transferred from the master plate to corresponding holes on the replica plate.

After each replication, the probes are sterilized by lowering the probes into a petri plate containing 95% ethanol. The plate containing the alcohol is removed, and the alcohol on the probes is ignited and allowed to burn for a few seconds. The flames are doused by lowering the plexiglass window and cutting off the supply of oxygen. The probes are cooled by lowering into agar medium and leaving for a few seconds. Using this method, about 10 plates can be replicated every 30 minutes.

d. Selection on Plates.   The possibility exists that the *Tetrahymena* derive most of their nutrients from the liquid medium in which they are added to the plate, not from the medium in the agar plate itself. This can occur because if cells in liquid complex medium are added to plates prepared in inorganic medium (2.75 gm of NaCl, 0.25gm of $MgSO_4 \cdot 7H_2O$ made up to 1 liter with 0.005 M potassium phosphate buffer, pH 6.5), near normal plating efficiency

TABLE VI

EFFICIENCY OF REPLICA PLATING ON SOLID MEDIUM CONTAINING "MICRO-HOLES"[a]

Strain	No. of colonies on master plate	No. of colonies on replica plate	No. of colonies not transferred from master to replica plate	No. of extra colonies on replica plate[b]
GL	29	26	3	0
	37	34	3	1
	38	37	1	0
	46	48	0	2
	37	34	3	0
	29	29	1	1
CHX F-3	27	27	0	0
	23	21	2	0
	32	28	4	0

[a]All experiments were at 28°C and plates were incubated for 24 hours after replica plating before colonies were counted.

[b]Possible reasons for the extra colonies are spreading of one colony to another hole during replica plating, or incomplete flaming of probes between replica platings.

and growth occur (Table VII). In this experiment, strain CA-103 resistant to 250 μg chloramphenicol per milliliter but sensitive to 10 μg cycloheximide per milliliter was used. However, data in Table VII also demonstrate that nutrients in the medium on the plates is utilized by the cells. Cells in liquid inorganic medium added to plates containing complex medium show growth and plating efficiency the same as when cells in liquid complex medium are added to plates containing complex medium. Furthermore, if 10 μg of cycloheximide per milliliter is included in agar in either inorganic medium or complex medium on plates, no growth occurs whether cells are added to the plates in liquid inorganic medium or liquid complex medium. Thus, this technique of manipulating *Tetrahymena* on solid medium can be used in steps of mutant isolations involving selection for mutants with an altered

TABLE VII

GROWTH IN VARIOUS MEDIA ON PLATES WITH "MICRO-HOLES"[a]

Strain	Medium on plates	Temperature (°C)	Liquid medium in which cells were added to plates	Number of colonies per plate	Mean
CA-103	Complex	28	Complex	47 30	39
	Complex	28	Inorganic	40 48	44
	Inorganic	28	Complex	25 30	28
	Inorganic	28	Inorganic	11 7	9[b]
	Inorganic plus cycloheximide (10 μg/ml)	28	Complex	0 0	0
	Complex plus cycloheximide (10 μg/ml)	28	Inorganic	0 0	0
	Complex plus cycloheximide (10 μg/ml)	28	Complex	0 0	0

[a]Approximately 60–65 *Tetrahymena* were added to each plate. Colonies on plates were counted after incubation for 48 hours.

[b]Colony size was much smaller after 3 days' incubation at 28°C than for the other samples. Some growth in inorganic medium is consistent with a 30–40% increase in cell number during starvation (Mowat *et al.*, 1974).

response to drugs or antimetabolites and possibly also in selection for mutants with altered nutritional requirements.

III.   Summary and Conclusions

Some of the interest of cell and molecular biologists in studying unicellular eukaryotic organisms is to study in organisms which can be easily manipulated fundamental biological mechanisms of potential general significance. *Tetrahymena pyriformis* grows axenically with a generation time of approximately 2.5 hours in complex medium and about 4–5 hours in completely defined medium, and it can be grown in large amounts with a high yield of cells. These and other characteristics make *Tetrahymena* a very interesting and valuable organism for many cell and molecular biological studies (Hill, 1972; Elliott, 1973). Although the genetic system of *Tetrahymena* is somewhat unique and complex (see Allen and Gibson, in Elliott, 1973), *Tetrahymena* is amenable to genetic study. Strains with combinations of available genetic markers can be obtained by crossing, and the possibility exists of studying cytoplasmic inheritance. Thus, this organism is amenable to the isolation and characterization of mutants affected in various biological mechanisms, particularly those of general interest in the study of the biology of eukaryotes. This approach has been very valuable in understanding the molecular biology of some prokaryotic microorganisms, particularly *Escherichia coli* and its viruses.

The methods described in this report were developed with the particular aim of being able to manipulate *Tetrahymena*, using procedures that have become routine with microorganisms, such as many bacteria, yeast, and cellular slime molds. In particular, growth and isolation of clones can now be performed in a single operation on solid medium which can contain selective agents to aid in mutant isolation. Previous methods for isolating clones following large-batch mutagenesis have generally involved mechanically picking single cells and distributing these in drops of nutrient medium for growth. "Replicating" colonies to selective media was accomplished by a mechanical procedure, very similar to the procedure we have adopted for replica plating from plates with "micro-holes" (Roberts and Orias, 1973a). Although these methods have not seriously limited manipulation of *Tetrahymena* for mutant isolations, it is hoped that the ability to easily manipulate *Tetrahymena* on solid medium containing selective agents will greatly facilitate mutant isolation and thus further genetic and biochemical studies with this organism. Although we have not manipulated all available strains of *Tetrahymena pyriformis* with our techniques, we have used enough strains, both micronucleate and amicronucleate to suggest that these methods will be generally applicable to *Tetrahymena.*

ACKNOWLEDGMENTS

This work was supported by Grant A-5395 from the National Research Council of Canada. We would like to thank our colleagues, particularly Dr. I. S. Villadsen, for advice and many helpful discussions during the course of the development of these methods. We would also like to thank Mr. Frank Burrows and others in the machine shop in the Biology Department, York University, for invaluable assistance in designing and constructing the apparatus for making holes in solid medium.

REFERENCES

Chu, E. H. Y. (1971). *In* "Environmental Chemical Mutagens" (A. Hollaender, ed.), pp. 411–444. Plenum, New York.

Elliott, A. M. (1973). "Biology of *Tetrahymena.*" Dowden, Hutchinson & Ross, Stroudsburg, Pennsylvania.

Elliott, A. M., Brownell, L. E., and Gross, J. A. (1954). *J. Protozool.* **1**, 193–199.

Gardonio, E., Crerar, M., and Pearlman, R. E. (1973). *J. Bacteriol.* **116**, 1170–1176.

Hill, D. L. (1972). "Biochemistry and Physiology of *Tetrahymena.*" Academic Press, New York.

Mowat, D., Pearlman, R. E., and Engberg, J. (1974). *Exp. Cell Res.* **84**, 282–286.

Roberts, C. T., Jr., and Orias, E. (1973a). *Genetics* **73**, 259–272.

Roberts, C. T., Jr., and Orias, E. (1973b). *Exp. Cell Res.* **81**, 312–316.

West, R. A., Jr., Barbera, P. W., Kolar, J. R., and Murrell, C. B. (1962). *J. Protozool.* **9**, 65–73.

Chapter 18

The Isolation of Nuclei with Citric Acid and the Analysis of Proteins by Two-Dimensional Polyacrylamide Gel Electrophoresis[1]

CHARLES W. TAYLOR, LYNN C. YEOMAN, AND HARRIS BUSCH

Nuclear Protein Labs, Department of Pharmacology, Baylor College of Medicine, Houston, Texas

I. Introduction

Recent studies that have highlighted the importance of nuclear and chromatin proteins include those of Spelsberg *et al.* (1971), Kostraba and

[1] This study was supported by the Cancer Research Center Grant CA 10893 and a generous gift from Mrs. Jack Hutchins.

Wang (1972), and LeStourgeon and Rusch (1973) on the changes in nuclear protein complement associated with differentiation; Church and McCarthy (1967), Bresnick (1971), Salas and Green (1971), Baserga (1972), and Kostraba and Wang (1973) on the amount and number of protein changes associated with cells in a growth state; Teng *et al.* (1971), Wilhelm *et al.* (1972), Richter and Sekeris (1972), MacGillivray *et al.* (1972), and Wu *et al.* (1973) on the specificity of nuclear proteins observed among different tissues; Patel (1972) and Wakabayashi *et al.* (1973) on the specificity of individual chromatin proteins in their affinity for unique DNA; and Calafat *et al.* (1970), Weisenthal and Ruddon (1972), and Stein and Thrall (1973) on differences in nuclear nonhistone proteins between normal and tumor cells.

A reexamination of nuclear isolation methods and protein analytical electrophoresis procedures was prompted by the need to compare nuclear proteins of normal cells and neoplastic cells. The ultimate goal was the examination of proteins from the whole nucleus and of interphase chromatin. Such a study was begun to determine whether there were in fact differences in proteins between normal and tumor cells and, in addition, to establish whether any of the protein differences could be responsible for altered gene expression or merely the result of it. Certainly, any approach would have to take into account a large number and many types of nuclear proteins that include the histones and a large number of nonhistone (acidic) nuclear proteins: cytonucleoproteins, enzymes, preribosomal proteins, nuclear particle proteins, nucleolar proteins, and gene regulatory proteins.

Methods available for the isolation of nuclei differ with respect to the medium employed for disruption of the cells. The procedure employing a hyperisomolar sucrose solution was introduced by Chauveau *et al.* (1956) and later modified to sucrose containing 3.3 mM calcium ion by Muramatsu *et al.* (1963). The use of citric acid was introduced by Crossman (1937) and was later shown by Gurr *et al.* (1963) to remove the outer layer of the nuclear envelope and its attached ribosomes. In addition, the use of detergents such as Triton X-100 by Hymer and Kuff (1964), Blobel and Potter (1966) and Tween 80 by Fisher and Harris (1962) after disruption in sucrose was shown to remove cytoplasmic enzyme activities from nuclear preparations. Finally, procedures were developed that employ organic solvents to attempt to circumvent the problems associated with hypotonic solutions (Behrens, 1932; Siebert, 1964).

Our recent studies compare the acid-soluble nuclear protein complement obtained from rat liver nuclei isolated in sucrose medium containing 3.3 mM Ca^{2+} and those isolated in 0.025 M citric acid. The citric acid procedure was chosen for the comparison studies on the 0.4 N sulfuric acid extracts and the chromatin fractions in normal and tumor cell nuclei. The absence of contaminating outer nuclear envelope elements, the low pH (2.5) of the 0.025 M citric acid medium, and the ease of cytoplasmic tag removal from the tumor cell nuclei were the reasons for the use of a citric acid procedure.

One way to determine the quality of nuclear isolation products is light microscope examination of the preparation. In addition, electron microscopic evaluation provides information that cannot be obtained by light microscopy.

Another level of study is evaluation of the proteins in the preparation. The application of high resolution and high sensitivity two-dimensional polyacrylamide gel electrophoresis provides further evaluation of nuclear isolation methods. This approach can supplement cytoplasmic enzyme assays (Siebert *et al.*, 1966) which would, however, not detect noncatalytic cytoplasmic contaminants or nuclear protein losses. With two-dimensional methods the protein constituents of nuclei and nucleoli can be evaluated at various stages of nuclear isolations and nuclear fractionations. This approach was employed along with electron microscopy in the evaluation of the 0.025 M citric acid method employing the Tissumizer® for cell disruption (Taylor *et al.*, 1973).

With the subsequent identification of preribosomal elements of the nucleolar and ribosomal elements of the cytoplasm by two-dimensional polyacrylamide gel electrophoresis (Prestayko *et al.*, 1974), it is now possible to define specific ribosomal protein contaminants in preparations of nuclei and to also identify preribosomal particle protein losses due to washout from the nucleolus. With the recent identification of cytonucleoproteins on two-dimensional gels (Kellermayer *et al.*, 1973), even the most subtle nucleoplasmic protein losses can be detected.

One of the virtues in the study of citric acid nuclei is the reported success in chromatin reconstitution by Paul and Gilmour (1968). In their work it was shown that chromatin from citric acid nuclei could be reconstituted to the same transcriptional template activity as native chromatin. In a study on the reconstitution of chromatin with a subfraction of the nonhistone proteins obtained from regenerating rat liver chromatin, Kostraba and Wang (1973) were able to demonstrate regenerating rat liver RNA transcripts when reconstitution was effected with regenerating rat liver nonhistone chromatin proteins. An application of two-dimensional polyacrylamide gel electrophoresis separations to nonhistone protein components is intended to aid the selection and identification of chromatin subfractions and single nonhistone proteins that can define specific transcripts.

II.   Methods

A.   Cell Preparation

All the methods for nuclear isolation must first take into account the extent of contamination of the starting sample. A consideration of equal importance is the particular experiment for which the nuclei and nuclear isolation procedure are needed.

When nuclei are isolated from solid tissues, it is important to remove the blood as well as other contaminating connective tissue. Removal of the blood from liver tissue is accomplished by perfusion with either 0.13 M NaCl, 0.005 M KCl, 0.008 M MgCl$_2$ (NKM) solution (Mauritzen *et al.*, 1971) or 0.25–0.34 M sucrose solution. The vena cava can be cut to give a more thorough perfusion.

To remove the covering tissue capsule and to excise large segments of connective tissue from the sample, physical maceration of the tissue is carried out in tissue presses. For solid tissues and tumors that are not amenable to perfusion, the macerated tissue can be washed with NKM immediately prior to dispersion in the homogenizing medium.

Cells grown in culture and ascites fluid cells can be washed by repeated suspension in NKM solution or 0.25 M sucrose and recovered by centrifugation. The cells can then be dispersed in the homogenization medium. This is an appropriate time in the nuclear isolation procedure to introduce antiproteolytic degradation agents, such as diisopropylfluorophosphate (DFP), phenylmethylsulfonyl fluoride, or sodium bisulfite. When DFP is used, care should be taken to avoid skin contact or inhalation because of its long-lasting anticholinesterase activity.

In the preparation of tissues for nonaqueous nuclear isolation procedures, the freeze-stop technique of Hohorst *et al.* (1959) has been applied. This treatment freezes the tissue sample *in situ* to suspend biological events as rapidly as possible. Subsequent steps involve the exhaustive removal of water from the sample prior to dispersion in nonaqueous media (Behrens, 1932; Siebert, 1964).

B.   Established Methods for Isolation of Nuclei

1.   Sucrose Ca^{2+} Methods for Isolation of Nuclei

The hypotonic sucrose solution which was introduced by Chauveau *et al.* (1956) together with homogenization has become one of the most widely used methods for nuclear isolation. The particular concentrations of sucrose employed have ranged from 0.25 to 2.4 M even though 0.25 M sucrose has been reported by Siebert (1967) to extract many nuclear proteins via the nuclear envelope. Siebert (1967a) claimed that the higher osmotic pressure within the nucleus caused a redistribution of many nuclear molecules.

This has led to the more general acceptance of sucrose concentrations in the range of 2.0–2.4 M for the cell disruption medium. Many investigators have included physiological concentrations of Ca^{2+} and Mg^{2+} ion in their sucrose solutions to preserve nuclear morphology (Muramatsu et al., 1963; Kellermayer et al., 1974). The addition of these divalent cations has proved to be particularly beneficial in the preparation of the nuclear membrane for subsequent steps leading to the isolation of nucleoli. Although the sucrose method has been very useful for preparation of nuclei from normal tissues, efforts to prepare nuclei from many types of tumor cells by this method have not generally been successful since most of the nuclei were surrounded by cytoplasmic tags. This problem has necessitated the development of alternative methods for isolation of nuclei from many types of tumors (Busch et al., 1974).

a. Continuous Homogenization Methods for the Isolation of Nuclei in Sucrose.   The need for large amounts of nuclear proteins and RNAs was generated by the amounts of material required for structural studies. For this reason a "continuous homogenizer" was developed. With this device, which consisted of a Lucite pestle screw turning in a stainless steel sleeve that could be cooled by a circulating refrigeration system, homogenate was produced at the flow rate of 2–4 liters per hour (Desjardins et al., 1965).

b. A Sucrose Method for Minimization of Contamination as Shown by Excess RNA in Rat Liver Nuclei.   The procedure introduced by Schneider (1948) which employed isosmotic sucrose produced rather contaminated nuclei with a 12–14% cellular RNA content. By the use of a more concentrated sucrose medium (2.2 M) Chauveau et al. (1956) were able to remove the microsomes and mitochondria; the RNA content of their nuclear preparations was 6–7%. Blobel and Potter (1966) disrupted rat liver tissue in 0.25 M sucrose containing 0.05 M Tris·HCl, pH 7.5, 0.025 M KCl, and 0.005 M $MgCl_2$ (TKM) at 20°. Sufficient sucrose was added to make the nuclear preparation 1.62 M with respect to sucrose containing TKM. The sample was underlayered with 2.3 M sucrose containing TKM (Blobel and Potter, 1966). The nuclear pellet obtained after centrifugation for 30 minutes at 124,000 g was found to have a high yield of nuclei and 4–5% cellular RNA content. However, electron microscopic examination revealed that half of the nuclei were damaged.

2.   CITRIC ACID PROCEDURES FOR THE ISOLATION OF NUCLEI

In 1941 Marshak reported the use of 0.25 M citric acid as the medium for the grinding of cells with a mortar and pestle to remove adherent cytoplasmic tags from nuclei. Successive resuspension was carried out in fresh 0.25 M citric acid 4–6 times. He noted that nuclei could also be prepared from tumor tissue by this method. Marshak (1941) indicated that the hydrogen ion concentration, not the action of an anion, was responsible for the separation of the nucleus from its cytoplasm.

Subsequent to Marshak's work, Arnesen et al. (1949) used 0.008 M citric acid in an 8.5% sucrose solution for the homogenization of leukemic mouse spleens. Since that time, Dounce et al. (1955), Higashi et al. (1966), Paul and Gilmour (1968), and Bornens (1968) have used 0.1 M citric acid in 0.44 M sucrose, 0.125 M citric acid, 0.025 M citric acid, and 0.005 M citric acid in 0.25 M sucrose, respectively, to prepare nuclei.

3. DETERGENT PROCEDURES FOR THE ISOLATION OF NUCLEI

Several methods for the preparation of nuclei have employed nonionic detergents such as Triton X-100 (Hymer and Kuff, 1964). This medium contained 0.5% (w/v) Triton X-100 in 0.25 M sucrose and yielded nuclei that were reported by Hubert et al. (1962) to have most of the outer layer of the nuclear membrane removed. Subsequently, Higashi et al. (1966) noted that this medium extracted much nuclear RNA.

The use of other nonionic detergents, including Cemusol NPT 6, Cemusol NPT 12, and Tween 80, in the isolation of nuclei was described by Fisher and Harris (1962) and Tsai and Green (1973) for liver cells, and by Woernley and Carruthers (1957) for tumor cells.

4. ORGANIC SOLVENT PROCEDURES FOR THE ISOLATION OF NUCLEI

One quite different approach to the isolation and purification of nuclei has involved nonaqueous conditions which employ organic solvents. Starting with the "freeze stop" method of Hohorst et al. (1959) a portion of the desired tissue is removed with liquid nitrogen cooled tongs and frozen in liquid nitrogen. After lyophilization until dry, the sample is then ground in a ball mill until a powder is obtained. Separation of nuclei is then obtained by centrifugation in mixtures of cyclohexane and carbon tetrachloride. Nuclei prepared by nonaqueous methods retain DNA in the native state and do not suffer from protein or ion washout to the extent that sucrose nuclei do. Solvent-isolated nuclei are undoubtedly depleted in their lipid content and may suffer from changes in the integrity of their lipoproteins.

Considerable criticism of this method has emerged from electron microscopic studies which show marked indentation of the nuclei. Cytoplasmic contamination is evident from these studies, and studies on RNA show marked degradation in such preparations by sucrose density gradient analyses.

B.   Recent Developments in Isolation of Nuclei

The need to make comparisons of the nuclear proteins in normal tissues versus those obtained from tumors prompted an investigation into suitable methods for nuclear isolation that could be used on both normal and neoplastic tissues. A new device for cell disruption, the Super Dispax or Tissumizer® (Fig. 1), along with the development of a high resolution method for the separation of nuclear proteins on a two-dimensional gel electrophoresis system (Orrick *et al.*, 1973) has facilitated this improved isolation of nuclei in 0.025 M citric acid (pH 2.5). This procedure is satisfactory for nuclei of both normal and tumor tissues. In turn, this method

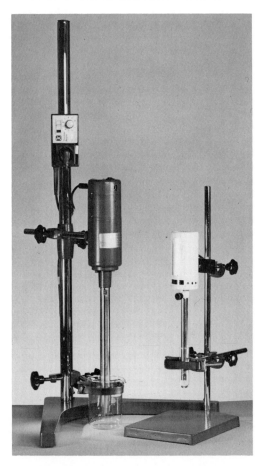

FIG. 1.   Photograph of the Tekmar Model SD-45K Super Dispax on the left and the Tekmar Model SDT-182N on the right. Both instruments can be obtained from Tekmar Company (Cincinnati, Ohio).

TABLE I

Protein, RNA, and DNA Composition of Rat Liver
Nuclei Prepared with the Sucrose-Ca^{2+} Procedure[a]

Protein	40%
RNA	3%
DNA	15%
DNA[b]	20–22%
RNA/DNA	0.19%
RNA/DNA[b]	0.18–0.21%

[a]Expressed as percentages of dry weight.
[b]Siebert (1967b).

has made possible direct comparisons of acid-soluble proteins of nuclei
and "Chromatin Fraction II." More recently, studies on phosphoproteins
have also been done on nuclei isolated by this method. The nuclei prepared
by this method show the protein, RNA, and DNA composition presented in
Table I and are free of cytoplasmic contamination (Taylor et al., 1973). Only
23% of the proteins are washed out (see Table II) by 0.025 M citric acid.
This probably is more due to removal of outer nuclear membrane associated
elements than nuclear protein washout.

1. Preparation of Tissue for Isolation of Nuclei

Male albino rats obtained from the Holtzman Rat Company (Madison,
Wisconsin) were fasted for 16 hours. The rats were killed and the livers
perfused with ice-cold NKM solution (Mauritzen et al., 1971) containing
0.13 M sodium chloride, 0.005 M potassium chloride, and 0.008 M magnesium
chloride. Tumor cells were maintained in male albino rats from the same
company. Six days after intraperitoneal transplantation of Novikoff hepatoma the rats were sacrificed and the ascites fluid was collected by abdominal
incision. The ascites cells were passed through cheesecloth and centrifuged

TABLE II

Protein, RNA, and DNA Released from Sucrose Rat Liver Nuclei by
Citric Acid

Nuclei	Released		
	% Protein	% RNA	% DNA
0.025 M Citric acid	23 ± 2.1[a]	15 ± 5.0	3 ± 3.5
0.25 M Citric acid	42 ± 4.0	19 ± 2.8	14 ± 2.1

[a]Standard deviation from the mean calculated by the following formula:
$S = [\Sigma x^2/(n-1)]^{1/2}$

at 3000 g for 20 minutes. The pellet was washed repeatedly with NKM and centrifuged at 3000 g until the tumor cells were almost white in appearance.

2. PREPARATION OF NUCLEI IN SUCROSE MEDIUM EMPLOYING THE TEFLON–GLASS HOMOGENIZER

The perfused rat liver tissue was chopped, passed through a tissue press, and homogenized in a loose-fitting Teflon–glass homogenizer (pestle clearance, 0.012 inch) with six strokes in 2.2 M sucrose containing 3.3 mM Ca^{2+} (1 gm of pressate to 10 volumes of sucrose solution). This suspension was layered over an equal volume of 2.2 M sucrose containing 3.3 mM Ca^{2+} and centrifuged at 53,000 g for 45 minutes. The pellet was resuspended in 1 volume of 0.34 M sucrose containing 3.3 mM Ca^{2+}, layered over 3 volumes of 0.88 M sucrose containing 3.3 mM Ca^{2+} and centrifuged at 1100 g for 20 minutes.

3. TISSUMIZER CITRIC ACID METHOD

Nuclei from normal rat liver pressates were prepared by homogenization in 0.025 M citric acid, pH 2.5 (Taylor et $al.$, 1973) or 2.2 M sucrose containing 3.3 mM Ca^{2+}, using a Tekmar SD-45K Super Dispax system (Tissumizer, Tekmar Company, Cincinnati, Ohio) for cell disruption at 0–4° with the G456 generator. Homogenization in 2.2 M sucrose required 3–5 minutes at a thyristor regulator setting of 7, whereas in citric acid only 15–30 seconds were required at the same thyristor setting. The nuclei obtained with sucrose using the Tissumizer did not differ by light microscopy from those obtained by hand homogenization. Nuclei obtained by 0.025 M citric acid were centrifuged for 20 minutes at 1100 g. The pellet was resuspended in 2.2 M sucrose containing 3.3 mM Ca^{2+}, layered over 2.2 M sucrose containing 3.3 mM Ca^{2+} and centrifuged for 60 minutes at 53,000 g. This nuclear pellet was resuspended in 0.34 M sucrose containing 3.3 mM Ca^{2+} and further purified by layering over 0.88 M sucrose containing 3.3 mM Ca^{2+} and centrifuging at 1100 g for 20 minutes.

Nuclei from washed Novikoff hepatoma ascites cells were isolated in media containing 0.025 M and 0.25 M citric acid with the aid of the Tissumizer. The time required for homogenization ranged from 3 to 5 minutes. The washed ascites cells were tissumized for 3 minutes at a regulator setting of 7 and centrifuged for 20 minutes at 1100 g. At this point no adherent cytoplasmic tags were visible. The nuclear pellet was further purified by resuspension in 2.2 M sucrose containing 3.3 mM Ca^{2+} and centrifugation for 60 minutes at 53,000 g. The pellet was layered in 0.34 M sucrose containing 3.3 mM Ca^{2+} over 0.88 M sucrose containing 3.3 mM Ca^{2+} and centrifuged for 20 minutes at 1100 g.

4. MASS ISOLATION PROCEDURE FOR THE PREPARATION OF 100 gm QUANTITIES OF NUCLEI

With the aid of the SD-45K Tissumizer, it has been possible to isolate 100-gm quantities of calf liver nuclei. One kilogram of the perfused tissue was cut into cubes, macerated, dispersed in 10 volumes of 1.6 M sucrose containing 3.3 mM Ca^{2+} and homogenized in a matter of minutes. When performed on 4-liter batches, the total homogenization was completed in less than 20 minutes. Continuous centrifugation of the homogenate was accomplished in the Sharples centrifuge over a 2.2 M sucrose, 3.3 mM Ca^{2+} cushion. Quantities of purified nuclei ranging from 60 to 90 gm were scraped directly from the rotor.

Such studies have not been carried out on nuclei obtained from 0.025 M citric acid medium, but the mass isolation approach will be tested in future experiments. Currently, 4–5 gm of nuclei are obtained in routine experiments.

III. Morphology of Sucrose and Citric Acid Nuclei

The morphology of nuclei treated with citric acid differs considerably from those of nuclei obtained by the sucrose Ca^{2+} method (Fig. 2). The differences are seen in the ultrastructural organization of the nuclear membrane, the chromatin, and the nucleolar components (Busch and Smetana, 1970). The outer layer of the nuclear envelope is lost when nuclei are homogenized with 0.025 M citric acid. The 0.025 M citric acid concentration was chosen after evaluating concentrations ranging from 0.012 M to 0.25 M. Using 0.025 M citric acid, nuclei of good quality were obtained from tumor tissues in homogenization times comparable to those required for homogenization of sucrose liver nuclei.

The most striking effect of the citric acid treatment is observed in the distribution and organization of the chromatin throughout the nucleoplasm as well as in the region of the nucleolus-associated chromatin. In "sucrose nuclei" the chromatin is dispersed throughout the nucleoplasm (Fig. 2), but in nuclei treated with 0.025 M citric acid the chromatin was collapsed and condensed into large clumps within the nucleus as well as at the nuclear periphery and around the nucleolus (Fig. 3A,B).

The increase in citric acid concentration to 0.25 M produced further condensation and clumping of the chromatin (Fig. 3C); particularly the perinucleolar, nucleolus-associated chromatin (Fig. 3C,D) was condensed.

Nucleoli exposed to 0.025 M citric acid retain much of their usual appear-

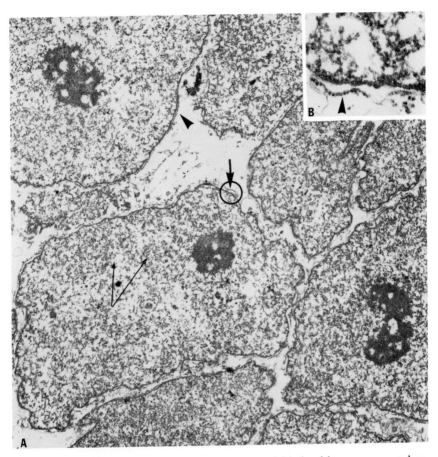

Fig. 2. Electron micrograph of normal rat liver nuclei isolated by sucrose procedure. Arrows indicate presence of outer nuclear membrane and ribosomes (inset). Chromatin is dispersed throughout the nucleoplasm (small arrows). (A) × 6000; inset (B) × 17,000.

ance with respect to the granular and fibrillar components (Fig. 3B) as well as the fine filaments in the nucleolar matrix while those treated with 0.25 M citric acid contained few granular elements and little if any fibrillar components or fine protein filaments (Fig. 3D).

To better visualize the fine ultrastructure of the nucleolar periphery that was partially obscured by the increased density of the stained chromatin due to the citric acid treatment, the chromatin structures were bleached with the EDTA procedure of Bernhard (1969), which is usually used for the preferential staining of RNA-containing structures. Figure 4 shows that

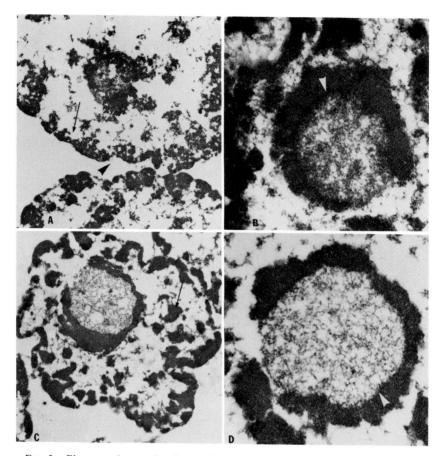

Fig. 3. Electron micrographs of normal rat liver nuclei isolated by citric acid procedure. (A) and (B) Nuclei treated with 0.5% citric acid; (C) and (D) 5% citric acid. (A) Note chromatin clumping around the periphery of nuclei (small arrow) and throughout the nucleoplasm. The outer layer of the nuclear envelope and ribosomes are absent, although some membranous vesicles are present (black pointer) × 7500. (B) An electron-dense area was found in the nucleolus-associated chromatin in nuclei treated with 0.5% citric acid. Note the distribution of granular and fibrillar components in the nucleolus. × 13,000. (D) A nucleolus of a preparation treated with 5% citric acid. The decrease of the granular and fibrillar components is clearly seen, as well as the decrease in the electron density of the proximal region of the nucleolus-associated chromatin (pointer). × 13,000. Electron micrographs are the courtesy of Drs. J. Daskal and K. Smetana.

Fig. 4. Electron micrographs of nucleoli of rat liver nuclei stained with the EDTA procedure. Nucleolus-associated (nac) and intranucleolar chromatin appear bleached. The fine proteinaceous filaments seen after treatment of nuclei with 0.5% citric acid (A) are mainly extracted at a concentration of 5% citric acid (B) (pointer). Some interchromatin granules (small arrows) resisted extraction even with 5% citric acid. × 24,000.

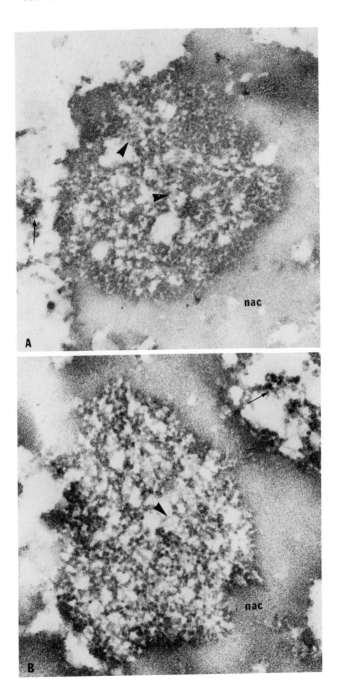

the fibrillar elements persist in the denser nucleoli of nuclei isolated with 0.025 M citric acid. In nucleoli of nuclei isolated with 0.25 M citric acid, the fibrillar elements were difficult to define and the nucleolus was much less dense.

IV.   Composition of Sucrose and Citric Acid Nuclei

Assays for protein, RNA, and DNA were performed on the sucrose nuclei and citric acid-washed sucrose nuclei. A summary of the absolute amounts of protein, RNA, and DNA in normal rat liver sucrose nuclei is presented in Table I. These values are based on dry nuclear weight. The ratio of RNA/DNA in the normal rat liver nucleus is in good agreement with the value reported by Siebert (1967b). It was found that 0.025 M citric acid released 23% of the proteins, 15% of the RNA, and 3% of the DNA from sucrose liver nuclei (Table II). The treatment with 0.25 M citric acid released 42% of the protein, 19% of the RNA, and 14% of the DNA from sucrose liver nuclei.

V.   Two-Dimensional Electrophoretic Analysis of Nuclear Proteins from Citric Acid Nuclei

The use of two-dimensional electrophoretic methods has greatly advanced the knowledge of the component protein parts that make up a ribosome and its subunits (Prestayko et al., 1974). This same clarification has been developed for the following components and fractions of the nucleus: 0.4 N sulfuric acid-soluble proteins (Yeoman et al., 1973a), "Chromatin Fraction II" proteins (Yeoman et al., 1973b), 0.4 N sulfuric acid soluble nucleolar proteins (Orrick et al., 1973), RNP-particle proteins (Prestayko et al., 1974), and dilute salt soluble proteins of the nuclear sap (Kellermayer et al., 1973). From 50 to 200 protein components can be displayed in a two-dimensional array by this method. The utility of this type of approach is apparent in nuclear protein comparisons. The dilute acid-soluble nuclear proteins and chromatin proteins from six rodent tissues have been compared (Yeoman et al., 1974) on two-dimensional polyacrylamide gels.

Three systems for polyacrylamide gel analysis of nuclear proteins have evolved. The first was developed by Orrick et al. (1973) and has a first-dimensional gel composition of 10% acrylamide, 0.35% bisacrylamide, 0.1% persulfate, 1% TEMED, 4.5 M urea, and 1.8 N acetic acid (pH 3.5). The second-dimensional slab gel with dimensions of 100 × 95 × 3 mm had a composition of 12% acrylamide, 0.31% bisacrylamide, 0.05% TEMED, 0.1% SDS and 0.1 M sodium phosphate buffer (pH 7.1). More recently the first-dimensional tube gel has been replaced with a gel of the following composition (Busch *et al.* 1974): 6% acrylamide, 0.21% bisacrylamide, 0.1% ammonium persulfate, 1% TEMED, 1.8 N acetic acid, and 4.5 M urea. In the second dimension, an 8% acrylamide gel slab was used with the following components: 8% acrylamide, 0.21% bisacrylamide, 0.05% TEMED, 0.1% SDS, 0.075% persulfate, 6 M urea and 0.1 M phosphate buffer (pH 7.1). The third system is an adaptation of the previous gel system. The 6% acrylamide first-dimensional tube gel is retained, but an 8 to 12% linear acrylamide gradient slab gel was substituted in the second dimension. The solutions A and B in Table III were mixed in the following manner to form the 8 to 12% acrylamide gradient gel. One volume of solution A was pumped at a flow rate of 1 ml/minute into a mixing chamber containing 1 volume of solution B. The mixing chamber solution was pumped at a flow rate of 2 ml/minute into two gel-forming cells at a time. Slab gels were stained in a 0.25% solution of Coomassie Brilliant Blue R in acetic acid–methanol–water (1:5:5) and destained in a solution that was 5% in methanol and 10% in acetic acid.

TABLE III

SOLUTIONS USED FOR THE 8–12% LINEAR GRADIENT
POLYACRYLAMIDE SLABS

	Solution[a]	
Reagents	A	B
Acrylamide	14.5%	8.0%
Bisacrylamide	0.36%	0.2%
Urea	6 M	6 M
Sodium dodecyl sulfate	0.1%	0.1%
Phosphate buffer (pH 7.1)	0.1 M	0.1 M
Sucrose	24.0%	5.0%
TEMED	0.05%	0.05%
Ammonium persulfate	0.019%	0.019%

[a]The solutions were prepared and stored for no more than 1 week as stock solutions without the ammonium persulfate. Ammonium persulfate was added to the mixing solution (A) and reservoir solution (B) just prior to the forming of the gradient gel. Gels were poured and overlayered with water within 45 minutes after mixing was initiated.

A.   Sulfuric Acid-Soluble Nuclear Proteins

The use of sulfuric acid in the extraction of whole nuclei was chosen as a starting point for several reasons: (1) earlier studies on histone extracts employing sulfuric and hydrochloric acid solutions, (2) the minimal degradation incurred in the presence of 0.4 N H_2SO_4, (3) the small amount of complicating nucleic acid extracted by this solution, and (4) the simplicity of this approach. Nuclei prepared by the 0.025 M citric acid method were suspended in 10 volumes of 0.4 N H_2SO_4 and given 10 strokes with a tight Teflon–glass (clearance: 0.004–0.005 inch) homogenizer. The supernatant was collected by centrifugation at 30,000 g for 20 minutes and dialyzed against distilled water. The resulting dialyzate was lyophilized and stored as a dry powder.

1.   DILUTE SULFURIC ACID-SOLUBLE NUCLEAR PROTEINS OF NORMAL RAT TISSUES

The two-dimensional polyacrylamide gel pattern obtained for the 0.4 N sulfuric acid-soluble nuclear proteins of normal rat liver citric acid nuclei is shown in Fig. 5. This gel has a 10% acrylamide first dimension and a 12%

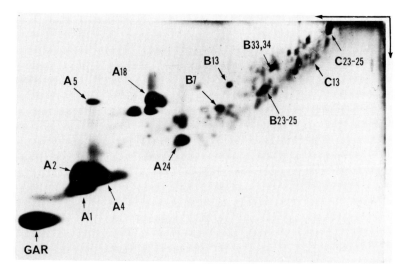

FIG. 5. Two-dimensional polyacrylamide gel electrophoretogram of sulfuric acid-extracted proteins from 0.025 M citric acid rat liver nuclei. The sulfuric acid-extracted proteins were layered on a 10% acrylamide, 6 M urea, 0.9 N acetic acid, 10 cm gel and run for 6 hours at 120 V. The second dimension was a 12% acrylamide, 0.1% SDS slab gel which was run for 14 hours at 50 mA/slab. Gels were stained with Coomassie Brilliant Blue R. Spots A1, A2, A4, and A18 were the F2a2 (AL), F2b (N-proline), F3 (AR5), and F1 (VLR) histones, respectively. The directions of electrophoresis are shown. The first dimension is represented by the horizontal arrow and the second dimension by the vertical arrow.

acrylamide second dimension. The pattern is divided into the A, B, and C regions along the horizontal and shows approximately 100 protein spots. The numbers designate spots seen in 0.4 N H_2SO_4 extracts of normal liver and Novikoff hepatoma nucleoli (Orrick et al., 1973) whereas lower-case letters indicate spots seen in whole nuclear acid extracts.

Valuable markers are the histone spots GAR, A1, and A4, which correspond in migration to the f2al, f2a2, and f3 histones, respectively. Suitable marker protein spots seen in many of the following 10%/12% acrylamide gell protein patterns are A24, B13, C23-24, and the histones.

The same extract from normal liver nuclei prepared by the 2.2 M sucrose method employing a Tissumizer is shown in the two-dimensional pattern of Figure 6. The pattern is virutally identical to that shown in Fig. 5 and constitutes strong support for the hypothesis that citric acid does not selectively extract particular nuclear proteins (Taylor et al., 1973).

The sulfuric acid-soluble nuclear proteins from regenerating rat liver are shown in Fig. 7 (Yeoman et al., 1974). The total number of nuclear proteins counted was approximately 90. A number of new protein spots were found in the A region of this gel that were not observed in normal liver.

The sulfuric acid-extractable nuclear proteins from the livers of rats that had received daily injections of thioacetamide for 9 days were analyzed on two-dimensional gels. A pattern is shown in Fig. 8, and approximately 90 protein spots are seen. Interestingly, no additional spots were found in this pattern when compared to the normal liver.

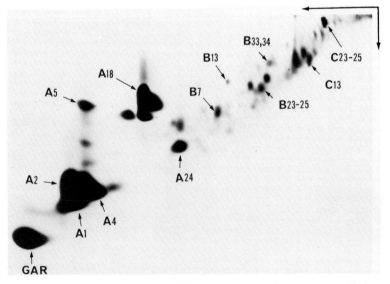

FIG. 6. Two-dimensional polyacrylamide gel electrophoresis pattern of sulfuric acid-extracted proteins from 2.2 M sucrose liver nuclei. See Fig. 5 for conditions.

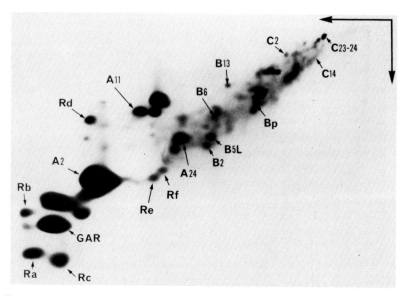

FIG. 7. Two-dimensional polyacrylamide gel electrophoresis of 250 μg of regenerating rat liver nuclear proteins. See Fig. 5 for conditions.

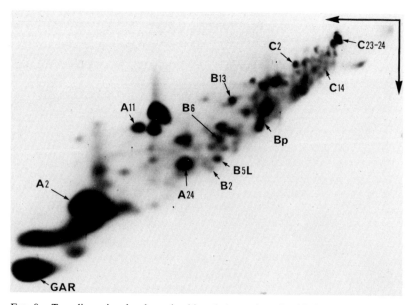

FIG. 8. Two-dimensional polyacrylamide gel electrophoresis of 250 μg of thioacetamide-treated rat liver nuclear proteins. See Fig. 5 for conditions.

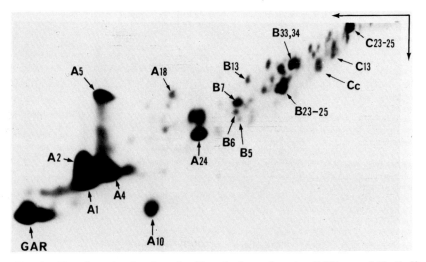

FIG. 9. Two-dimensional polyacrylamide gel electrophoresis of 250 μg of Novikoff hepatoma nuclear proteins. See Fig. 5 for conditions.

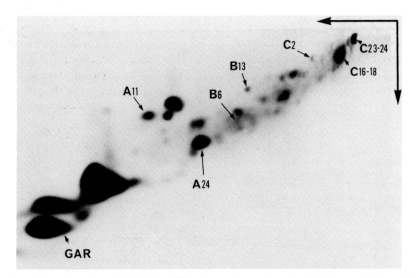

FIG. 10. Two-dimensional polyacrylamide gel electrophoresis of 250 μg of Walker 256 carcinosarcoma nuclear proteins. See Fig. 5 for conditions.

2. Dilute Sulfuric Acid-Soluble Nuclear Proteins of Rat Tumors

The pattern shown in Fig. 9 is that of Novikoff hepatoma 0.4 *N* sulfuric acid-soluble nuclear proteins. This pattern demonstrates approximately 100 protein spots, 6 of which were not found in any of the liver patterns. A high density was noted in the vicinity of spots C16-C18.

The 0.4 *N* sulfuric acid-soluble nuclear proteins of Walker 256 carcinosarcoma nuclei are shown in Fig. 10. Of the 90 spots counted in this pattern, those in the vicinity of C16-18 were noted to be more intense than in the liver, as was the case in the Novikoff hepatoma (Yeoman *et al.*, 1974).

The acid-soluble nuclear proteins from the Morris 9618A hepatoma are shown in Fig. 11. This pattern continues the trend of high density for spots in the area of C16-18 in rat tumor patterns.

B. "Chromatin Fraction II" Proteins

After isolation of 0.025 *M* citric acid nuclei, chromatin was prepared by a modification of the method of Marushige and Bonner (1966). After extraction of this material with 0.4 *N* H_2SO_4, the DNA was digested with DNase I (Yeoman *et al.*, 1973b). The protein that had remained soluble in 0.9 *N* citric acid, 10 *M* urea, and 1% β-mercaptoethanol was "Chromatin Fraction II." This approach to the survey of nonhistone nuclear proteins offered several advantages; (a) a minimum number of fractions, (b) drastic conditions to stop changes in the sample due to metabolism or degradation and (c) minimum time and manipulation of the sample. Although this

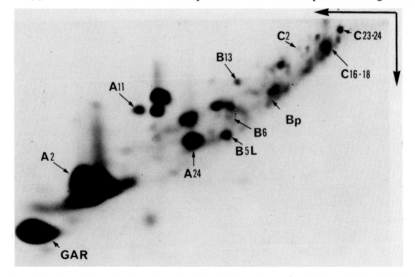

Fig. 11.   Two-dimensional polyacrylamide gel electrophoresis of 250 μg of Morris 9618A hepatoma nuclear proteins. See Fig. 5 for conditions.

approach was not designed to yield native, biologically active molecules, enzymes treated with SDS have been reconstituted to biological activity after the removal of SDS (Weber and Kuter, 1971).

1. "CHROMATIN FRACTION II" PROTEINS OF NORMAL RAT TISSUES

Isolation of "Chromatin Fraction II" from normal rat liver nuclei and electrophoresis on the 6%/8% two-dimensional polyacrylamide gel electrophoresis system produced the pattern of proteins shown in Fig. 12. In this case the left half of the pattern was the B region and the right half the C region with capital letters designating protein spots found in chromatin. In this particular pattern 60 protein spots were visualized.

A comparable gel pattern for the chromatin of thioacetamide treated rat liver is shown in Fig. 13. It was interesting to note that spots C18, C21, and CQ were found to be more dense in a nongrowing tissue with a large nucleolus.

The "Chromatin Fraction II" from rat kidney showed the pattern in Fig. 14. Interestingly, spot BJ′ was observed and appears to be kidney specific. Other than BJ′, there were no additional spots. In fact, the C region appeared to contain fewer and less densely stained protein spots than in the previous rat liver patterns (Figs. 12 and 13).

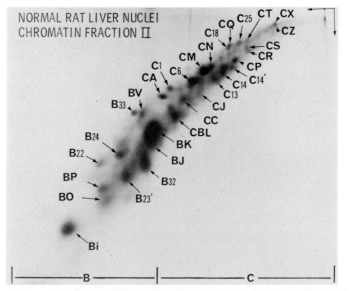

FIG. 12. Two-dimensional gel electrophoretogram of normal rat liver "Chromatin Fraction II." A 500-μg sample was run on the first-dimensional gel of 6% acrylamide, 6 M urea, and 0.9 N acetic acid at 120 V for 4.5 hours. The second dimension was an 8% acrylamide, 0.1% sodium dodecyl sulfate, 0.1 M phosphate (pH 7.1) slab gel which was run at 50 mA per slab for 15 hours. The gels were stained for 6 hours with 0.2% Coomassie Brilliant Blue R in acetic acid–methanol–water (1:5:5).

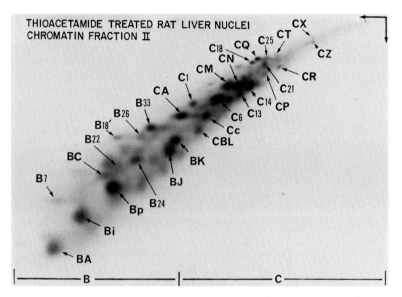

FIG. 13. Two-dimensional gel electrophoretogram of thioacetamide-treated rat liver "Chromatin Fraction II." See Fig. 12 for conditions.

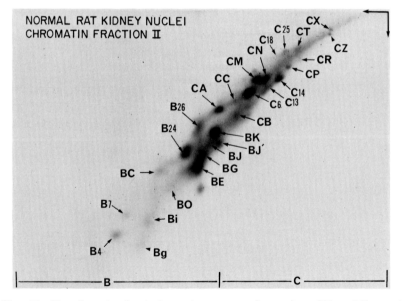

FIG. 14. Two-dimensional gel electrophoretogram of normal rat kidney "Chromatin Fraction II." See Fig. 12 for conditions.

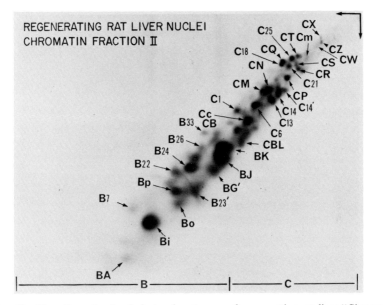

FIG. 15. Two-dimensional gel electrophoretogram of regenerating rat liver "Chromatin Fraction II." See Fig. 12 for conditions.

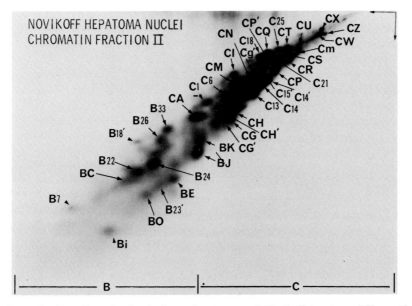

FIG. 16. Two-dimensional gel electrophoretogram of Novikoff hepatoma "Chromatin Fraction II." See Fig. 12 for conditions.

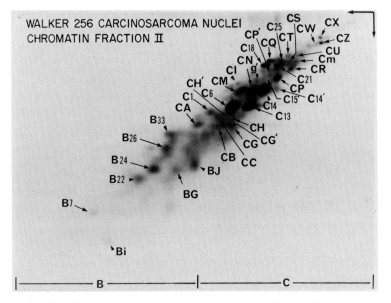

Fig. 17. Two-dimensional gel electrophoretogram of Walker 256 carcinosarcoma "Chromatin Fraction II." See Fig. 12 for conditions.

2. "Chromatin Fraction II" of Growing Rat Tissues

The "Chromatin Fraction II" patterns for regenerating rat liver, Novikoff hepatoma, and Walker 256 carcinosarcoma are shown in Figs. 15, 16, and 17, respectively. It appeared that spots CP, C25, CR, CS, and C8 in the growing tissues were much larger and more dense than in the nongrowing liver tissues and kidney tissue. In addition, spots BG', BH', and BP' were apparently tumor specific.

C.   Reconstitution of Chromatin from Citric Acid Nuclei

In the studies by Paul and Gilmour (1968), chromatin isolated from citric acid nuclei of several tissues was extracted and reconstituted. They found that the reconstituted chromatin exhibited the same template activity as the original chromatin. In a following study by Gilmour and Paul (1969), it was further shown that the chromatin transcripts from reconstituted chromatin were quantitatively the same as those observed from natural chromatin. These impressive results, together with the application of the Tissumizer to the citric acid method for nuclear isolation provide a basis for studies on reconstitution in additional normal and tumor tissues.

With two-dimensional methods for the analysis of the protein fractions used in reconstitution, the ability to identify specific protein factors responsible for the observed RNA transcripts reaches a greater level of sensitivity. In addition, the two-dimensional gel provides a protein spot assay for the integrity of the reconstituted chromatin product when compared to the native chromatin.

D.   Chromatin-Associated RNA from Citric Acid Nuclei

Reports concerning small amounts of RNA associated with chromatin have appeared from Shih and Bonner (1969), Wilt and Ekenberg (1971), Kanehisa *et al.* (1972), and Getz and Saunders (1973). Opinions regarding the origin of this RNA have ranged from the opinion that these are breakdown products of either ribosomal or messenger RNAs to the proposal that chromatin-associated RNA represents a unique RNA species. In light of the use of tissue as a starting material for the chromatin by such workers as Shih and Bonner (1969), the contamination by ribosomal and cytoplasmic elements is a potential source of degradation since neither citric acid nor Triton X-100 were used to remove the outer layer of the nuclear envelope and its associated ribosomal components. The employment of citric acid in the preparation of nuclei prior to the isolation of chromatin greatly reduces this risk of ribosomal type RNA contamination (Tolstoshev and Wells, 1974; Holmes and Bonner, 1974). This further points to the potential usefulness of citric acid-isolated nuclei in the resolution of this problem.

VI.   Isolation of Nucleoli

A.   Isolation of Nucleoli in Sucrose Medium

The disruption of purified nuceli in 0.34 *M* sucrose has been described earlier (Busch, 1967). Clean nuclei are suspended in 5–10 volumes of 0.34 *M* sucrose and sonicated until only a few whole nuclei remain. For rat liver nuclei, this is usually accomplished within 30 to 60 seconds. The resulting sonicate is layered over at least 2 volumes of 0.88 *M* sucrose and centrifuged at 1100 *g* for 30 minutes. The resulting pellet is examined by phase-contrast microscopy with 0.5% azure C in 0.25 *M* sucrose for an acceptable nucleolar product (Busch and Smetana, 1970).

B.   Isolation of Nucleoli from Citric Acid Nuclei

A procedure has been described for the preparation of nuclei and sonication to nucleoli in 0.25 *M* citric acid by Ro-Choi *et al.* (1973). Recently, it has been found that nuclei prepared by the 0.25 *M* citric acid procedure can be sonicated to provide nucleoli; such nuclei are dispersed and sonicated in 0.34 *M* sucrose. The sonicated nucleoli are obtained by layering over 2 volumes of 0.88 M sucrose and centrifuging at 1100 g for 30 minutes. The advantage of using 0.025 M citric acid nuclei is the reduced likelihood of adsorption of cytoplasmic and ribosomal type elements when the nucleolus is exposed to the sonication medium without the more extensive washout of nuclear protein elements observed when 5% citric acid is employed.

C. Nucleoli Obtained by Detergent Methods

A procedure for the isolation of nucleoli was developed by Penman *et al.* (1966), who employ Tween 40 and sodium deoxycholate treatment of the nuclei from a hypotonic medium. After centrifugation the nuclei are suspended in 0.5 M NaCl, 0.05 M MgCl$_2$, and 0.01 M Tris (pH 7.4) and treated with DNase. The pellet resulting from centrifugation at 10,000 g for 5 minutes is the nucleolar preparation. In place of the hypotonic medium and ball mill for homogenization, 0.025 M citric acid medium and the Tissumizer could be employed.

VII.   Summary

The use of citric acid in 0.025 M concentration with a Tissumizer provides a very satisfactory method for the preparation of nuclei from a large number of normal and tumor cells. This has proved to be of particular value in nuclear protein comparison studies. The advantages of two-dimensional polyacrylamide gel electrophoresis results from the number of protein components resolved by this method. It has been particularly useful as an analytical approach in the comparison of normal and tumor cell nuclear proteins.

REFERENCES

Arnesen, K., Goldsmith, Y., and Dulaney, A. D. (1949). *Cancer Res.* **9**, 669.
Baserga, R. (1972). *J. Invest. Dermatol.* **59**, 21.
Behrens, M. (1932). *Hoppe-Seyler's Z. Physiol. Chem.* **209**, 59.
Bernhard, W. (1969). *J. Ultrastruct. Res.* **27**, 250.
Blobel, G., and Potter, V. R. (1966). *Science* **154**, 1662.
Bornens, M. (1968). *C. R. Acad. Sci., Ser. D* **266**, 596.
Bresnick, E. (1971). *Methods Cancer Res.* **6**, 347.
Busch, G. I., Yeoman, L. C., Taylor, C. W., and Busch, H. (1974). *Physiol. Chem. Phys.* **6**, 1.
Busch, H. (1967). *In* "Methods in Enzymology" (L. Grossman and K. Moldave, eds.), Vol. 12, Part A. p. 448. Academic Press, New York.
Busch, H., and Smetana, K. (1970). "The Nucleolus." Academic Press, New York.

Calafat, J., Den Engelse, L., and Emmelot, P. (1970). *Chem-Biol. Interact.* **2**, 309.

Chauveau, J., Moule, Y., and Rouiller, C. (1956). *Exp. Cell Res.* **11**, 317.

Church, R. B., and McCarthy, B. J. (1967). *J. Mol. Biol.* **23**, 459.

Crossman, G. (1937). *Science* **85**, 250.

Desjardins, R., Smetana, K., and Busch, H. (1965). *Exp. Cell Res.* **40**, 127.

Dounce, A. L., Witler, R. F., Monty, K. J., Pate, S., and Cottine, M. A. (1955). *J. Biophys. Biochem. Cytol.* **1**, 139.

Fisher, H. W., and Harris H. (1962). *Proc. Roy. Soc., Ser. B* **156**, 521.

Getz, M. J., and Saunders, G. F. (1973). *Biochim. Biophys. Acta* **312**, 555.

Gilmour, R. S., and Paul, J. (1969). *J. Mol. Biol.* **40**, 137.

Gurr, M. I., Finean, J. B., and Hawthorne, J. N. (1963). *Biochim. Biophys. Acta* **70**, 406.

Higashi, K., Narayan, K. S., Adams, H. R., and Busch, H. (1966). *Cancer Res.* **26**, 1582.

Hohorst, H. J., Kreutz, F. H., and Bucher, T. (1959). *Biochem. Z.* **332**, 18.

Holmes, D. S., and Bonner, J. (1974). *Biochemistry* **13**, 841.

Hubert, M. T., Fanard, P., Carasso, N., Rosencuiajg, R., and Zalta, J. P. (1962). *J. Microsc. (Paris)* **1**, 435.

Hymer, W. C., and Kuff, E. L. (1964). *J. Histochem. Cytochem.* **12**, 359.

Kanehisa, T., Tanaka, T., and Kano, Y. (1972). *Biochim. Biophys. Acta* **277**, 584.

Kellermayer, M., Lehane, D. E., Lane, M., and Busch, H. (1973). *Physiol. Chem. Phys.* **5**, 503.

Kellermayer, M., Olson, M. O. J., Smetana, K., Daskal, I., and Busch, H. (1974). *Exp. Cell Res.* **85**, 191.

Kostraba, W. C., and Wang, T. Y. (1972). *Biochim. Biophys. Acta* **262**, 169.

Kostraba, W. C., and Wang, T. Y. (1973). *Exp. Cell Res.* **80**, 291.

LeStourgeon, W. M., and Rusch, H. P. (1973). *Arch. Biochem. Biophys.* **155**, 144.

MacGillivray, A. J., Camerson, A., Krauze, R. J., Rickwood, D., and Paul, J. (1972). *Biochim. Biophys. Acta* **277**, 384.

Marshak, A. (1941). *J. Gen. Physiol.* **25**, 275.

Marushige, K., and Bonner, J. (1966). *J. Mol. Biol.* **15**, 160.

Mauritzen, C. M., Choi, Y. C., and Busch, H. (1971). *Methods Cancer Res.* **6**, 253.

Muramatsu, M., Smetana, K., and Busch, H. (1963). *Cancer Res.* **23**, 510.

Orrick, L. R., Olson, M. O. J., and Busch, H. (1973). *Proc. Nat. Acad. Sci. U.S.* **70**, 1316.

Patel, G. L., (1972). *Life Sci.* **11**, 1135.

Paul, J., and Gilmour, R. S. (1966). *J. Mol. Biol.* **16**, 242.

Paul, J., and Gilmour, R. S. (1968). *J. Mol. Biol.* **34**, 305.

Penman, S., Smith, J., and Holtzman, E. (1966). *Science* **154**, 786.

Prestayko, A. W., Klomp, G. R., Schmoll, D. J., and Busch, H. (1974). *Biochemistry* **13**, 1945.

Richter, K. H., and Sekeris, C. E. (1972). *Arch. Biochem. Biophys.* **148**, 44.

Ro-Choi, T. S., Smetana, K., and Busch, H. (1973). *Exp. Cell Res.* **79**, 43.

Salas, J., and Green, H. (1971). *Nature (London), New Biol.* **229**, 165.

Schneider, W. C. (1948). *J. Biol. Chem.* **176**, 259.

Shih, T. Y., and Bonner, J. (1969). *Biochim. Biophys. Acta* **182**, 30.

Siebert, G. (1964). *In* "Biochemisches Taschenbuch" (H. M. Rauen, ed.), 2nd ed., Vol. II, p. 541. Springer-Verlag, Berlin and New York.

Siebert, G. (1967a). *Methods Cancer Res.* **2**, 287.

Siebert, G. (1967b). *Methods Cancer Res.* **2**, 290.

Siebert, G., Villalobos, J., Ro., T. S., Steele, W. J., Lindenmayer, G., Adams, H., and Busch, H. (1966). *J. Biol. Chem.* **241**, 71.

Spelsberg, T. C., Steggles, A. W., and O'Malley, B. W. (1971). *Biochim. Biophys. Acta* **254**, 129.

Stein, G. S., and Thrall, C. L. (1973). *FEBS (Fed. Eur. Biochem. Soc.) Lett.* **32**, 41.
Taylor, C. W., Yeoman, L. C., Daskal, I., and Busch, H. (1973). *Exp. Cell Res.* **82**, 215.
Teng, C. S., Teng, C. T., and Allfrey, V. G. (1971). *J. Biol. Chem.* **246**, 3597.
Tolstoshev, P., and Wells, J. R. E. (1974). *Biochemistry* **13**, 103.
Tsai, R. L., and Green, A. (1973). *Nature (London), New Biol.* **243**, 168.
Wakabayashi, K., Wang, S., Hard, G., and Hnilica, L. S. (1973). *FEBS (Fed. Eur. Biochem. Soc.) Lett.* **32**, 46.
Weber, K., and Kuter, D. J. (1971). *J. Biol. Chem.* **246**, 4504.
Weisenthal, L. M., and Ruddon, R. W., (1972). *Cancer Res.* **32**, 1009.
Wilhelm, J. A., Ansevin, A. T., Johnson, A. W., and Hnilica, L. S. (1972). *Biochim. Biophys. Acta* **272**, 220.
Wilt, F. H., and Ekenberg, E. (1971). *Biochem. Biophys. Res. Commun.* **44**, 831.
Woernley, D. L., and Carruthers, C. (1957). *Arch. Biochem. Biophys.* **67**, 493.
Wu, F. C., Elgin, S. C. R., and Hood, L. E. (1973). *Biochemistry* **12**, 2792.
Yeoman, L. C., Taylor, C. W., and Busch, H. (1973a). *Biochem. Biophys. Res. Commun.* **51**, 956.
Yeoman, L. C., Taylor, C. W., Jordan, J. J., and Busch, H. (1973b). *Biochem. Biophys. Res. Commun.* **53**, 1067.
Yeoman, L. C., Taylor, C. W., and Busch, H. (1974). *Cancer Res.* **34**, 424.

Chapter 19

Isolation and Manipulation of Salivary Gland Nuclei and Chromosomes

M. ROBERT

Institut für Genetik, Universität des Saarlandes, Saarbrücken, West Germany, and Department of Biology, Johns Hopkins University, Baltimore, Maryland

I. Introduction

Chromosomes, as reaction products between polyanionic and polycationic macromolecules, undergo structural changes in response to variations in their environment of physicochemical parameters like pH and ionic strength. The polytene chromosomes of Diptera, owing to their size and their longitudinal differentiation, represent a most useful experimental material for the study of these changes. However, any attempt to characterize the direct effects of chemical agents and to understand their mechanism of action requires the availability of isolated systems. A method is described here which allows the semi-mass isolation of nuclei and chromosomes from salivary glands of *Chironomus*.

Different techniques for the isolation of nuclei with polytene chromosomes (Kroeger, 1966; Lezzi, 1966; Ristow and Arends, 1968; Boyd *et al.*, 1968; Cohen and Gotchel, 1971; Zweidler and Cohen, 1971), of nuclear contents (Kroeger *et al.*, 1973) and of polytene chromosomes (D'Angelo, 1946; Karlson and Löffler, 1962; Lezzi, 1965; Lezzi and Gilbert, 1970) have been described. These techniques include the use of detergents such as Triton X-100 (Boyd *et al.*, 1968; Cohen and Gotchel, 1971), sodium deoxycholate (Ristow and Arends, 1968; Cohen and Gotchel, 1971) and Tween 80 (Ristow and Arends, 1968), of enzymes, e.g., chitinase (Ristow and Arends, 1968) and pronase (Karlson and Löffler, 1962; Ristow and Arends, 1968), or of micrurgical operations (D'Angelo, 1946; Lezzi, 1965, 1966; Kroeger, 1966; Gruzdev and Belaja, 1968; Lezzi and Gilbert, 1970). The author's intention is not to present here a comparative review of these different techniques, but to describe in more detail than previously (Robert, 1971), together with some modifications and new results, the technique he has developed for the isolation of salivary gland nuclei and chromosomes of *Chironomus*. The guidelines presented here should allow any one with a minimum of manual dexterity to use the method successfully.

Any method for the isolation of chromosomes should include as the first step the separation of clean nuclei in order to avoid contamination by cytoplasmic and extracellular substances. The method presented here consists of the following steps: (a) soaking of explanted salivary glands in a Ringer solution containing digitonin, a nonionic detergent, (b) homogenization by gentle pipetting, (c) collecting and washing the nuclei by means of micropipettes, and (d) breaking the nuclear membrane and freeing the chromosomes by forcing the nuclear suspension through micropipettes of specific diameter.

Nuclei and chromosomes isolated according to this procedure have made possible the study of the structural and functional modifications induced by physicochemical parameters like pH, ionic strength, and relative concentrations of different types of ions (Robert, 1971; Lezzi and Robert, 1972; Robert and Gopalan, 1974). A brief account of the results obtained is given in Section IV.

II. Material

A. Larvae

Fourth-instar larvae of a laboratory-bred strain of *Chironomus thummi* were used in this study. The method of culture has been described in detail by Kroeger (1973).

B. Special Instruments

1. MICROPIPETTES

Glass micropipettes with inner diameters at the opening ranging from 0.1 to 0.5 mm were made by elongating melting point capillaries of 0.9 to 1.8-mm diameter over the flame. Since standardization of this technique is rather difficult, it is advisable to prepare large quantities of micropipettes and to select the suitable ones. The micropipettes used for the isolation of nuclei and chromosomes should have a short tip (ca. 0.5 cm) with an inner diameter of 0.4–0.5 mm and 0.25–0.35 mm, respectively. The micropipettes used for the transfer of isolated nuclei and chromosomes should present a long tip (ca. 3 cm) of 0.1–0.15 mm inner diameter.

The micropipettes are siliconized by immersing in a 1% aqueous solution of Siliclad (Clay Adams, Division of Becton, Dickinson and Company, New Jersey).

The micropipettes are connected to mouthpieces by means of soft rubber tubing. It is recommended to have the micropipettes, the connecting tubing, and the mouthpieces affixed to the stereomicroscope in order to give more freedom of movement of the operator's hands.

2. INCUBATION CHAMBERS

Incubation chambers for isolated nuclei and chromosomes are prepared by a technique modified from Lezzi and Gilbert (1970). Pieces of parafilm pierced with three holes (5 mm in diameter) situated on a line 1.5 cm apart are heat sealed onto nonsiliconized microscope slides. After cooling, two narrow channels connecting the holes are cut out with a razor blade. The middle hole serves as an incubation chamber, while the other two function as reservoirs for exchanging the medium. If necessary, the middle hole can be covered, after depositing the chromosomes or the nuclei, with a cover slip securely affixed onto the parafilm with silicone grease.

This type of incubation chamber is particularly suitable for experiments requiring repeated changes of the medium. Other types of incubation chambers are used when the volume of the medium has to be kept small. These chambers consist of a single hole of arbitrary diameter in a small piece of parafilm. Small rings of parafilm are particularly useful for autoradiographic studies as they can easily be detached from the slide with thin forceps.

C. Isolation Medium

The modified *Chironomus* Ringer solution (CR medium) is composed of 87 mM NaCl, 3.2 mM KCl, 1.3 mM CaCl$_2$, and 1 mM MgCl$_2$. It is buffered at pH 7.3 with 10 mM Tris·HCl (CR$_{7.3}$ medium) or at pH 6.3 with 10 mM Tris-maleate (CR$_{6.3}$ medium).

III.   Method

A.   Isolation of Salivary Gland Nuclei

Thirty to forty salivary glands are explanted into a drop of ice-cold $CR_{7.3}$ medium (for composition see Section II,C). The explanted salivary glands are then washed by transfer into two drops of $CR_{7.3}$ medium and two drops of $CR_{6.3}$ medium. Alternatively, the glands can be washed by low speed centrifugation in conical polypropylene tubes of about 1 ml capacity. During the washing procedure a large portion of the gland secretion is released into the solution. This release facilitates the subsequent isolation of the nuclei, since the presence of large quantities of secretion in the homogenate would make the solution too viscous for easy pipetting of the nuclei.

After washing, the glands are suspended in 0.5 ml of ice-cold $CR_{6.3}$ medium containing 10 mg of digitonin per milliliter, a nonionic detergent of the steroid type. Since the solubility of digitonin in water is very low, the medium is prepared as follows: 10 mg of digitonin are dissolved by rapid boiling in 0.5 ml of distilled water and added, after cooling, to 0.5 ml of double-concentrated $CR_{6.3}$ medium. This solution is stable for many hours at 2°C. Certain commercially available digitonin preparations have been found difficult to solubilize completely. In such cases, the insoluble material should be removed by centrifugation.

The glands are incubated for 40 minutes at 2°C in covered polypropylene tubes. The suspension is then homogenized by gently passing it through a siliconized micropipette of 0.4 to 0.45 mm inner diameter at the opening. The pipetting is repeated until all visible tissue pieces have been broken down. This generally requires twenty to thirty passages through the pipette. Two factors are critical for the success of this operation: the inner diameter of the micropipette and the strength of suction. Pipettes of smaller diameter or too vigorous suction create shearing forces which may rupture the nuclear membrane. On the other hand, pipettes of larger diameter or too gentle suction yield nuclei not completely separated from adjacent cytoplasmic layers.

The homogenate, which consists of intact nuclei plus cellular and glandular debris, is transfered onto a siliconized microscope slide, and subsequent operations are performed under the stereomicroscope at magnifications of 50 to 100 ×.

In order to facilitate the collecting of the nuclei it is convenient first to pipette off the gross glandular debris which tend to float in the solution. The nuclei, which are generally found on the surface of the glass, are then easily collected with siliconized micropipettes of 0.1 mm inner diameter. Care is

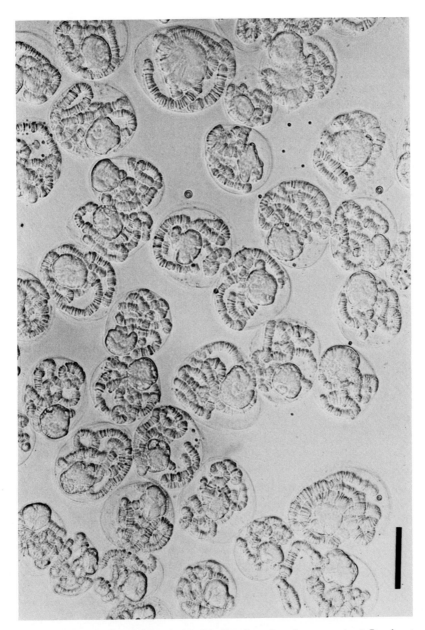

FIG. 1. Isolated salivary gland nuclei. $CR_{7,3}$ medium. Interference contrast. Bar denotes 50 μm.

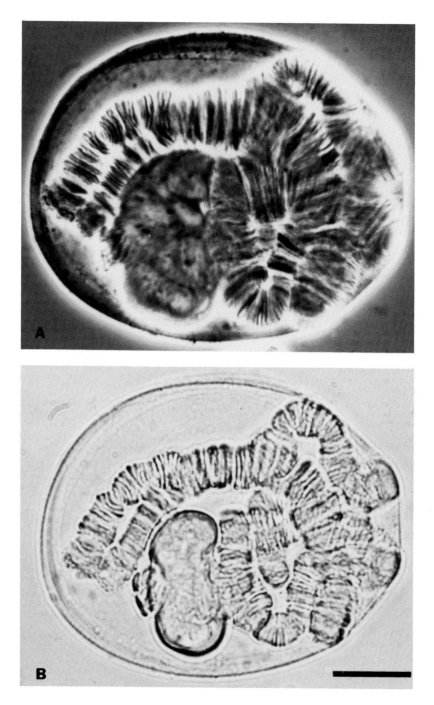

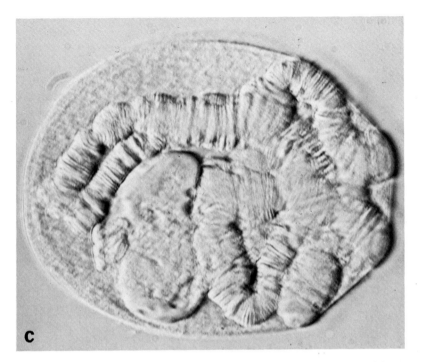

FIG. 2. Isolated salivary gland nucleus. CR$_{7.3}$ medium. (A) Phase contrast, (B) bright field, and (C) oblique illumination. A–C represent the same nucleus. Bar denotes 20 μm.

taken to collect only clean nuclei with as little as possible of the surrounding medium. The nuclei are then washed by five successive transfers into drops of ice-cold CR$_{6.3}$ medium. Attempts to separate and purify the nuclei by sedimentation through sucrose gradients gave unsatisfactory results. The yield of microscopically clean nuclei (Figs. 1 and 2) can vary between 80 and 100%. This slight variation is principally due to the difficulty encountered in making micropipettes exactly similar in size and in shape.

The CR$_{6.3}$ medium was selected because it gives the best yield of clean and structurally intact nuclei. The presence in this medium of divalent ions and the slightly acidic pH help preserve the structural integrity of the nuclei (see Section IV). Equally important, at pH 6.3 the nuclei show no or little tendency to aggregate or to adhere on glass surfaces, two properties that are very pronounced at pH values over 7. Media containing sucrose do not allow a good separation of the nuclei from the cytoplasm. The presence of the sugar seems to cause a hardening of the cytoplasm which impairs the liberation of the nuclei.

B.   Observation and Manipulation of Isolated Nuclei

At low and high magnifications, the best way of observing isolated nuclei is to use bright field illumination with slightly closed condenser diaphragm (Fig. 2B). Oblique illumination (Fig. 2C) as well as interference contrast optics (Fig. 1) allow the visualization of the three-dimensional surface structure of the nuclei. The utilization of phase contrast is limited because it requires a flattening of the preparation, which can be achieved only by gentle squashing with a cover slip. This renders any subsequent change of medium difficult.

In order to test the effects of different environments, isolated nuclei can be transfered into drops of new media by means of micropipettes. However, this operation is easy to execute only as long as the pH is kept at values below 7. If the pH is raised over 7, the transfer is complicated by the tendency of the nuclei to adhere on glass surfaces, even if siliconized. Another difficulty comes from the fact that, upon transfer in a solution of higher osmolarity, the nuclei float toward the surface of the liquid, where they can hardly be observed and picked up with micropipettes. For experiments in which the medium has to be repeatedly changed, it is more convenient to deposit the nuclei on the bottom of incubation chambers of the three-hole type, as described in Section II,B,2. A cover slip is placed on the top of the chamber, and the medium can be changed by aspiration through one channel while new medium is being added from the other side. This technique has also the advantage of allowing the direct observation of the nuclei while the medium is being changed.

C.   Isolation of Polytene Chromosomes

Micropipetting is also the method of choice for rupturing the nuclear envelope and freeing the chromosomes. Other methods including sonication, pressurization followed by rapid depressurization, and the use of detergents or enzymes gave unsatisfactory results.

The nuclear suspension in $CR_{6.3}$ medium is transferred in a conical polypropylene tube and repeatedly forced through a siliconized micropipette of ca. 0.3 mm inner diameter at the opening. Here again, the diameter of the micropipette and the strength of suction are critical. Too narrow pipettes do not allow a suction vigorous enough to rupture the nuclear envelope. On the other hand, in pipettes of too large diameter, even a vigorous suction cannot produce the necessary shearing forces. The optimum diameter lies in the range 0.25 to 0.35 mm. However, even within these limits, the strength of suction has to be adapted to the diameter of the pipette; a less vigorous suction being required in pipettes of 0.25 mm diameter than in pipettes of

0.35 mm diameter. Otherwise, the isolated chromosomes will be distorted. It should be noted that a certain percentage of the isolated chromosomes are regularly found to be stretched. The largest chromosome (I) is particularly susceptible to shearing forces produced during the pipetting (Fig. 3). However, the stretching of the chromosomes is not necessarily a disadvantage and may in certain cases be desirable; stretched chromosomal regions allowing a better resolution of the banding pattern.

Free chromosomes can be separated from remaining intact nuclei by means of micropipettes of ca. 0.15 mm diameter and transferred into drops of $CR_{6.3}$ medium for washing. However, their high tendency to adhere on glass surfaces, even siliconized, and in spite of the slightly acidic pH renders this operation hazardous. It is therefore recommended to transfer the chromosomes immediately after isolation into chambers of the types described in Section II,B,2.

The yield of isolated chromosomes can vary considerably. This variation is principally due to the impossibility of making micropipettes exactly similar

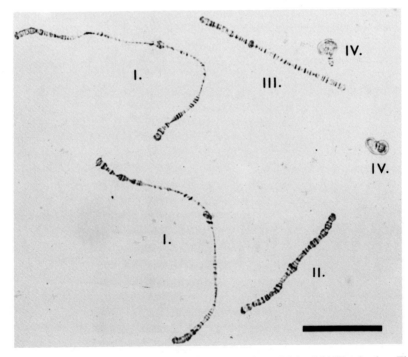

FIG. 3. Isolated polytene chromosomes. $CR_{7.3}$ medium. Bright field illumination. The chromosome numbers are indicated. Bar denotes 100 μm.

in size and in shape. In addition to the inner diameter of the pipette and the strength of suction, factors like shape and sharpness of the edge at the opening of the pipette seem to contribute to the efficiency of this technique.

D.   Observation and Manipulation of Isolated Chromosomes

At low and high magnifications, isolated chromosomes are best observed using bright field illumination with slightly closed condenser diaphragm (Figs. 3 and 4). At high magnification, phase contrast optics give a clear picture of the chromosomes (see Robert, 1971). However, it requires a reduction of the thickness of the preparation, which can be achieved only by gentle squashing under a cover slip. This type of observation is consequently not recommended for experiments in which the medium has to be changed repeatedly. For such experiments, bright-field illumination is more convenient. At high magnification, the objectives of the microscope can be immersed directly into the incubation medium (see, e.g., Fig. 5).

As stated in Section III,C, isolated chromosomes have a high tendency to adhere on glass surfaces. This renders hazardous any transfer by pipetting. It is best to transfer the chromosomes immediately after isolation into incubation chambers of the types described in Section II,B,2. Displacements of chromosomes within a drop of medium can be controlled by producing currents with a micropipette. Using the same technique, chromosomes can

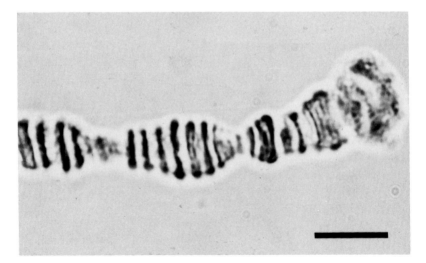

FIG. 4.   Banding pattern in the right arm of an isolated chromosome I. $CR_{7,3}$ medium. Bright field illumination. Bar denotes 10 μm.

be pressed against the bottom of the chamber, where they rapidly adhere. Repeated changes of medium generally do not detach the chromosomes from the glass. The use of incubation chambers described in Section II,B,2 allows the direct observation of isolated chromosomes while the medium is being changed (see, e.g., Fig. 5 A–C).

IV.   Dependence of Chromosome Structure on Ionic Environment

The isolation technique described above has made possible the study of the direct action of physicochemical parameters like pH, ionic strength and relative concentrations of different types of ions on the structure and function of polytene chromosomes (Robert, 1971; Lezzi and Robert, 1972; Robert and Gopalan, 1974). In brief, low pH values, intermediate ionic strengths, and the presence of divalent ions like Mg^{2+} and Ca^{2+} contribute to keep the chromosomal bands in a condensed state. On the other hand, high pH values, low and high ionic strengths, and the absence of divalent ions promote a decondensation of the bands. In addition to these general effects, ionic conditions have been found which cause a differential decondensation of the bands: in the presence of divalent ions or at acidic pH the bands decondense in a specific sequence as the ionic strength increases. Furthermore, the pattern of decondensation along the chromosomes can be modified to some extent by varying the relative concentrations of ions, e.g., Na^+, K^+, Mg^{2+}, and Ca^{2+}.

Not only inorganic ions are able to affect the structural state of the chromosomes; such organic ions as the amino acids are also effective structure modifiers (Robert and Gopalan, 1974). Increasing concentrations of amino acids lead to a decondensation of the bands (Fig. 5 A and B) and to an expansion of the chromosomes. This type of decondensation is reversible; however, if the decondensation has proceeded too far, the chromosomes are reduced in diameter upon recondensation as compared with the original (Fig. 5C). In addition to these general effects, certain amino acids, e.g., tryptophan and phenylalanine, are able to elicit specific responses on individual bands.

It has been suggested (Kroeger, 1963; Kroeger and Lezzi, 1966; Lezzi and Robert, 1972) that changes in the ionic environment of chromosomes may play a role in the physiological control of gene activity patterns. The results obtained using isolated chromosomes support this hypothesis in that (1) specific patterns of structural changes can be elicited in isolated chromosomes by manipulating the ionic environment, (2) these structural changes

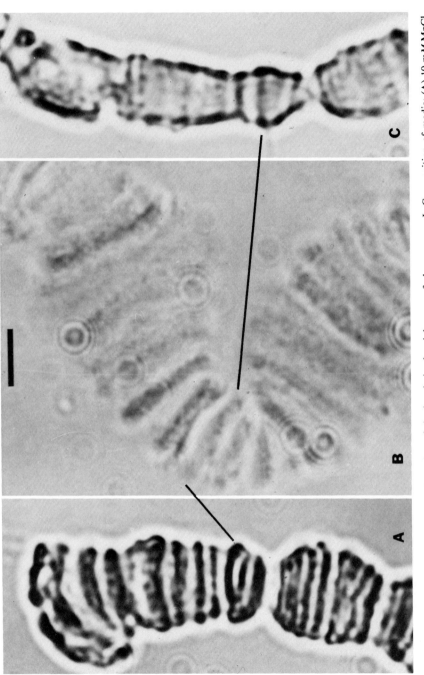

FIG. 5. Amino acid-mediated decondensation of the bands in the right arm of chromosome I. Composition of media: (A) 10 mM MgCl$_2$, (B) 10 mM MgCl$_2$ + 500 mM L-lysine-HCl, and (C) 10 mM MgCl$_2$. All solutions at pH 7.5 with 50 mM Tris · HCl. Bar denotes 5 μm.

seem to be correlated with an increased template activity of the chromosomes (Lezzi and Gilbert, 1970; Robert, 1971), and (3) the ionic concentrations required to induce a partial decondensation of the chromosomes at physiological pH are close to those encountered in salivary gland nuclei (Kroeger *et al.*, 1973).

The isolation techniques described here also make it possible to investigate the biochemical and biophysical events involved in the response of chromosomes to changes in their ionic environment. Such investigations are presently underway. For instance, the nuclear proteins of *Chironomus* salivary glands have been characterized by gel electrophoresis (Fig. 6). The electro-

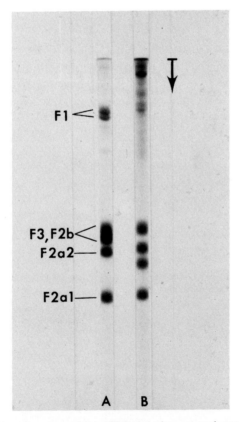

FIG. 6.   Electrophoretic pattern of (A) calf thymus chromosomal proteins, and (B) *Chironomus* salivary gland nuclear proteins. Sodium dodecyl sulfate (SDS) polyacrylamide gel electrophoresis. The samples were solubilized in a mixture of SDS, urea, and mercaptoethanol. The direction of migration is indicated. F1, F2a1, F2a2, F2b, and F3: calf thymus histone fractions.

phoretic pattern of these proteins differ from that of the calf thymus chromosomal proteins (Fig. 6). Preliminary results indicate that, as postulated earlier (Robert, 1971), some of the structural transitions induced in chromosomes by changing ionic environment coincide with a release of certain protein fractions or subfractions.

ACKNOWLEDGMENTS

This study was supported in part by a Public Health Service International Research Fellowship (No. 1 FO5 TWO 1898). The author wishes to thank Dr. H. Kroeger, Saarland's University, Saarbrücken, West Germany, and Dr. E. N. Moudrianakis, Johns Hopkins University, Baltimore, Maryland, for providing the facilities and help that have made this work possible, as well as Mr. P. Kurth, Johns Hopkins University, for help during the preparation of the English version of this chapter.

REFERENCES

Boyd, J. B., Berendes, H. D., and Boyd, H. (1968). *J. Cell Biol.* **38**, 369–376.
Cohen, L. H., and Gotchel, B. V. (1971). *J. Biol. Chem.* **246**, 1841–1848.
D'Angelo, E. G. (1946). *Biol. Bull.* **90**, 71–87.
Gruzdev, A. D., and Belaja, A. N. (1968). *Tsitologiya* **10**, 297–305.
Karlson, P., and Löffler, U. (1962). *Hoppe-Seyler's Z. Physiol. Chem.* **327**, 286–288.
Kroeger, H. (1963). *Nature (London)* **200**, 1234.
Kroeger, H. (1966). *In* "Methods in Cell Physiology" (D. M. Prescott, ed.), Vol. 2, pp. 61–92. Academic Press, New York.
Kroeger, H. (1973). *Z. Morphol. Tiere* **74**, 65–88.
Kroeger, H., and Lezzi, M. (1966). *Annu. Rev. Entomol.* **11**, 1–22.
Kroeger, H., Trösch, W., and Müller, G. (1973). *Exp. Cell Res.* **80**, 329–339.
Lezzi, M. (1965). *Exp. Cell Res.* **39**, 289–292.
Lezzi, M. (1966). *Exp. Cell Res.* **43**, 571–577.
Lezzi, M., and Gilbert, L. I. (1970). *J. Cell Sci.* **6**, 615–628.
Lezzi, M., and Robert, M. (1972). *In* "Results and Problems in Cell Differentiation" (W. Beermann, J. Reinert, and H. Ursprung, eds.), Vol. 4, pp. 35–57. Springer-Verlag, Berlin and New York.
Ristow, H., and Arends, S. (1968). *Biochim. Biophys. Acta* **157**, 178–186.
Robert, M. (1971). *Chromosoma* **36**, 1–33.
Robert, M., and Gopalan (1974). In preparation.
Zweidler, A., and Cohen, L. H. (1971). *J. Cell Biol.* **51**, 240–248.

Subject Index

A 5
B 6
C 7
D 8
E 9
F 0
G 1
H 2
I 3
J 4